Structural Foundations Manual for Low-Rise Buildings

Second Edition

Structural Foundations Manual for Low-Rise Buildings

Second Edition

M.F. Atkinson

CEng, FIStructE, MICE, HND Arb

Spon Press
Taylor & Francis Group

LONDON AND NEW YORK

First published 2004 by Spon Press
11 New Fetter Lane, London EC4P 4EE

Simultaneously published in the USA and Canada
by Spon Press
29 West 35th Street, New York, NY 10001

Spon Press is an imprint of the Taylor & Francis Group

© 2003 M. F. Atkinson

Typeset in Times by Steven Gardiner Ltd, Cambridge
Printed and bound in Great Britain by TJ International Ltd, Padstow, Cornwall

British Library Cataloguing in Publication Data
A catalogue record for this book is available from the British Library

Library of Congress Cataloging in Publication Data
Atkinson, M. F.
Structural foundations manual for low-rise buildings / M. F. Atkinson –
2nd ed.
p.   cm.
Includes bibliographical references and index.
ISBN 0-415-26643-2 (alk. paper)
1. Foundations. 2. Structural engineering. I. Title
TH2101.A88 2003
624.1'5 – dc21   2003053943

ISBN 0-415-26643-2

Contents

Preface viii
Examples ix

1 Site investigations 1

1.1 Walk-over survey 1
1.2 Desk study 2
1.3 Site investigation: field work 4
    1.3.1 Trial pit logs 4
    1.3.2 Borehole record 4
1.4 Site investigation procedure 4
    1.4.1 Borehole logs 4
    1.4.2 Trial pit logs 4
    1.4.3 Groundwater 4
    1.4.4 Standard penetration tests 4
1.5 Interpretation of laboratory testing 9
    1.5.1 Chemical tests 9
1.6 Solution features 11
    1.6.1 Limestones 11
    1.6.2 Chalk 11
    1.6.3 Salt 13
    1.6.4 Gypsum 14
Case study 1.1 Investigation of former mining site, Sheffield 14
Bibliography 21

2 Foundation design 23

2.1 Introduction 23
    2.1.1 Width of footing 23
    2.1.2 Soft spots 23
    2.1.3 Stratum variation in excavation 23
    2.1.4 Firm clays overlying soft strata 25
    2.1.5 Depth of footings 26
2.2 Widened reinforced strip footings 29
2.3 Reinforced strip footings on replacement granular fill 32
2.4 Trench fill foundations 33
2.5 Raft foundations 34
2.6 Pad and pier foundation 44
    2.6.1 Disused wells 44
2.7 Piled foundations 47
    2.7.1 Bored piles 47
    2.7.2 Design of a bored pile 48
    2.7.3 Design of bored and driven piles 48

    2.7.4 Driven piles 50
    2.7.5 Driving precast piles 53
    2.7.6 Test loading 53
Bibliography 66

3 Foundations in cohesive soils 67

3.1 Introduction 67
3.2 Settlements in cohesive soils 67
3.3 Consolidation settlement 68
    3.3.1 Bearing capacity of cohesive soils 68
    3.3.2 Vertical stress distribution 70
    3.3.3 Construction problems on clay sites 70
    3.3.4 Foundation designs on clay soils 71
    3.3.5 Settlements in clay soils 72
3.4 Moisture movements 74
    3.4.1 Liquid limit test 76
    3.4.2 Plastic limit test 77
Bibliography 79

4 Foundations in sands and gravels 81

4.1 Classification of sands and gravels 81
    4.1.1 Composite sands and gravels 81
    4.1.2 Dilatant sands 82
    4.1.3 Calcareous sands 82
4.2 Relative densities of granular soils 82
    4.2.1 Field density assessment 82
    4.2.2 Visual observations 82
    4.2.3 Groundwater levels 83
    4.2.4 The standard penetration test 83
    4.2.5 Interpretation of SPT results 84
    4.2.6 Ultimate bearing capacities 84
4.3 Construction problems in granular soils 86
4.4 Foundation design in granular soils 87
4.5 Plate bearing tests 89
4.6 Piling into sands and gravel strata 89
    4.6.1 Bored piles 90
    4.6.2 Continuous flight auger piles 90
    4.6.3 Design of bored piles 90
    4.6.4 Set calculations 92
    4.6.5 Dynamic pile formula 92
    4.6.6 Re-drive tests 93
    4.6.7 Base-driven steel tube piles 93
    4.6.8 Top-driven steel piles 93
Bibliography 94

5   Building in mining localities   95

5.1   Coal mining, past and present   95
5.2   Coal shafts   96
5.3   Shallow mineworkings   97
5.4   Drilling investigations   99
5.5   Stabilizing old workings   99
      5.5.1   Collapsed workings   99
      5.5.2   Special conditions   100
5.6   Foundations in areas with shallow workings   100
5.7   Active mining   100
5.8   Future mining   100
5.9   Mitigating the effects of mining subsidence   101
      5.9.1   Longwall mining (advancing system)   101
      5.9.2   Designing buildings for future mining
              subsidence   101
      5.9.3   CLASP system of construction   102
      5.9.4   Mining rafts   103
      5.9.5   Irregular-shaped units   104
      5.9.6   Designing strip footings in active
              mining areas   107
      5.9.7   Movement joints   109
Bibliography   112

6   Sites with trees   113

6.1   Foundation design   113
      6.1.1   Climatic variation   119
      6.1.2   Distances between trees and
              foundations   119
      6.1.3   Foundation depths related to
              proposed tree and shrub planting   121
      6.1.4   Measurement of foundation depths   121
6.2   Building on wooded sites   122
      6.2.1   Piled foundations   122
      6.2.2   Deep trench-fill concrete foundations   123
      6.2.3   Deep strip footings with loose stone
              backfill   124
      6.2.4   Stiff raft foundations on a thick
              cushion of granular fill   124
      6.2.5   Deep pad and stem foundations   125
6.3   Precautions to take when there is evidence of
      clay desiccation   125
      6.3.1   Suspended floors   126
      6.3.2   Drainage and services   126
      6.3.3   Protection to drainage   126
      6.3.4   Precautions against clay heave   127
6.4   Foundations in granular strata overlying
      shrinkage clays   134
Bibliography   136

7   Developing on sloping sites   137
7.1   Stability of slopes   137
Case study 7.1   Sloping site with clay fills over
                 boulder clay   141
7.2   Developing on sloping sites   146
      7.2.1   Additional weight of dwellings   146

      7.2.2   Additional weight of regrade fill   147
      7.2.3   Changes in the groundwater level or
              surface runoff   147
      7.2.4   Excavations for deep drainage   147
      7.2.5   Removal of trees and vegetation   147
      7.2.6   Split-level housing   148
7.3   Retaining systems   148
      7.3.1   Gravity type retaining systems   148
      7.3.2   Cantilever walls: reinforced concrete
              or brickwork   148
      7.3.3   Gabions, crib walling, reinforced earth 149
      7.3.4   Steel sheet piling   150
7.4   Designing retaining walls   150
      7.4.1   Active pressure on walls   150
      7.4.2   Surcharge loading   151
      7.4.3   Passive resistance (granular soils)   152
7.5   Cantilevered retaining walls   152
      7.5.1   Mass brick or block walls   152
      7.5.2   Reinforced cavity walls   156
      7.5.3   Pocket-type walls   156
7.6   Damp-proofing to retaining walls   169
      7.6.1   Type A structures: tanked protection   169
      7.6.2   Type C structures: drained cavity
              construction   172
Bibliography   174

8   Building on filled ground   175

8.1   Opencast coal workings   177
8.2   Foundations   178
      8.2.1   Stiff raft foundations   178
      8.2.2   Piled foundations   178
8.3   Suspended ground-floor construction   179
8.4   Compaction of fills to an engineered
      specification   181
      8.4.1   Procedure   181
      8.4.2   Site testing before backfilling   181
      8.4.3   Foundations   182
      8.4.4   Roads and drainage   182
      8.4.5   Groundwater   182
8.5   Ground improvement techniques   182
      8.5.1   Dynamic consolidation   182
      8.5.2   Surcharge loading   182
8.6   Compaction of structural fills   182
      8.6.1   Materials specification   183
      8.6.2   Definitions   183
      8.6.3   Suitable fill materials   183
      8.6.4   Unsuitable materials   183
      8.6.5   Compaction   183
      8.6.6   Testing on site   184
Bibliography   184

9   Ground improvement   185

9.1   Vibro-compaction techniques   186
      9.1.1   Types of treatment   186
      9.1.2   Ground conditions   187

9.1.3	Engineering supervision	190
9.1.4	Design of vibro-compaction stone columns	193
9.1.5.	Foundations on vibro-compaction sites	197
9.2	Dynamic consolidation	198
9.2.1	Testing	201
9.3	Preloading using surcharge materials	201
9.4	Improving soils by chemical or grout injection	202
Bibliography		203

10   Building up to existing buildings — 205

10.1	Site investigation	205
10.2	Foundation types	205
10.3	Underpinning	212
10.3.1	Beam and pad solution	216
10.3.2	Pile and needle beam solution	216
Case study 10.1	Investigation and underpinning of detached house on made ground, York	218
Case study 10.2	Differential settlement, Leyburn	219
Case study 10.3	House on made ground, Beverley	222
Case study 10.4	House founded on sloping rock formation, Scarborough	223
10.4	Shoring	226
Bibliography		227

11   Contaminated land — 229

11.1	Contaminated sites	229
11.2	UK policy on contaminated land	229
11.3	Risk assessment	230
11.4	Industrial processes	233

11.4.1	Asbestos	235
11.4.2	Scrap yards	235
11.4.3	Sewage treatment works	235
11.4.4	Timber manufacturing and timber treatment works	235
11.4.5	Railway land	236
11.4.6	Petrol stations and garage sites	236
11.4.7	Gasworks sites	236
11.4.8	Metal smelting works	236
11.4.9	Old mineral workings	236
11.4.10	Toxicological effects of contaminants	238
11.5	Landfill sites	239
11.5.1	Gas migration	240
11.5.2	Gas monitoring	241
11.5.3	Carbon dioxide	241
11.5.4	External measures	242
11.6	Desk study	244
11.6.1	Local geological study	244
11.6.2	Industrial history of the site	244
11.6.3	Mining investigation	244
11.6.4	Site reconnaissance	244
11.7	Site investigation	245
11.7.1	Trial pits	245
11.7.2	Boreholes	245
11.7.3	Testing for toxic gases	245
11.7.4	Chemical analysis	245
11.7.5	Safety	245
11.7.6	Conclusion	245
Case study 11.1		246
Bibliography		248

Index — 251

Preface

All structures resting on the earth must be carried by a foundation. The foundations are therefore the most important part of the structure as they are the element which transfers the superstructure loads into the soil mass. Foundation engineering is not an exact science, and engineers involved in foundation design have to use their training, experience and engineering judgement in formulating economic foundation solutions.

The design of foundations for low-rise buildings is just as important as that required for larger buildings, the only difference being one of scale. The correct engineering solution must be economic and practical if the builders are to maintain profit margins.

This book is about providing suitable foundations for low-rise buildings where ground conditions are classed as hazardous. It deals primarily with housing construction, but much of the information is relevant to the design and construction of other categories of building. It has been written for those professionals involved in building design and is intended to be a source of guidance for builders, architects, engineers, building inspectors and site agents for dealing with foundation problems in which they themselves are not specialists.

The book includes state-of-practice methods together with a mixture of 'how to' with worked examples and case histories covering a broad range of foundation problems. Its special feature is the interfacing of foundation designs and their 'buildability'. The realistic examples included are drawn from the author's extensive experience in civil and structural engineering in dealing with hazardous and marginal ground conditions.

I am indebted to Mr G. Studds for his assistance in carrying out the revisions to Chapter 11 on contaminated land.

In preparing this book I am particularly indebted to my colleague Mr J. Hine for his help in checking the calculations. I am also grateful to the following for permission to reproduce diagrams and other material from their publications: the National House-Building Council (NHBC) and the Construction Industry Research and Information Association (CIRIA). Extracts from British Standards and Codes of Practice are reproduced by permission of the British Standards Institution, 2 Park Street, London W1, from whom copies of the complete Standards and Codes of Practice can be obtained.

Examples

		Page
2.1	Semi-rigid raft design	35
2.2	Structural calculations for three-storey flats	40
2.3	Disused well	45
2.4	Bored piles	49
2.5	Detached house: pile and ground beam design	53
3.1	Strip footing on clay soil	71
3.2	Settlements on clay soil	73
3.3	Triaxial test	78
4.1	Strip footing on granular soil	87
4.2	Pad foundation on sand	88
4.3	Strip footing on sand, high water table	88
4.4	Bearing pressure of granular soil	88
4.5	Bored piles	91
4.6	Working load of precast concrete piles	92
4.7	Steel piles	93
4.8	Driving precast concrete piles	93
5.1	Calculating the reinforcement in a mining raft	104
5.2	Designing the reinforcement in an irregular shaped dwelling	107
5.3	Reinforcement mesh	108
5.4	Calculating movement joint sizes	109
6.1	Foundations on clay soil with young poplar trees	116
6.2	Strip foundations on clay soil with mature and semi-mature trees	127
6.3	Foundations on a site with mature trees, subject to settlement	128
6.4	Site investigation and foundation design, heavily wooded site	129
6.5	Foundation design on a site with mature trees, subject to heave	134
7.1	Thrust on a retaining wall	155
7.2	Pressures and bending moments on a retaining wall	156
7.3	Thickness of retaining wall	157
7.4	Pocket-type retaining wall	160
7.5	Reinforced cavity retaining wall	161
7.6	Concrete filled cavity retaining wall	162
7.7	Brick retaining wall	165
7.8	Reinforced concrete retaining wall to BS 8002	167
8.1	Suspended floor slab over filled ground	180
9.1	Design of stone columns	194
9.2	Improving mixed clay fills	194
9.3	Partial depth treatment of filled ground	196
9.4	Dynamic consolidation	200
10.1	Foundations beside existing building	205

Chapter 1
Site investigations

Before purchasing a site a builder must establish whether there are any hazards on or below the site which could result in expensive house foundations.

Remember: you pay for a site investigation whether you have one done or not. In other words, the consequences of inadequate site investigation could be at least as expensive, and most probably much more so, than having one carried out.

The builder should visit the site and make enquiries. The most important thing he should do is carry out a pre-liminary ground investigation before making an offer for the land. This investigation may be no more than simple trial pits, mechanically excavated to depths of approximately 3 m.

If there are hazardous ground conditions the builder may have to consult a chartered structural or chartered civil engineer to have special foundations designed and costed before purchasing the land. Not all ground hazards require engineer-designed foundations. Sometimes it may just be a simple case of placing the strip footings deeper into firmer strata or catering for the effect of existing trees by implementing the requirements of the *NHBC Standards* Chapter 4.2. These are decisions which any experienced builder should be able to make. However, the builder must recognize when a specialist needs to be consulted.

Site investigation should be split up into three sections:

- a walk-over survey (very important);
- an initial desktop study;
- field investigations using trial pits or boreholes.

1.1 WALK-OVER SURVEY

This is an essential part of a site investigation and it should always be carried out. The Technical Questionnaire Survey Sheet (Fig. 1.1) should be used as an *aide-mémoire* when carrying out the walk-over survey.

The object of the survey is to check and supplement information gleaned from the desktop study and the field

TECHNICAL QUESTIONNAIRE SURVEY SHEET

Site address *Green Lane, York* Date visited *16:1:92*

1. Give description of site. — *Pasture land*

2. Give direction and gradient of slopes. Are there any indications of hill creep? — *Site fairly level No*

3. Presence of trees, streams, marshy areas etc. — *Yes. A copse of beech trees in SE corner*

4. Are there any soil exposures in trenches, nearby quarries etc.? — *No*

5. Is there any indication of fill being present? — *No*

6. Has the site been previously developed? Give any indications of depth and type of existing foundations. — *No*

7. Are any old basements present? If so, give details of extent and depth. — *No*

8. Have you taken ground levels and obtained the location of boreholes and trial pits? — *Yes*

9. Are there nearby buildings? Any signs of cracking? Any information on foundations of such buildings? — *Yes. No visible signs of damage*

10. Is the site in a mining area? — *No*

11. Have you obtained all information on soil conditions from the trial pits? Depth and description of strata. Level to the groundwater. — *Yes No groundwater*

12. Have you obtained shear strengths using the shear vane? — *Yes*

13. What samples have been taken? — *4 No. bulk samples*

14. Have ground water samples been taken? From which pits? — *None encountered*

Fig. 1.1 Technical questionnaire survey sheet.

investigation by observing the site topography. Information can be gleaned from local inhabitants and people working in the area, i.e. for statutory authorities such as gas, water and electricity. Any evidence of likely ground instability should be noted, such as the following.

• Surface depressions	Past mining or quarries (Fig. 1.2)
• Hummocky terraced ground	Slope instability, hill creep (Fig. 1.3)
• Existing vegetation and trees	Clay desiccation
• Railway cuttings	Exposed strata (Fig. 1.4)
• Spoil heaps	Possible mine shafts or bell pits (Fig. 1.5)
• Existing structures	Record any visible damage
• Slope of site	Will retaining walls be required? Signs of landslip
• Groundwater, springs etc.	Could construction problems arise?
• Surface hollows	May be solution features in chalk (Fig. 1.6) May be collapsed mine workings
• Reeds, bullrushes etc.	Generally indicate peaty or wet ground
• Manholes	Old sewer lines, infilled watercourse
• Existing dwellings	Check if there are any cellars
• Surface cracking	Indicative of soils plasticity
• Gaps in existing houses	Why? Could be sewers, mineshaft (Fig. 1.7)
• Street names	Quarry Lane, Coal Pit Lane?
• Flooded areas	Construction problems; expensive land drainage
• Site bouncy under foot	Indicative of high water table

1.2 DESK STUDY

This involves collecting as much information as possible about the site. Sources include geological maps, Ordnance

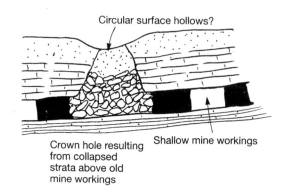

Fig. 1.2 Crown holes.

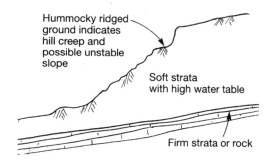

Fig. 1.3 Slope stability.

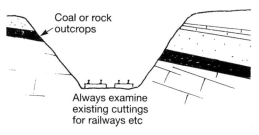

Fig. 1.4 Coal outcrops.

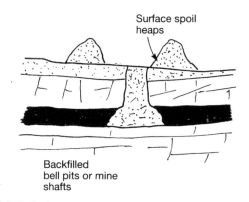

Fig. 1.5 Bell pits.

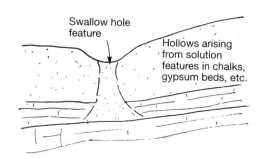

Fig. 1.6 Solution features.

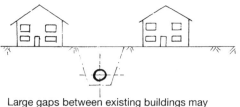

Fig. 1.7 Gaps in existing development.

Survey maps, geological memoirs, aerial photographs, mining records, and previous site investigations for the site or adjacent sites. Most of this information can be obtained at a relatively low cost from public libraries and local sources.

- Old Ordnance plans indicate old ponds, old hedge lines, and old watercourses. Evidence of previous buildings — old mill buildings for example — can be defined. All these features could have an effect on the proposed development.
- Geological memoirs and maps give information on mine shafts, coal outcrops and strata succession (the different levels and types of strata).
- Aerial photographs give valuable information relating to previous trees or hedges on the site. They can also reveal

old quarries, old landslips, or depressions caused by old mine shafts or infilled quarries.

- Mining reports: Always consult British Coal who have detailed plans and knowledge of the local coalfield together with mine shaft records.
- British Rail: Where a railway tunnel passes below the site, construction shafts may be present. Details of these may be held by British Rail engineers.
- District mineral valuers: They have useful information on old mines, quarries and mineral workings.

Always consult with the NHBC regional engineers and local authority building inspectors for information on and adjacent to the site. Consult with all the statutory service undertakings in respect of any existing services which may be on the site to be investigated.

DESK STUDY CHECKLIST

1. **Site topography**, salient vegetation and drainage
   (a) Are there any springs, ponds, or watercourses on or near the site?
   (b) Is the site steeply sloping?
   (c) Are there any signs of previous tree growth on site?
2. **Ground conditions**
   (a) Is the site in a known mining area for coal or other minerals?
   (b) What is the geological strata succession below the site?
   (c) Are the clays in the high plasticity range?
   (d) Is there any evidence of slope instability?
   (e) What information is available on the ground strengths?
3. **Proposed housing**
   (a) What type of development is intended?
   (b) What foundation loads are expected?
   (c) What soils investigation is required to enable special foundations to be designed, e.g. piling, dewatering?
4. **Identification of ground conditions**. By consulting geological records and maps a lot of information can be obtained about the site conditions below the surface.
   (a) Type of drift material, e.g. sands, clays, shales.
   (b) Thickness of various strata bands. Usually indicated on borehole records and geological maps.
   (c) Positions of old coal pits, old mine shafts, geological faults, whinstone dykes, buried glacial valleys.

For low-rise housing (two to three storeys) the foundations generally require ground bearing capacities in the 50–100 kN/m^2 range when using strip footings. Settlements under such footings are generally of a low order and less than 25 mm.

In general, it is only when soft alluvium, peaty strata and filled ground are encountered that settlements can reach unacceptable magnitudes. Alluvial soils, peats and glacial head deposits often have considerable variations in stratum thickness over relatively short distances. Peat beds are often in a lenticular form or can be found in old infilled valley formations.

Sands, gravels and rock strata usually provide excellent bearing capacities with settlements taking place over a short timescale. The effect of groundwater on sands and gravels is to reduce the ground bearing capacity and allowance should be made for variations likely because of seasonal variations.

Site investigations carried out in the summer months may not reveal water problems encountered in the period from October to February periods.

Where clay strata are known to be present the plasticity index of the clays needs to be determined, especially if there are trees on site or trees have been removed from the site. This information is required in order that foundations can be designed in accordance with the NHBC Chapter 4.2, formerly NHBC Practice Note 3.

Information about the site and its underlying stability may have an influence on the type of housing layout and the construction programme: avoiding long terrace blocks in mining areas, provision of public open space in any no-build zones, and providing movement joints at changes in storey heights, for example.

Where British Coal opencast plans or old quarry record plans show no-build zones at the high wall batter planes the housing layout may need to be modified to avoid placing houses in these areas if the site is to be developed economically.

1.3 SITE INVESTIGATION: FIELD WORK

For low-rise housing it is generally considered that simple trial pits are the most useful and economic way of determining ground conditions. When excavated, they should be deep enough to confirm that the strata below the proposed foundation level remain competent and adequate for at least a further metre below a standard-width footing of 600 mm. The minimum depth of the trial pits should range from 2.5 to 3.0 m and where possible they should be at least 3.0 m from proposed dwellings. If the position of dwellings has not been finalized then the trial pits should be accurately located for the site survey.

If an adequate formation depth is not encountered within the reach of a mechanical excavator then a larger machine may be required. If this is likely to be too disruptive, and *in-situ* field tests are needed, then shell and auger boreholes will need to be drilled.

If poor ground conditions are evident during the drilling of boreholes then the boreholes should be taken well below any suitable strata which may have to support piles.

1.3.1 Trial pit logs

The following data should be recorded in the trial pit logs together with a site plan showing an accurate location of the trial pits (Fig. 1.8):

- depth and nature of the strata encountered: e.g. stony clays, wet loose sands;
- description of the various strata in regard to their strength: e.g. firm to stiff clays, loose, medium dense or dense sands;
- depths at which groundwater entered the trial pit and the final depth at which it stabilized;
- whether the sides of the trial pit collapsed, either during excavation or shortly after excavation (in sands and gravels this can be a good indication of the state of compaction).

During the excavation of the trial pit the following information should be taken.

- Penetrometer or shear vane results enable the undrained shear values for clay strata to be determined. These values when doubled will give an approximate allowable bearing pressure allowing for a factor of safety of 3.0.
- Clay samples can be obtained for laboratory testing (1.0 kg bulk samples). These can be tested for natural moisture content, plastic limit, liquid limit, plasticity index, soluble sulphates and pH (acidity).
- If filled ground is evident, this should be described: e.g. topsoil, vegetation, organic refuse, rags, metal, timber. If possible the percentage of organic matter such as timber should be indicated.

1.3.2 Borehole record

This should indicate the following information (Fig. 1.9).

- Ground level at the borehole position. This is very important, but regretably it is often omitted. When the investigation is being done to assess the site stability for possible shallow coal workings, it is essential that ground levels are known if accurate strike lines of the seam are to be drawn.
- Depths and description of the various strata encountered.
- Depths of any water entries and the final standing level.
- Piezometer readings if applicable.
- Depths at which various samples or field tests are taken, e.g. standard penetration tests or U-100 mm samples (samples obtained undisturbed from the borehole).

Figure 1.10 shows the standard key symbols for annotating strata variations.

1.4 SITE INVESTIGATION PROCEDURE

Site investigations should be carried out in accordance with BS 5930 (1999). The tests carried out on samples obtained in the fieldwork are to be in accordance with the procedures laid down in BS 1377 (BSI, 1990).

1.4.1 Borehole logs

These show the ground conditions only at the position of the borehole. Ground conditions between boreholes can be correlated but the accuracy cannot be guaranteed.

1.4.2 Trial pit logs

These enable soil conditions to be examined in more detail and bulk sampling to be carried out. The stability of the trench sides and details of groundwater ingress provide useful indicators of the soil strengths.

1.4.3 Groundwater

Groundwater levels can vary because of seasonal effects and some allowance may be required depending on the time of year when the works are carried out. The use of temporary casings may seal off seepage into the borehole.

1.4.4 Standard penetration tests

These are generally used in sands and gravel strata but can also be taken in cohesive soils and made ground. They can also be performed in soft or weathered rock strata. (See Chapter 4 for a description of the standard penetration test.) Because of the difficulty in obtaining accurate N values especially in filled strata containing large stones, the results should only be used as a guide when assessing the ground strength. Tables 1.1 and 1.2 give an approximate guide to soil strengths based on SPT values.

	From	To	Key	Description of strata	
	GL	0.30		MADE GROUND. comprising loose bricks, lumps of concrete, soil stones and fragments of timber	
	0.30	1.70		MADE GROUND, comprising moderately compact dark black-brown clayey fine to coarse-grained stony ashes, with stones and firm brown clay. Brick footings encountered down to 0.80 m.	
	1.70	2.10		MADE GROUND, comprising firm medium grey brown silty clay with black ashes and stones	
	2.10	2.50		Soft to firm medium to dark grey silty CLAY.	
	2.50	2.70		Firm medium grey-brown silty CLAY with friable mudstone fragments	
	2.70	3.00		Firm to stiff medium grey-brown CLAY with sandstone fragments	
		3.00		End of trial pit	

SITE SURVEYS LTD

TRIAL PIT RECORD

No. 3.

Date: 16 MARCH 1991

Client: C.M. JACKSON (BUILDERS) LTD

Site Address: Rowan Avenue, Leeds, Yorkshire

Location: Phase 4

Test results, water observations and remarks

Slow ground water ingress at 2.00 m
Bulk samples taken at 0.75m & 1.50m
Vane test readings at 2.20 m & 2.50m from bulk sample.

Fig. 1.8 Typical trial pit log.

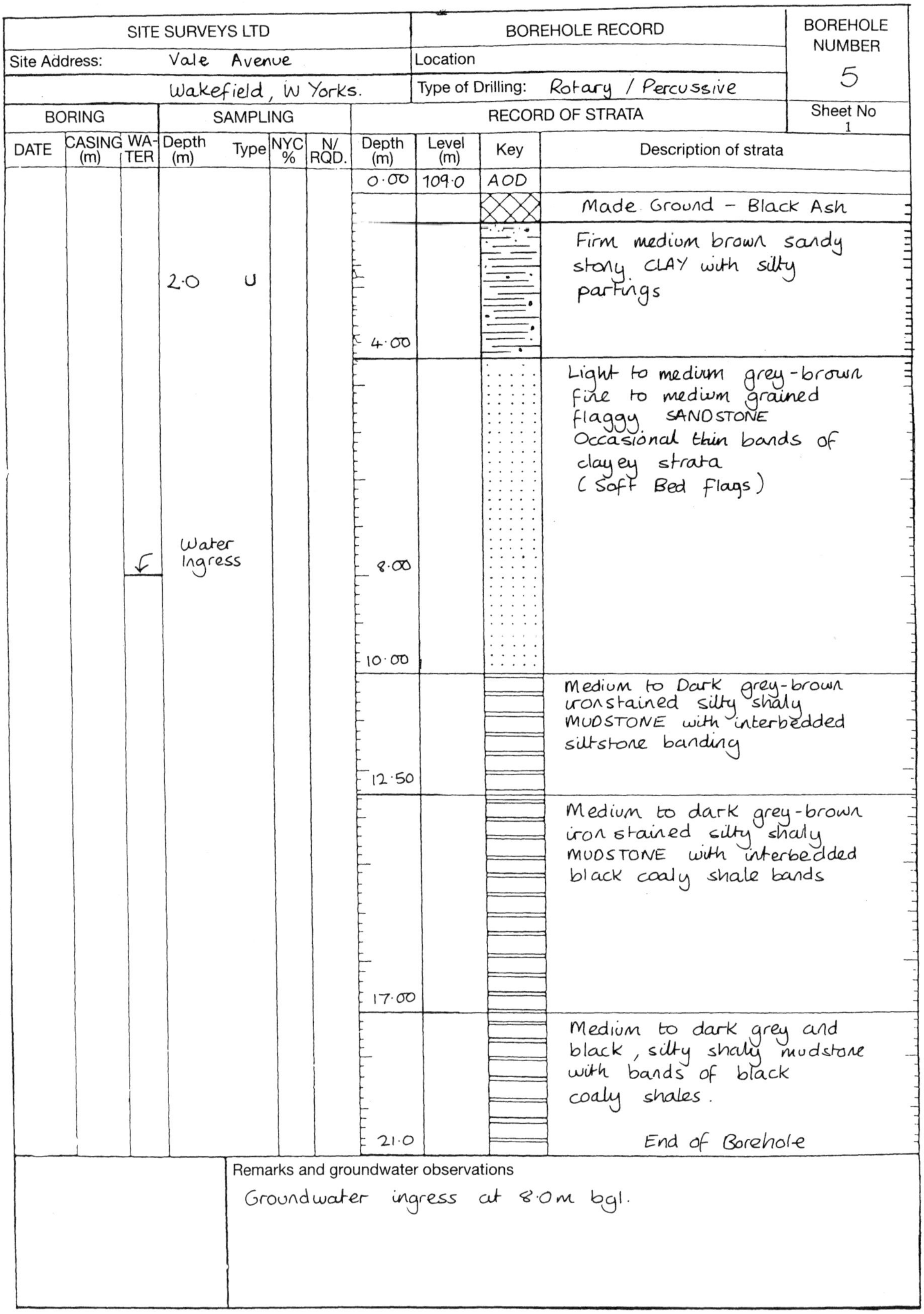

SITE SURVEYS LTD							BOREHOLE RECORD			BOREHOLE NUMBER
Site Address:			Vale Avenue				Location			5
			Wakefield, W Yorks.				Type of Drilling:	Rotary / Percussive		
BORING			SAMPLING				RECORD OF STRATA			Sheet No 1
DATE	CASING (m)	WA-TER	Depth (m)	Type	NYC %	N/RQD.	Depth (m)	Level (m)	Key	Description of strata
							0·00	109·0	AOD	
										Made Ground – Black Ash
			2·0	U						Firm medium brown sandy stony CLAY with silty partings
							4·00			
										Light to medium grey-brown fine to medium grained flaggy SANDSTONE Occasional thin bands of clayey strata (Soft Bed Flags)
		⌄	Water Ingress				8·00			
							10·00			
										Medium to Dark grey-brown iron stained silty shaly MUDSTONE with interbedded siltstone banding
							12·50			
										Medium to dark grey-brown iron stained silty shaly MUDSTONE with interbedded black coaly shale bands
							17·00			
										Medium to dark grey and black, silty shaly mudstone with bands of black coaly shales.
							21·0			End of Borehole

	Remarks and groundwater observations
	Groundwater ingress at 8·0 m bgl.

Fig. 1.9 Typical borehole log.

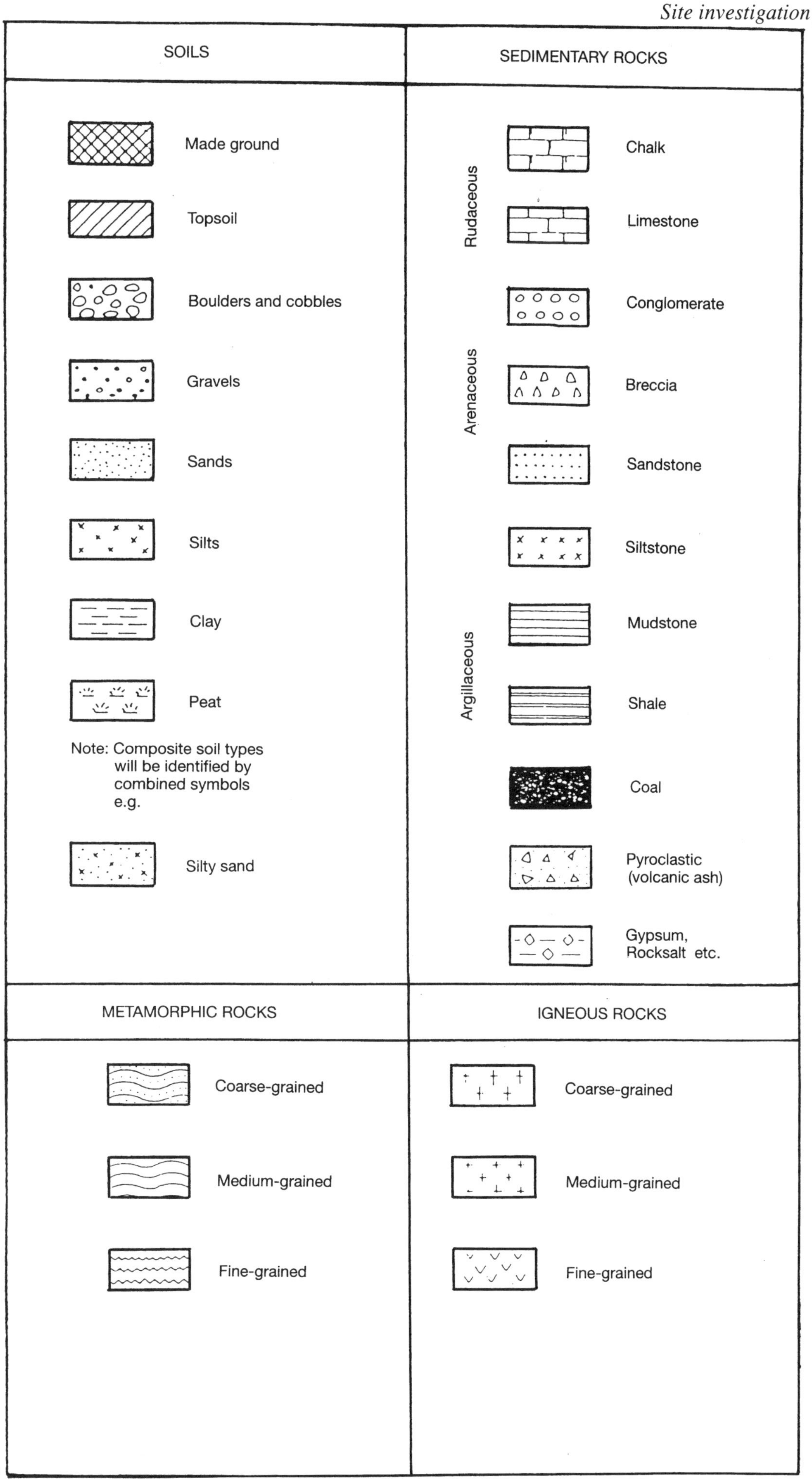

SOILS	SEDIMENTARY ROCKS

SOILS

Made ground

Topsoil

Boulders and cobbles

Gravels

Sands

Silts

Clay

Peat

Note: Composite soil types will be identified by combined symbols e.g.

Silty sand

SEDIMENTARY ROCKS

Rudaceous

Chalk

Limestone

Arenaceous

Conglomerate

Breccia

Sandstone

Argillaceous

Siltstone

Mudstone

Shale

Coal

Pyroclastic (volcanic ash)

Gypsum, Rocksalt etc.

METAMORPHIC ROCKS

Coarse-grained

Medium-grained

Fine-grained

IGNEOUS ROCKS

Coarse-grained

Medium-grained

Fine-grained

Fig. 1.10 Key to soil symbols in trial pit logs.

LABORATORY RESULTS: SYMBOLS

B	Bulk sample: disturbed	SO_3	Soluble sulphate content
c	Cohesion (KN/m^2)	U	Undisturbed 100 mm dia. samples
c'	Effective cohesion intercept (kN/m^2)	U	Undrained triaxial compression test
CBR	California Bearing Ratio	W	Water sample
D	Jar sample: disturbed	w	Natural moisture content (%)
LVT	Laboratory vane test	W_L	Liquid limit (%)
m_v	Coefficient of volume decrease (m^2/kN)	W_p	Plastic limit (%)
N	Standard penetration test value (blows / 300 mm)	γ_b	Bulk density (kg/m^3)
		γ_d	Dry density (kg/m^3)
pH	Acidity / alkalinity index	ϕ	Angle of shearing resistance (degrees)
PI	Plasticity index (%)	ϕ'	Effective angle of shearing resistance (degrees)

Soil strengths. Soil tests used on cohesive strata fall into two main categories. The undrained triaxial test is carried out on a soil which is stressed under conditions such that no changes occur in the moisture content. This reproduces the

Table 1.1. SPT values for cohesive soils

Consistency	Undrained shear strength (kN/m^2)	N value
Very stiff	> 150	> 20
Stiff	75–150	10–20
Firm	40–75	4–10
Soft	20–40	2–4
Very soft	< 20	< 2

Table 1.2. SPT values for sands and gravels

Consistency	N value
Very dense	> 50
Dense	30–50
Medium dense or compact	10–30
Loose	4–10
Very loose	< 4

Table 1.3. Field assessment of soil strengths

Consistency of soil	Method of testing	Approximate undrained shear strength (kN/m^2)
Very soft	Exudes between fingers when squeezed in one's hand	< 20
Soft	Moulded by light finger pressure	20–40
Firm	Moulded by strong finger pressure	40–80
Stiff	Indented by thumb pressure	80–150
Very stiff	Indented by thumbnail	150–300
Hard	Difficult to indent with thumbnail	> 300

Table 1.4. Typical ground bearing capacities

Types of rock and soil	Maximum safe bearing (kN/m^2) capacity	
Rocks		
Igneous and gneissic rocks in sound condition	10 700	
Massively bedded limestones and hard sandstones	4300	
Schists and slates	3200	
Hard shales, mudstones and soft sandstones	2200	
Clay shales	1100	
Hard solid chalk	650	
Thinly bedded limestones and sandstones	[a]	
Heavily shattered rocks	[a]	
Non-cohesive soils[b]	Dry	Submerged
Compact, well-graded sands and gravel–sand mixtures	430–650	220–320
Loose, well-graded sands and gravel–sand mixtures	220–430	110–220
Compact uniform sands	220–430	110–220
Loose uniform sands	110–220	55–110
Cohesive soils[c]		
Very stiff boulder clays and hard clays with a shaly structure	430–650	
Stiff clays and sandy clays	220–430	
Firm clays and sandy clays	110–220	
Soft clays and silts	55–110	
Very soft clays and silts[d]	55–nil	
Peats and made ground[d]		

[a] To be assessed after inspection
[b] With granular soils the width of foundation is to be not less than 900 mm. 'Dry' means that the goundwater level is at a depth not less than 900 mm below the foundation base
[c] Cohesive soils are susceptible to long-term consolidation
[d] To be determined after investigation
Source: BSI (1986) BS 8004.

conditions most likely to occur under the actual foundations. The undrained test gives the apparent cohesion C_u and the angle of shearing resistance ϕ_u. With saturated non-fissured clays ϕ_u tends to zero and the apparent cohesion C_u is equal to one-half of the unconfined compression strength q_u. The drained triaxial test can also be carried out using the triaxial compression apparatus and is known as the slow test.

The approximate consistency of a clay soil can be assessed in the field by handling the sample. Table 1.3 lists the criteria used in arriving at approximate values for the undrained shear strength.

Table 1.4 gives typical ground bearing capacities for various soils; these figures should only be used as a guide.

1.5 INTERPRETATION OF LABORATORY TESTING

Once the results of the laboratory tests are determined they can be used in conjunction with field tests to determine the strength of the soil strata below the site. By examining the borehole logs it is possible to build up a picture of the stratification of soils below the foundations.

Fig. 1.10 shows the standard symbols adopted in soil reports carried out in accordance with BS 5930 (BSI, 1981) and tested in accordance with BS 1377 (BSI, 1990).

Generally, suitable 100 mm samples are subjected to quick undrained triaxial compression testing using three specimens, 38 mm dia by 76 mm long, taken from each sample. The samples are tested under lateral pressures of 70, 140 and 210 kN/m² at a constant rate of strain of 2% per minute. The results of these tests, comprising shear strength parameters, apparent cohesions, angle of internal friction ϕ, natural moisture contents, field dry density and Atterberg limits can then be determined.

Table 1.5 illustrates the results of several quick undrained compression tests. When the angles of internal friction are indicated by the test results then **Mohr circle stress diagrams** are drawn (Fig. 1.11). From these values the shear strength parameters can be determined. The allowable bearing capacities can then be determined using the appropriate factors of safety.

Table 1.5. Compression test results

Borehole no.	Depth (m)	Cohesion (kN/m²)	Angle of internal friction (degrees)	Natural moisture content	Bulk density
5	0.50*	45	4	30	2.018
5	1.50*	20	6	24	2.050
5	2.0	51	11	11	2.218
5	3.0	100	13	8	2.103
7	0.30	46	9	22	2.035
7	0.80	50	4	18	2.090
7	1.40*	112	10	21	2.070
7	2.30*	138	0	14	2.195

*Indicates test results used to produce the Mohr stress circles in Fig. 1.11.

Using the bulk samples the Atterberg limits can be determined and the values obtained for liquid limit, and plasticity index can be plotted on the standard **Casagrande plasticity chart** (Fig. 1.12) to confirm the clay classification.

If the bulk samples are sands or gravel mixtures, grading tests will be carried out to determine the **particle size distribution**. The results are plotted on a chart (Fig. 1.13) which indicates which zone the materials fall into.

These tests will not reveal the bearing strength of the materials but they are of particular value when considering ground improvement techniques such as vibro-compaction, dewatering or grouting operations.

1.5.1 Chemical tests

The tests generally carried out are those which determine whether there are any substances in the ground which could be aggressive to Ordinary Portland Cement. The most common are the **soluble sulphur trioxide** (SO_3) and the **hydrogen ion pH** tests. Table 1.6 lists the classification of sulphates in soils and groundwater.

BRE Digest 363 (BRE, 1991) gives various recommendations for the design of suitable concrete mixes to resist sulphate attack. These are listed in Table 1.7. In addition to using the required minimum cement contents and water cement ratios the concrete should be dense, well cured and of low permeability to provide the maximum resistance to chemical attack. These recommendations only apply to concrete with a low workability.

When examining a number of tests carried out on a site, the greater emphasis should be placed on those samples

Table 1.6. Classification of sulphates in soils and groundwater

Class	Total SO_3 (%)	SO_3 in 2:1 aqueous extract (g/litre)	In groundwater (g/litre)
1	< 0.20	< 1.0	< 0.30
2	0.20–0.50	1.0–1.90	0.30–1.20
3	0.50–1.0	1.90–3.10	1.20–2.50
4	1.0–2.0	3.10–5.60	2.50–5.0
5	> 2.0	> 5.60	> 5.0

which fall into the higher classification. For example, if six out of ten samples are found to be within the Class 1 range and are hence non-aggressive, while the other four fall into Class 2, then the site should be placed into the Class 2 category of risk.

The pH values of soils and groundwater can be critical when determining concrete mix designs. Table 1.8 gives the acidity classification for various pH values. Any foundations placed in soils with organic acids of pH values less than 6 will need to have special precautions taken to ensure the durability of the concrete used below ground.

Table 1.7. Requirements for well compacted cast-*in-situ* concrete 140–450 mm thickness exposed on all vertical faces to a permeable sulphated soil or fill materials containing sulphates

	Concentration of sulphate and magnesium							
	In soil or fill							
Class	By acid extraction (%)	By 2:1 water/soil extract		In groundwater (g/l)		Cement type	Minimum cement content (kg/m³) Notes 1 and 2	Maximum free water/cement ratio Note 1
	SO_4	SO_4	Mg	SO_4	Mg			
1	<0.24	<1.2		<0.4		A–L	*Note 3*	0.65
2		1.2–2.3		0.4–1.4		A–G	330	0.50
						H	280	0.55
						I–L	300	0.55
3	If >0.24 classify on basis of 2:1 extract	2.3–3.7		1.4–3.0		H	320	0.50
						I–L	340	0.50
4		3.7–6.7	<1.2	3.0–6.0	<1.0	H	360	0.45
						I–L	380	0.45
		3.7–6.7	>1.2	3.0–6.0	>1.0	H	360	0.45
5		>6.7	>1.2	>6.0	>1.0	As for Class 4 plus surface protection		
		>6.7	>1.2	>6.0	>1.0			

Note 1 Cement content includes pfa and slag.
Note 2 Cement contents relate to 20 mm nominal maximum size aggregate. In order to maintain the cement content of the mortar fraction at similar values, the minimum cement contents given should be increased by 40 kg/m³ for 10 mm nominal maximum size aggregate and may be decreased by 30 kg/m³ for 40 mm nominal maximum size aggregate as described in Table 8 of BS 5328: Part 1.
Note 3 The minimum value required in BS 8110: 1985 and BS 5328: Part 1: 1990 is 275 kg/m³ for unreinforced structural concrete in contact with non-aggressive soil. A minimum cement content of 300 kg/m³ (BS 8110) and maximum free water/cement ratio of 0.60 is required for reinforced concrete. A minimum cement content of 220 kg/m³ and maximum free water/cement ratio of 0.80 is permissible for C20 grade concrete when using unreinforced strip foundations and trench fill for low-rise buildings in Class 1.
Source: Based on *BRE Digest 363*, July 1991.

Table 1.8 pH Values

Classification of acidity	pH value
Negligible	> 6.0
Moderate	5.0–6.0
High	3.50–5.0
Very high	< 3.50

The pH values should also be taken into account when the sulphate contents are borderline between the various classes. If the pH is less than 6 the sample should be placed in the more severe class of sulphate classification. Table 1.9 relates the acids in groundwater to the probable aggressiveness to ordinary Portland cement.

Table 1.9. Acids in groundwater: probable aggressiveness to ordinary Portland cement concretes
(This table is intended as a broad guide only)

Category	pH range	Comments
(a)	7.0–6.5	*Attack probably unlikely*
(b)	6.5–5.0	*Slight attack probable* Where the pH and chemical analysis of the groundwater suggests that some slight attack may occur, and if this can be tolerated, then the concrete should be fully compacted, made with either ordinary Portland cement or Portland blast furnace cement and aggregates complying with BS 882 (BSI,1973). The maximum free water/cement ratio should be 0.50 by weight with a minimum cement content of 330 kg/m³.
(c)	5.0–4.50	*Appreciable attack probable* For conditions where appreciable attack is probable, the concrete should be fully compacted, made with either ordinary Portland cement or Portland blast furnace cement and aggregates complying with BS 882. The maximum free water/cement ratio to be 0.45 by weight with a minimum cement content of 370 kg/m³.
(d)	< 4.50	*Severe attack probable* Where severe attack is likely, the concrete should comply with the requirements of category (b) but should have the lower water/cement ratio of 0.45. In addition the concrete should be physically protected by coatings of bitumen, asphalt, or other inert materials reinforced with a glass-fibre membrane. High alumina and supersulphated cements are generally recognized to be more acid-resisting than concretes made with ordinary Portland cements.

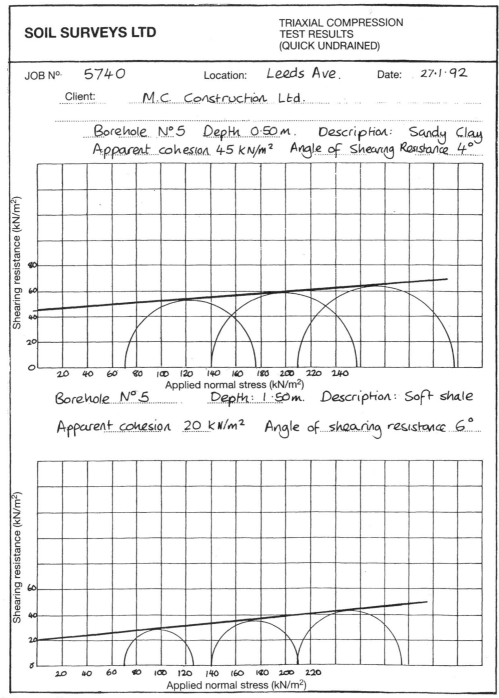

SOIL SURVEYS LTD

TRIAXIAL COMPRESSION
TEST RESULTS
(QUICK UNDRAINED)

JOB N°. 5740   Location: Leeds Ave.   Date: 27·1·92

Client: M.C. Construction Ltd.

Borehole N°5   Depth 0·50m.   Description: Sandy Clay
Apparent cohesion 45 kN/m²   Angle of Shearing Resistance 4°

Borehole N°5   Depth: 1·50m.   Description: Soft shale

Apparent cohesion 20 kN/m²   Angle of shearing resistance 6°

Fig. 1.11 Mohr circle stress diagram.

1.6 SOLUTION FEATURES

Damage to structures can often occur because of severe subsidence caused when certain types of strata below ground are affected by water and become soluble. Materials such as salt, gypsum, chalk and certain limestones can all be affected in this way.

1.6.1 Limestones

Limestones are generally very jointed and the action of acidic groundwaters can produce solution features known as

sink-holes. These sink-holes generally develop where joints in the limestone intersect, and they can result in large open galleries. The size of these galleries and sink-holes depends on the geological structure and the existence of layers of impervious strata.

1.6.2 Chalk

Chalk is a pure soft limestone, which usually contains approximately 95% calcium carbonate. Solution features do not generally form underground in the form of caverns as the chalk being softer than limestone collapses as it goes into

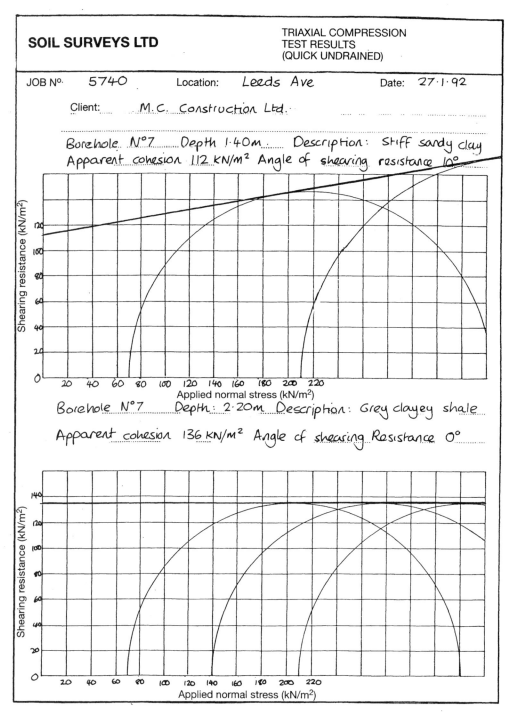

SOIL SURVEYS LTD

TRIAXIAL COMPRESSION
TEST RESULTS
(QUICK UNDRAINED)

JOB N°· 5740   Location: Leeds Ave   Date: 27·1·92

Client: M.C. Construction Ltd.

Borehole N°7   Depth 1·40m   Description: Stiff sandy clay
Apparent cohesion 112 kN/m² Angle of shearing resistance 10°

Borehole N°7   Depth: 2·20m Description: Grey clayey shale
Apparent cohesion 136 kN/M² Angle of shearing Resistance 0°

Fig. 1.11 continued.

solution. However, solution piping and swallow holes are often found in chalk deposits. Where chalk mining has taken place, in areas such as Norwich and Bury St Edmunds, subsidence in the form of sink-holes is common if the deposits are relatively shallow.

Swallow holes and sink-holes in carbonate rock strata can be very difficult to determine using standard site investigation methods. In known areas of subsidence, detailed research of the available geological maps is recommended and this may need to be followed up with specialized site investigation. If this investigation reveals unstable strata it may be necessary to carry out a check drill and grouting scheme on a 3.0 m grid spacing. If no evidence of unstable strata is found in the site investigation and there are carbonate strata still intact, it is generally prudent to provide a stiff raft foundation.

Figure 1.14 provides a decision flow chart for sites on chalk.

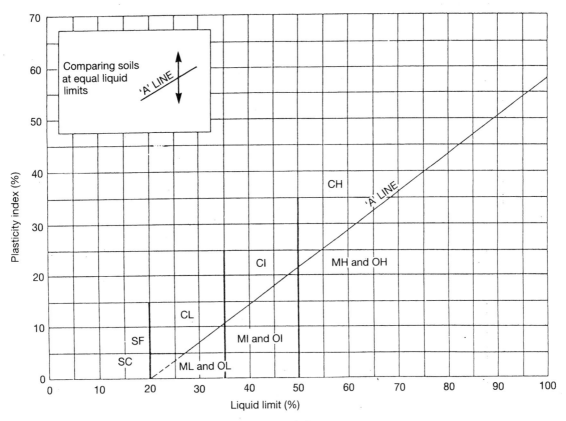

Fig. 1.12 Casagrande plasticity chart (for key to abbreviations see Table 3.6).

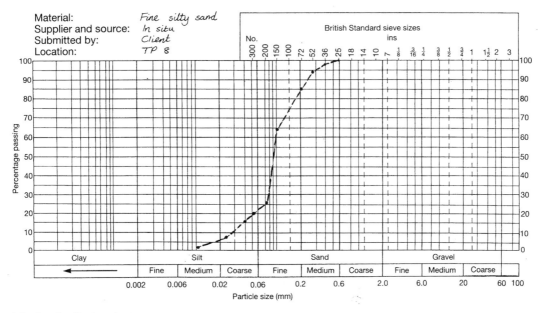

Fig. 1.13 Particle size distribution chart.

1.6.3 Salt

Salt is extracted from below ground using solution mining techniques and brine pumping is carried out in several parts of the UK, particularly Cheshire and Cleveland. Subsidence arising from brine extraction can be very unpredictable.

Wild brine extraction takes place around Northwich, Middlewich, Sandbach, and Lymm and accounts for about 10% of the brine extracted in Cheshire. The remaining brine production is from controlled solution mining from cavities formed in the salt strata. The effects of this type of extraction have not resulted in major subsidence.

Buildings constructed in the areas covered by the Brine Extraction Board are generally placed on raft foundations and it is usual to discuss the foundation proposals with the Board.

1.6.4 Gypsum

Gypsum is more readily soluble than soft limestone, and sink-holes and large caverns can therefore develop in thick beds, especially when water is a dominant factor. The most significant areas are in Ripon in Yorkshire and the surrounding areas north of Ripon where subsidence hollows have been recorded along the outcrop of the gypsum beds within the marl strata. Extensive geological mapping of these solution features has been carried out by the British Geological Survey. There is evidence that shows that a lot of subsidence took place many thousands of years ago, but there are areas where gypsum is still present and could be a problem in the future.

From the statistical evidence available it appears that one building in a hundred may be at risk in the Ripon area and it is considered prudent to provide stiff raft foundations in the floundering area designated by the British Geological Survey. Old subsidence hollows have often been filled with organic silts and peats and in these situations the only solution is to use a piled raft with the piles taken below the gypsum.

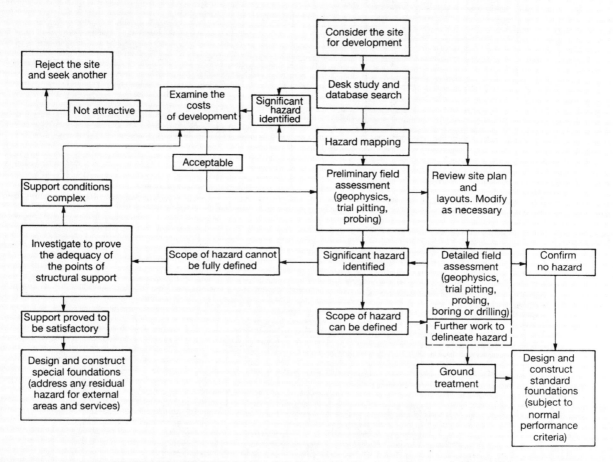

Fig. 1.14 Decision flowchart for sites on chalk (Edmonds and Kirkwood, 1989).

Case study 1.1   Investigation of former mining site, Sheffield

INTRODUCTION

A large tract of land is to be offered for sale by treaty to a major housebuilder and developer for future residential development. The site is 11 km north of Sheffield city centre, South Yorkshire, and the site location north of Wortley Road is shown in Fig. 1.15.

Preliminary enquiries with British Coal mining surveyors have revealed that there may be unrecorded mine shafts and possible shallow mine workings over part of the site. Other shafts and spoil resulting from ironstone workings could also be present on the site.

A desk study report on the geological and past mining situation has been commissioned by the developers. This report is to outline the geological, geotechnical and past mining problems that may be encountered in the development of the site. In addition the report should include a detailed soils investigation using trial pits and recommendations as to the need for a borehole investigation to establish the nature and extent of any shallow coal workings beneath the site.

Case study 1.1 *(contd.)*

RESEARCH SOURCES

In compiling this report we have examined information and records and made enquiries from the following sources:

1. British Coal Mining Surveyors at Technical Headquarters. Burton upon Trent to examine the shaft register held in the archives;

2. The British Coal Opencast Division at Burton upon Trent;

3. Ordnance Survey County Series plans 1850 edition and 1905 edition;

4. The County Series geological maps and memoirs including consultations with the British Geological Survey at Keyworth, Nottingham;

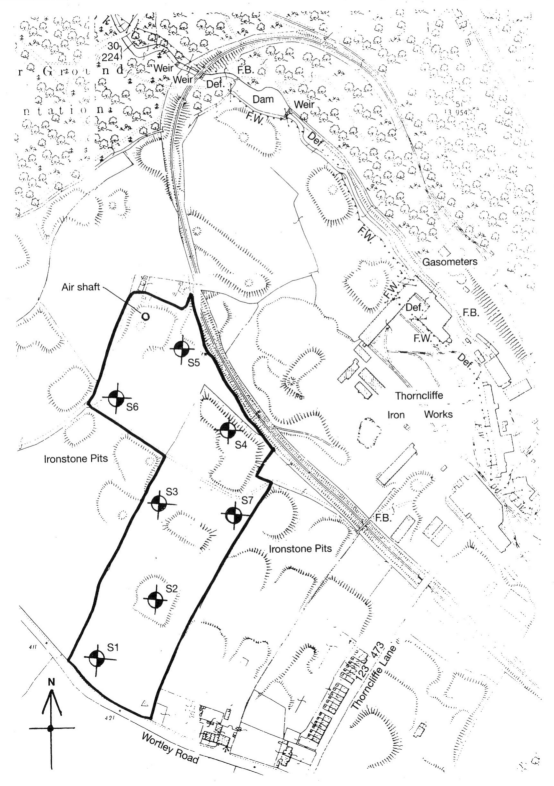

Fig. 1.15 Case study 1.1: site location and layout.

Case study 1.1   (*contd.*)

5. Sheffield City Technical Services Dept;
6. Mining Records Office at Rawmarsh, Rotherham, South Yorkshire;
7. Plans of abandoned mines and quarries contained in the archives of the Health and Safety Executive, London;
8. Mineral Valuer District Offices; Sheffield.

SITE GEOLOGY

The 1:10 560 scale County Series geological map NZ282 SE published by the British Geological Survey shows most of the site to be overlain by deposits of boulder clay of glacial origin. The thickness of these drift deposits is not recorded on the geological maps but from information gleaned from

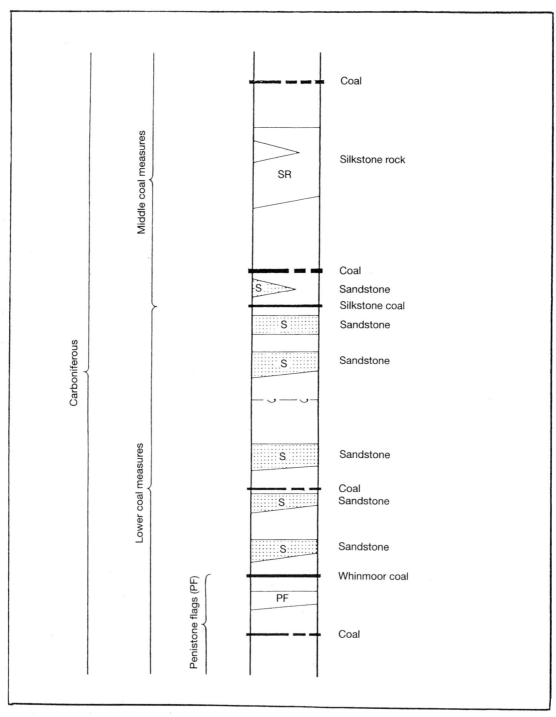

Fig. 1.16 Case study 1.1: generalized vertical section of strata below the superficial deposits. Unlabelled areas are shales or mudstones.

Case study 1.1 (*contd.*)

previous site investigations in the site locality, the boulder clay has been proven up to depths of 8 m beneath the existing surface levels.

Research into the history of the site, described later in this report, also established the existence of previous colliery spoil heaps within the site boundaries. There is a possibility therefore that some areas of the site could be overlain by deposits of colliery spoil. In addition there may be backfilled pits on site as a result of ironstone extraction;

these were indicated on the 1905 Ordnance Survey map. The old Ordnance plans show the Thorncliffe Ironworks to be close to the northern boundary of the site. The boundary is in fact marked by the old mineral line which served the ironworks and is now disused. A local shaft is noted to have been worked close to the ironworks from Thorncliffe Colliery and a generalized vertical section of the site is shown in Figs 1.16 and 1.17 taken from the Thorncliffe Colliery shaft records.

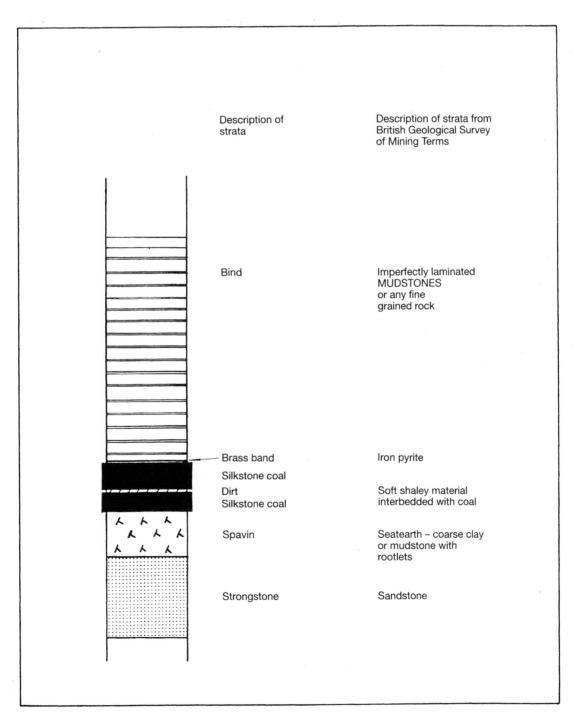

Fig. 1.17 Case study 1.1: vertical section of strata at Thorncliffe colliery.

Case study 1.1 (*contd.*)

SOLID GEOLOGY

The published geological maps show that the drift measures are underlain by Middle and Lower Coal Measure strata of the Carboniferous Age. These consist of sandstones, siltstones and mudstones with interspaced coal and fireclay horizons. Several collieries existed close to the site, notably the Thorncliffe Colliery about 0.50 miles east of the site.

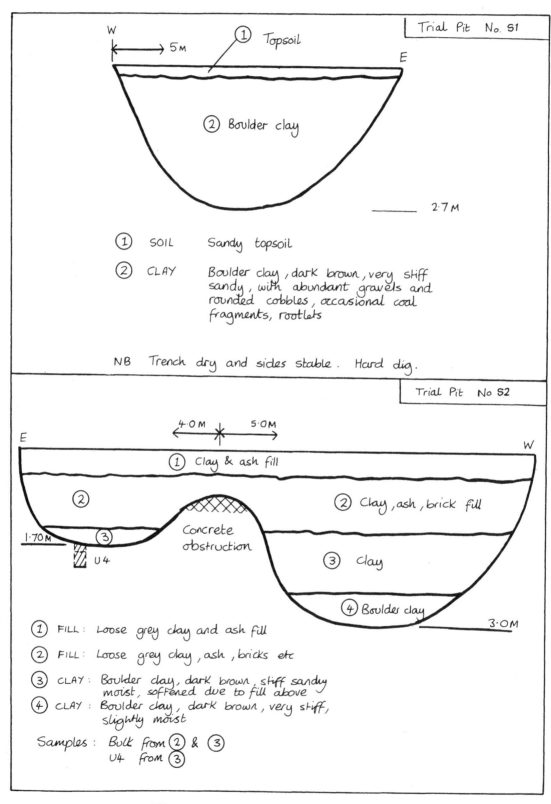

Fig. 1.18 Case study 1.1: trial pit logs S1 and S2.

Case study 1.1   (contd.)

There was also an ironworks close to the site referred to as the Thorncliffe Iron Works.

The Silkstone coal seam, average thickness 1.80 m, is conjectured to outcrop south of Wortley Road with the Whinmoor coal seam in excess of 40 m below the site. Other workable coal seams are the Hard Bed coal seam (Ganister), the Coking coal seam and the Pot Clay coal seam. However, it is considered that if these seams have been worked in the

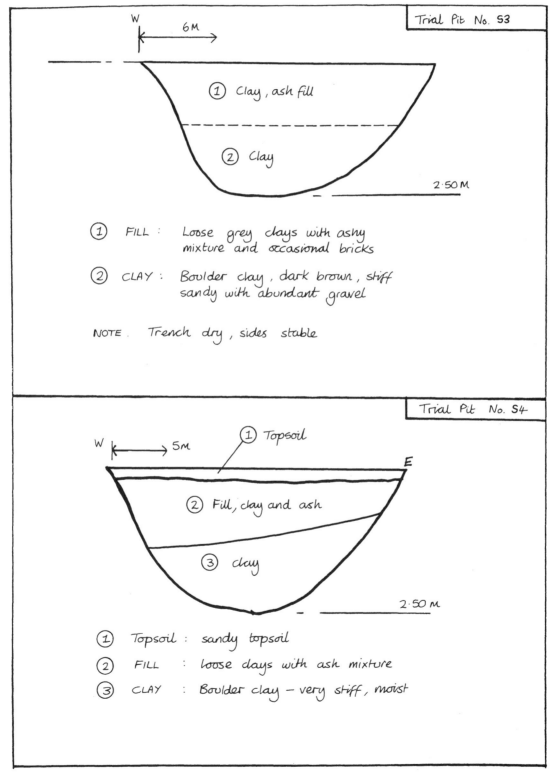

Fig. 1.19 Case study 1.1: trial pit logs S3 and S4.

Case study 1.1   *(contd.)*

past then any surface subsidence associated with the workings will have long since ceased. British Coal have stated in their mining report that no future mining is expected to take place.

The coal seams are known to dip to the north at a gradient varying between 1 in 5 and 1 in 15 and it is therefore considered that the Silkstone seam will be at such a depth below the site that there will be sufficient rock cover over any possible workings.

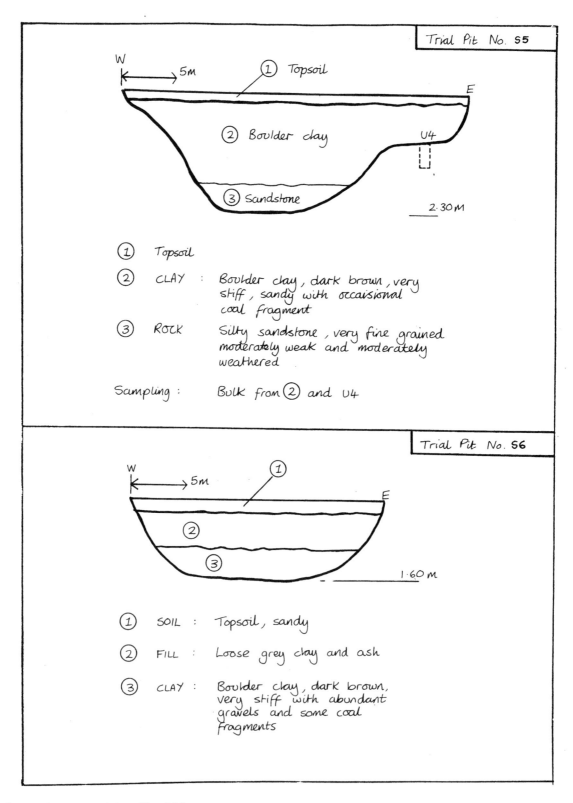

Fig. 1.20 Case study 1.1: trial pit logs S5 and S6.

Case study 1.1   (*contd.*)

Geological faults

The geological maps also show that the site locality is traversed by numerous geological faults which have resulted in localized variations in the stratification of the coal seams in regard to their depth beneath the site and degree of dip. The Silkstone Coal is present at depths which should provide an adequate cover of competent rock over any possible old workings.

SITE HISTORY

The various editions of the Ordnance Survey County Series were examined. The 1850 and 1905 publications record the position of the old Thorncliffe Colliery northeast of the site, and the Thorncliffe Iron Works.

The 1905 Ordnance plans show several ironstone pits on the site and an air shaft in the northwest corner.

The later Ordnance Survey plan also shows ponds together with colliery spoil heaps in the northern area adjacent to the mineral line. An inspection of the site reveals that the spoil heaps have been removed or regraded.

PAST MINING ACTIVITY

The available records show that coalmining has taken place beneath the site in several main seams. The depth of the Silkstone seam dips from approximately 16 m south of Wortley Road away to the north. It is likely that the Silkstone coal seam has been mined by pillar and stall methods in the past but it is considered that they present no risk to stability on this site.

Researches of mine records and topographical plans have confirmed the positions of one air shaft within the site area. The recorded position of this shaft has been investigated and identified using a JCB backacting excavator to carry out slit trenching under our supervision. Detailed records of the shaft do not exist; it will require capping off at rockhead.

Coalmining has ceased in the locality, the main collieries having been closed. The possibility of future underground mining for coal or any other mineral can reasonably be discounted.

TRIAL PITS

The trial pit investigation proved up to 1.50 m of colliery spoil and ironstone debris in various parts of the site, underlain by natural brown boulder clays. The maximum thickness of boulder clay recorded in the trial pits was 3.0 m.

There may be other areas of the site which are overlain by colliery spoil and which have not been revealed in this investigation.

The trial pit records are included in the appendix to this report (see Figs 1.18–1.20).

CONCLUSIONS AND RECOMMENDATIONS

The site is underlain by superficial deposits which are variable in their thickness and lateral extent. These deposits are generally natural boulder clays but parts of the site are shown to be overlain by approximately 1.50 m thickness of colliery spoil and ironstone minerals. These fills could be a result of bell pit workings for ironstone.

In those areas of the site underlain directly by natural boulder clays it is recommended that standard-width strip footings be used for dwellings of 2–3 storeys. These foundations should be reinforced with a nominal layer of mesh type B283 top and bottom.

Where colliery spoil is evident, foundations will need to be taken down below the fill into the natural boulder clays for a minimum distance of 300 mm.

Though the soluble sulphate content of the colliery spoils was within the Class 1 range it is recommended that a Class 3 concrete mix be adopted owing to the low pH values. All mortar below ground should use sulphate-resisting cement.

Ground floor construction should be a voided precast beam and block system.

We are of the opinion that the Silkstone coal seam is at such a depth beneath the site that any abandoned mine workings within these seams would not present a source of potential surface instability.

It is therefore recommended that excavations during the site strip are examined to ensure there are no other mine shafts present on the site. Should any be encountered the local British Coal mining surveyor should be informed.

It is recommended that the old air shaft be capped off at rockhead level. Should the depth to rockhead be excessive the shaft infill should be grouted down to rockhead prior to capping at the surface.

BIBLIOGRAPHY

BRE (1991) BRE Digest 363: *Concrete in sulphate-bearing soils and groundwater*, Building Research Establishment.

BSI (1992) BS 882: Part 2. *Specification for aggregates from natural sources for concrete*, British Standards Institution.

BSI (1999) BS 5930: *Code of practice for site investigations*, British Standards Institution.

BSI (1986) BS 8004: *British Standard code of practice for foundations*, British Standards Institution.

BSI (1990) BS 1377: *Methods of testing for civil engineering purposes*, British Standards Institution.

Edmonds, C.N. and Kirkwood, J.P. (1989) Suggested approach to ground investigation and the determination of suitable substructure solutions for sites underlain by chalk. *Proc. International Chalk Symposium*, paper 12, Thomas Telford, London.

Joyce, M.D. (1980) *Site Investigation Practice*, E. & F.N. Spon, London.

NHBC (1977) *NHBC Foundation Manual: Preventing Foundation Failures in New Buildings*, National House-Building Council, London (now rewritten as *NHBC Standards* Chapter 4.1).

Tomlinson, M.J. (1980) *Foundation Design and Construction*, 4th edn, Pitman.

Chapter 2
Foundation design

2.1 INTRODUCTION

Once all the relevant desk study and fieldwork data have been collected and collated the type of foundation most suitable for the site conditions can be established. The most simple foundation for low-rise dwellings is a strip footing or trenchfill concrete. Generally, in good strata, construction is a straightforward operation but in some situations where hazardous ground conditions are encountered special care must be taken. This chapter points out where this special care is needed and offers some practical advice on economic design and construction techniques.

The following foundation types can be classified based on where the load is carried by the ground:

- **Shallow foundations**: strip footings or pad foundations, raft foundations etc. Generally the depth over width (ratio of foundation depth to foundation width) is less than 1.0.
- **Deep foundations**: piles, pier and pad foundations, deep trench fill. Generally the depth over width exceeds 4.0.

If we consider shallow foundations in natural clay or granular strata the following conditions apply.

2.1.1 Width of footing

This is governed by the allowable bearing capacity of the ground at the foundation depth. In clay soils with plasticity index greater than 20% the minimum recommended foundation depth is 900mm. In areas of the UK where clay soils have a plasticity index less than 20%, foundations are traditionally placed at a minimum depth of 750 mm below final ground level.

For most two-storey housing, wall-line loads of 35–45kN/m run are achieved using traditional brick-and-block construction. Where the allowable bearing pressure of the stratum exceeds 75 kN/m^2, and the stratum remains competent for approximately 1.0 m below the footing, then a 600mm wide footing will suffice and will be well within allowable settlements. These footings can be unreinforced except when building in a mining area.

Where ground bearing capacities of between 50 and 75kN/m^2 exist, the strip footings may need to be widened, and they usually require transverse reinforcement in the bottom of the footing. It is often good practice to use a structural mesh with the main wires spanning across the footing width (Fig. 2.1).

Where ground conditions result in bearing capacities between 25 and 50 kN/m^2 then the use of strip footings should be avoided because of the excessive widths required.

To keep settlements within a tolerable magnitude the use of **'pseudo'-raft foundations** is recommended. Such rafts are based on empirical rules and usually take the form of a reinforced concrete slab within downstand edge beams and internal thickenings designed to span 3m and cantilever 1.5m at the corners (Fig. 2.2).

2.1.2 Soft spots

Where soft spots are encountered in a footing excavation and firm ground exists at a lower level the soft stratum should be dug out and replaced with a lean-mix concrete, of C7.5 or C10.0 strength.

Where the soft spot is found to be too deep to dig out then it may be practical to reinforce the foundation to span across the weak section of ground. When this is done the foundation at each end of the reinforced section will need to be checked for the additional loading condition and may need to be widened out locally.

2.1.3 Stratum variation in excavation

Where strip footing excavations reveal a variation in soil strata from firm clays to compact sands then a layer of light fabric mesh should be added in the bottom of the footing. B283 spanning along the length of the footing is the most suitable reinforcement.

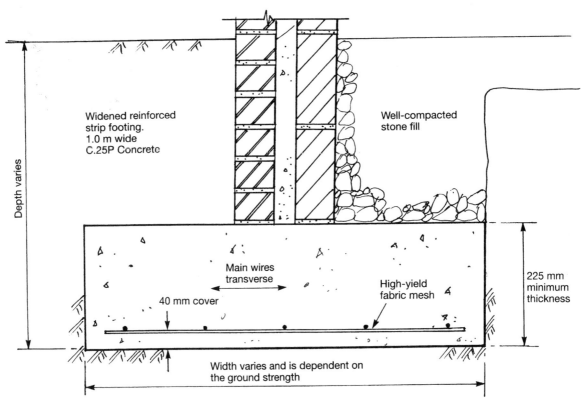

Fig. 2.1 Widened reinforced strip footing.

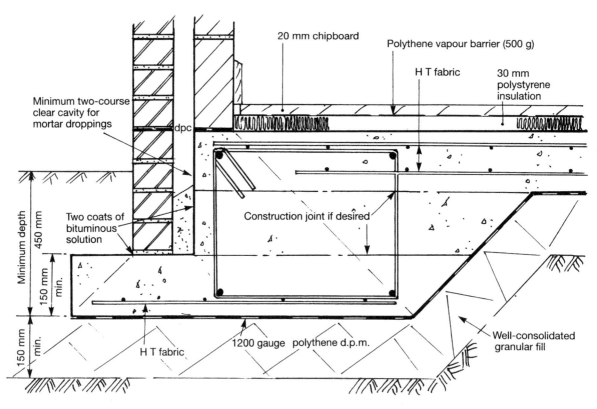

Fig. 2.2 Pseudo-raft foundation.

Where rock is encountered it is essential that the house is placed *wholly* on the rock stratum. If the rock stratum cannot be excavated at a reasonable depth, i.e. less than 3 m, then a different foundation solution should be considered. Figure 2.3 indicates the use of a raft slab on a thick cushion of crushed stone fill. Alternatively, piles and ring beams or pier and pad with ring beams are the usual options worth considering when these conditions are encountered. The use of concrete manhole rings to form mass concrete piers is a practical, economic and safe construction method and, where

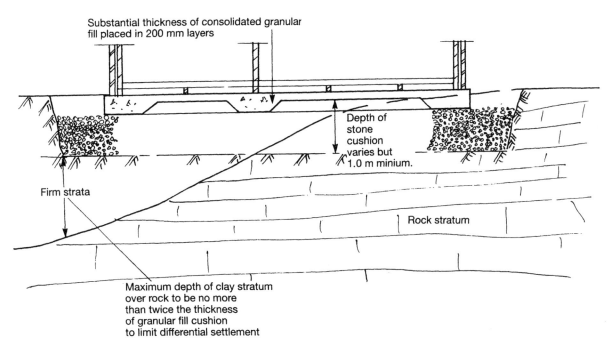

Fig. 2.3 Sloping rock formation.

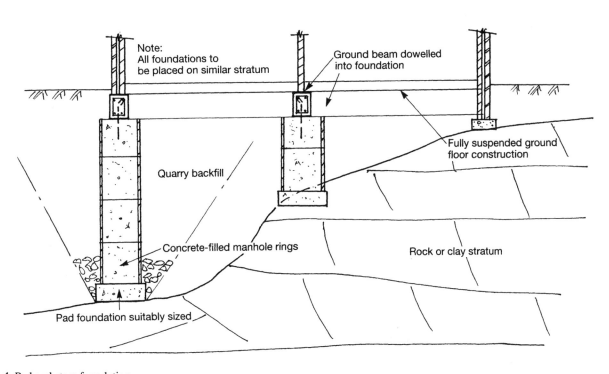

Fig. 2.4 Pad and stem foundation.

only a few dwellings are affected, avoids the expense of piling and avoids delays in the construction programme (Fig. 2.4).

2.1.4 Firm clays overlying soft strata

Avoid deep footings where a firm clay stratum overlies a soft stratum which reduces in strength with depth. Where footings have to be deepened to cater for existing trees or existing deep drainage then it is preferable to adopt a pseudo-raft foundation on a thick cushion of granular fill (Fig. 2.5). If a strip footing is deepened it is important to check that the footing load does not overstress the softer strata at depth if excessive settlements are to be avoided. In such ground conditions the width of the footing should be kept as narrow as possible so keeping the pressure bulb within the firm stratum.

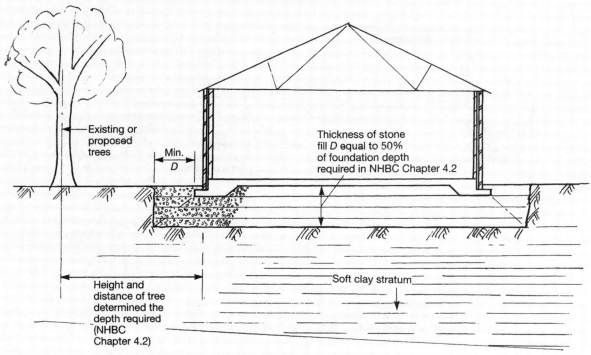

Fig. 2.5 Pseudo-raft on stone fill over soft clay stratum.

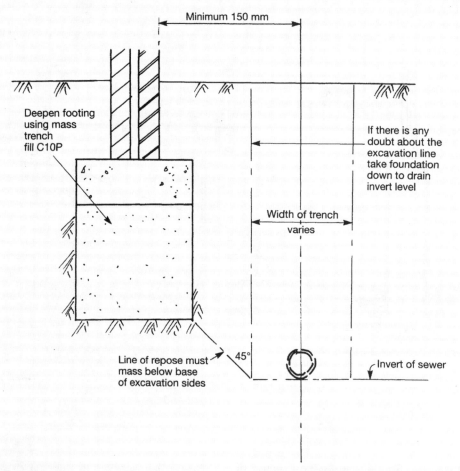

Fig. 2.6 Existing or proposed sewers.

2.1.5 Depth of footings

The depth of footings is governed by a number of factors:

1. depth at which a suitable bearing stratum exists;
2. depths of adjacent drainage runs or other existing

services, and existing basements adjoining the new development (Fig. 2.6);
3. distance from existing trees or proposed tree-planting scheme (Fig. 2.7);
4. depth of water table;
5. frost susceptibility of the stratum;

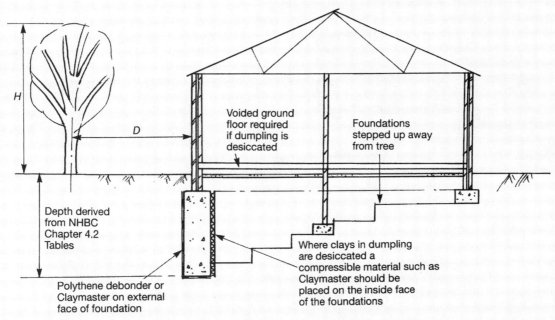

Fig. 2.7 Existing or proposed trees: *H*, mature height of tree; *D*, distance from dwelling to tree.

6. shrinkability characteristics of any clay soils which may exist and which could exhibit large volumetric changes because of seasonal variations.

If adequate formation cannot be obtained at relatively shallow depths, i.e. 1.2 m, then a trial pit should be dug about 3 m away from the dwelling to determine whether a good stratum exists at depths less than 3 m below ground level. If a good stratum is not encountered within 3 metres the following alternative foundations can be considered.

- **Widened** reinforced strip footings.
- **Reinforced strip** footings on a wide cushion of well-consolidated stone fill. This fill must be of constant depth under all load-bearing walls to avoid any differential settlement (Fig. 2.8).
- **Deep trench fill** (Fig. 2.9).
- **Pseudo-raft foundation**. Designed as an edge beam raft to span 3 m and cantilever 1.5 m at the corners. Can be used with a thick cushion of granular fill as replacement for weak strata such as peats and fills.

SYMBOLS

BS 8110: part 1: 1997
Structural use of concrete

A_s	Area of tension reinforcement (mm^2)	M_{sx}	Maximum design ultimate moment on strip of unit width and span L_x. $M_{sx} = \alpha_{sx}\, n\, L_x^2$
b	Width or effective width of the section or flange in the compression zone (mm)	M_{sy}	Maximum design ultimate moment on strip of unit width and span L_y. $M_{sy} = \alpha_{sy}\, n\, L_x^2$
b_v	Breadth of section	α_{sx}	Bending moment coefficient from Table 3.14 in BS 8110: Part 1: 1985
d	Effective depth of the tension reinforcement (mm)	α_{sy}	Bending moment coefficient from Table 3.14 in BS 8110: Part 1: 1985
f_y	Specified characteristic strength of the reinforcement (N/mm^2)	n	Total design ultimate load per unit area = (1.4 g_k + 1.6 q_k) (kN/m^2)
f_{yv}	Characteristic strength of links (not to be taken as more than 460 N/mm^2)	q_k	Characteristic imposed load (unit) (kN/m^2)
f_{cu}	Characteristic strength of concrete (N/mm^2)	Q_k	Characteristic imposed load (total) (kN)
g_k	Characteristic dead load (unit) (kN/m^2)	V	Design shear force due to ultimate load (kN)
G_k	Characteristic dead load (total) (kN)	v	Design concrete shear stress at cross-section (N/mm^2)
k'	Lever arm factor	v_c	Design concrete shear stress (Table 3.9) (N/mm^2)
L_x	Length of shorter side (m)		
L_y	Length of longer side (m)	z	Lever arm (= lever arm factor × d) (mm)
M_{ult}	Design ultimate moment at the section considered (kNm)	τ_f	Partial safety factor for load
		τ_m	Partial safety factor for strengths of materials

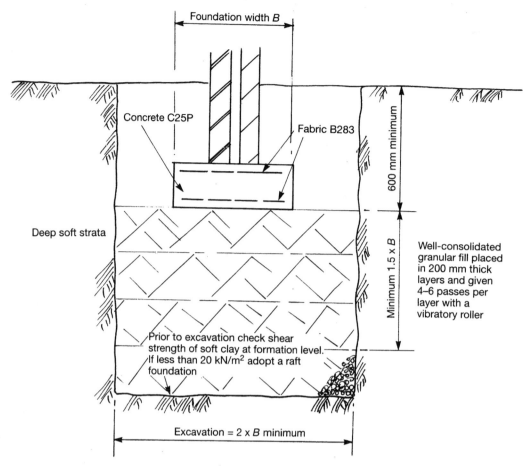

Fig. 2.8 Double reinforced footing on consolidated granular fill.

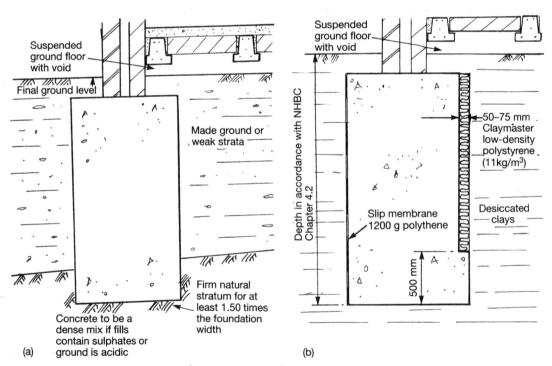

Fig. 2.9 Trench-fill foundations: (a) filled ground; (b) catering for existing trees and vegetation.

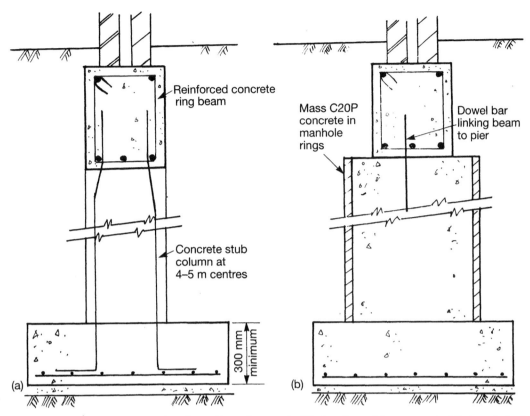

Fig. 2.10 Pad and stem foundations: (a) piers formed from formwork; (b) piers formed from concrete-filled manhole rings. Size of pad foundation to suit bearing capacity of ground.

- **Pad and stem foundations.** Designed using ground beams spanning between piers formed from concrete-filled manhole rings or formwork (Fig. 2.10).
- **Bored or driven piling** with reinforced ring beams at ground level.

2.2 WIDENED REINFORCED STRIP FOOTINGS

Designed in accordance with BS 8110.

Unfactored line load including self-weight of footing = 40 kN/m

Allowable bearing capacity of ground taken as 40 kN/m²

Therefore footing width = 40/40 = 1.0 m.

Net uplift pressure = 40 – (0.225 × 24) = 34.6 kN/m²

Load factor = 1.50

Ultimate transverse moment = $1.50 \times 34.60 \times \dfrac{0.375^2}{2} = 3.65\,\text{kN m}$.

Using the Design Chart No. 1 (Fig. 2.11):

$$\frac{M}{bd^2 f_{cu}} = \frac{3.65 \times 10^6}{10^3 \times 170^2 \times 25} = 0.005$$

Therefore lever arm, $l_a = 0.95d$

From BS 8110: Part 3:

$f_{cu} = 25$ N/mm $d = 170$ mm

$$A_s = \frac{3.65 \times 10^6}{0.87 \times 460 \times 0.95 \times 170} = 56\,\text{mm}^2$$

Minimum percentage $\dfrac{0.13 \times 10^3 \times 225}{100} = 292\,\text{mm}^2$

Use A393 fabric mesh (which has the required minimum percentage of steel in both directions) in bottom. The finished foundation is shown in Fig. 2.13.

29

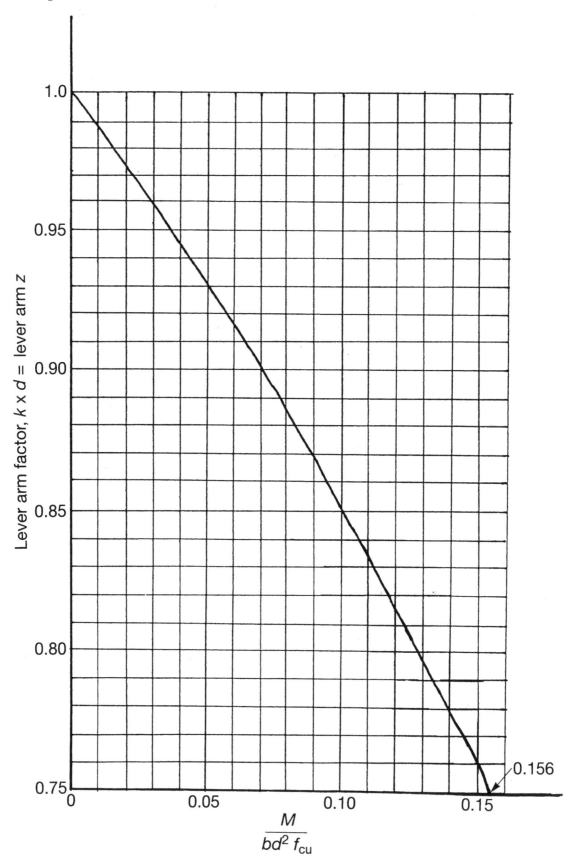

When M/bd^2f_{cu} exceeds 0.156, compression steel is required.
$A_s = M/0.87f_yk_fd.f_y = 460$ up to 16 mm dia; $f_y = 425$ over 16 mm dia; $f_y = 485$ wire

Fig. 2.11 Design Chart No. 1: lever arm curve for limit state design.

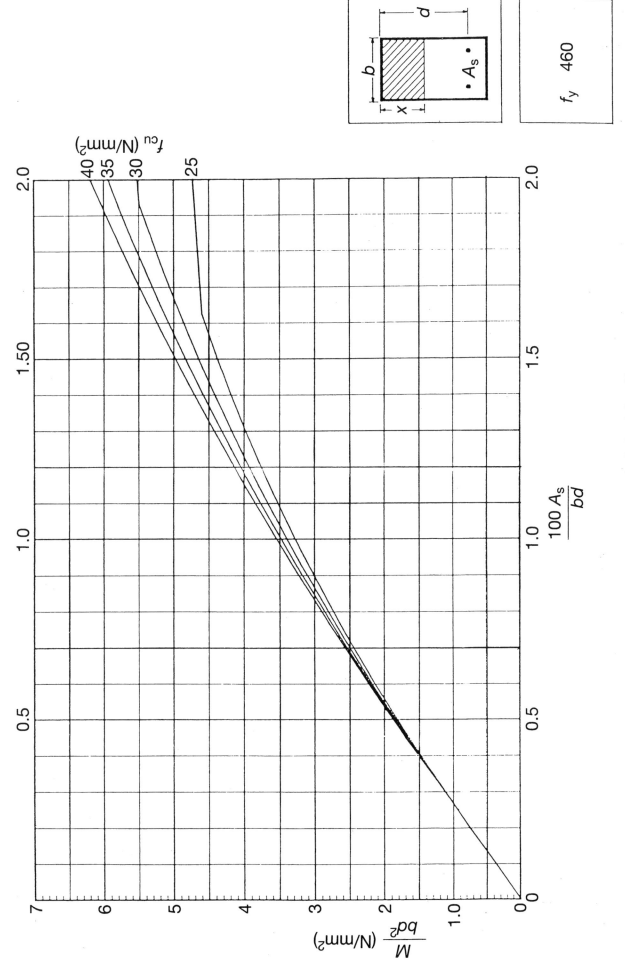

Fig. 2.12 Design Chart No. 2: singly reinforced beams.

2.3 REINFORCED STRIP FOOTINGS ON REPLACEMENT GRANULAR FILL

Where a weak stratum exists with allowable bearing capacities between 25 and 50 kN/m² then a reinforced footing can be used placed on a thick mat of replacement stone filling (Fig. 2.8). The bulb of pressure remains within the stone depth and the underlying soft clays are not stressed. This method can often be utilized instead of a raft foundation. It is important to ensure that the thickness of stone fill below the

Table 2.1. Design Chart No 3. Reinforcement: areas of groups of bars

Diameter (mm)	Area (mm²) for numbers of bars											
	1	2	3	4	5	6	7	8	9	10	11	12
6	28	57	85	113	142	170	198	226	255	283	311	340
8	50	101	151	201	252	302	352	402	453	503	553	604
10	79	157	236	314	392	471	550	628	707	785	864	942
12	113	226	339	452	566	679	792	905	1020	1130	1240	1360
16	201	402	603	804	1010	1210	1410	1610	1810	2010	2210	2410
20	314	628	943	1260	1570	1890	2200	2510	2830	3140	3460	3770
25	491	982	1470	1960	2450	2950	3440	3930	4420	4910	5400	5890
32	804	1610	2410	3220	4020	4830	5630	6430	7240	8040	8850	9650
40	1260	2510	3770	5030	6280	7540	8800	10100	11300	12600	13800	15100
50	1960	3930	5890	7850	9820	11800	13700	15700	17700	19600	21600	23600

Diameter (mm)	Area (mm²) for spacings in mm								
	50	75	100	125	150	175	200	250	300
6	566	377	283	226	189	162	142	113	94
8	1010	671	503	402	335	287	252	210	168
10	1570	1050	785	628	523	449	393	314	262
12	2260	1510	1130	905	745	646	566	452	377
16	4020	2680	2010	1610	1340	1150	1010	804	670
20	6280	4190	3140	2510	2090	1800	1570	1260	1050
25	9820	6550	4910	3930	3270	2810	2450	1960	1640
32	16100	10700	8040	6430	5360	4600	4020	3220	2680
40	25100	16800	12600	10100	8380	7180	6280	5030	4190
50	39200	26200	19600	15700	13100	11200	9800	7850	6550

Diameter (mm)	6	8	10	12	16	20	25	32	40	50
Mass (kg/m)	0.222	0.395	0.616	0.888	1.579	2.466	3.854	6.313	9.864	15.413

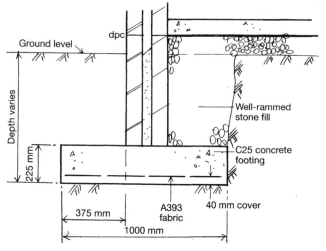

Fig. 2.13 Widened reinforced footing.

footings remains fairly constant. Shear vane results are 40 kN/m² to depths of 3.0 m.

At a depth of 450 mm the line load of 40 kN/m run imposes a ground pressure of 40/0.60 = 66 kN/m². For an increased line load of 60 kN/m run, it will be necessary to check the ground pressures at critical levels.

Checking at underside of footing

$q_m = 60/0.60 = 100$ kN/m²

This exceeds the allowable and will require the weak clays to be replaced by consolidated granular fill. 100 kN/m² is acceptable for a well-compacted granular fill placed in 200 mm layers.

Table 2.2. Design Chart No. 4

BS reference	Mesh sizes (nominal pitch of wires)		Wire sizes		Cross-sectional area per metre width		Nominal mass per square metre
	Main (mm)	Cross (mm)	Main (mm)	Cross (mm)	Main (mm)	Cross (mm)	(kg)
Square mesh fabric							
A 393	200	200	10	10	393	393	6.16
A 252	200	200	8	8	252	252	3.95
A 193	200	200	7	7	193	193	3.02
A 142	200	200	6	6	142	142	2.22
A 98	200	200	5	5	98.0	98.0	1.54
Structural mesh fabric							
B 1131	100	200	12	8	1131	252	10.9
B 785	100	200	10	8	785	252	8.14
B 503	100	200	8	8	503	252	5.93
B 385	100	200	7	7	385	193	4.53
B 283	100	200	6	7	283	193	3.73
B 196	100	200	5	7	196	193	3.05
Long mesh fabric							
C 785	100	400	10	6	785	70.8	6.72
C 636	100	400	9	6	636	70.8	5.55
C 503	100	400	8	5	503	49.0	4.34
C 385	100	400	7	5	385	49.0	3.41
C 283	100	400	6	5	283	49.0	2.61
Wrapping fabric							
D 49	100	100	2.5	2.5	49.1	49.1	0.770

Checking 1.35 m below ground level

Vertical pressure factor = 0.386 (pressure bulb, Fig. 2.14)

$$q_{all} = \frac{5.70s\mu}{3.0}$$

where s = undrained shear strength,

μ = plasticity index correction factor

Allowable bearing pressure = $\frac{5.7 \times 40 \times 0.8}{3}$ = 45.60 kN/m^2

Actual pressure at 1.35 m depth = 0.386×100 = 38.60 kN/m^2

< 45.60

Use a 600 mm wide by 225 mm thick double reinforced concrete footing reinforced with fabric mesh B283 top and bottom and specify a concrete mix of C25P.

2.4   TRENCH FILL FOUNDATIONS

These are only suitable if a good bearing stratum is known to be present at an economic and practical depth. The stratum below the base of the trench fill must remain competent for at least 1.5 times the foundation width.

This foundation is useful for sites where deep fills, thick soft strata or peat beds overlie firm natural strata. Due regard must be taken of the possibility of soluble sulphates in the fill materials or in acidic materials such as peats. In these situations the concrete should have a minimum cement content based on *BRE Digest 363* (1991 edition).

This type of foundation is also useful for meeting the requirements of NHBC Chapter 4.2 where houses are built close up to existing mature trees or landscape planting. In these situations it must be used with caution especially if the ground within and adjacent to the house walls is in a desiccated state. If the clays are desiccated the following recommendations should be adopted to prevent foundation heave should the clays rehydrate.

1. Provide a fully suspended voided ground-floor construction using full-span timber joists or joists on sleeper walls on their own foundation. In deep foundations the use of sleeper walls can be uneconomic. Alternatively a precast beam and block floor can be used.

2. Provide a low-density polystyrene material of suitable thickness on the internal face of the trench fill to all foundation walls affected by desiccation of clays. The density of the polystyrene should be no greater than 11 kg/m^3. A product such as Claymaster or Claylite would be suitable.

3. Provide a slip membrane by using 1200 g polythene sheets on the external face of the trench fill. If the existing trees are within a distance less than twice the foundation depth then a suitable thickness of low-density polystyrene should be provided on both sides of the trench fill. The density of the polystyrene should not exceed 11 kg/m^3.

4. Ensure that the width of the trench fill is maintained without any projections occurring at higher levels. If the clays rehydrate, any projections are going to be subjected to

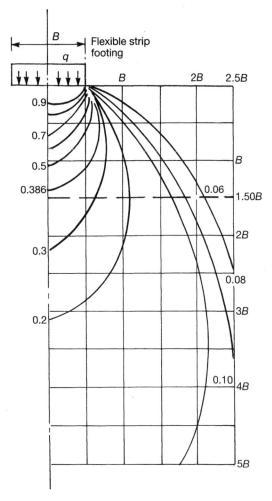

Fig. 2.14 Vertical pressure under a rectangular footing.

uplift pressures; these have been known to separate the upper wider sections of concrete from the lower sections (Fig. 2.15).

2.5 RAFT FOUNDATIONS

Raft foundations are most suitable for use in ground conditions such as soft clays and filled ground, in old mining areas which may have a potential for instability, and in active mining localities. The raft foundation has that inherent stiffness, not available in strip footings, which by virtue of its load-spreading capacity is more resistant to differential settlements.

If the fill is old, well layered and does not contain significant voids or organic matter, a pseudo edge beam raft can be used placed on a layer of granular fill which should extend under and beyond the raft edge for about 0.5 m (Fig. 2.16). The raft should have edge beams about 450 mm deep with sufficient width under the load-bearing walls to suit the ground conditions and be capable of spanning 3.0 m and cantilevering 1.5 m at the corners.

If the fills are variable and contain mixed materials it is advisable to remove the poor fill and proof-roll the formation prior to either replacing the fill if it is suitable for recompaction or using imported granular materials of a suitable grading. This method is obviously only economic when depths of replacement fill are less than 1.5 m.

This foundation method is most suitable for use on sites where old cellars are encountered and the minimum thickness of stone fill over the old cellar walls should be 1.0 m to prevent a hard spot from developing.

The pseudo edge beam raft is also suitable for sites where trees and heavy vegetation have been removed below potential

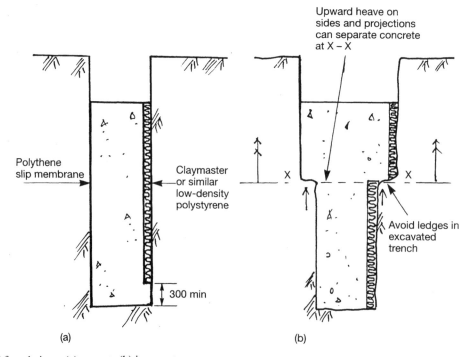

Fig. 2.15 Trench-fill foundations: (a) correct; (b) incorrect.

house plots and where the upper clays have become desiccated and could be affected by trees which are to remain. NHBC Chapter 4.2 recommends that the thickness of stone fill below the underside of the raft slab should be 50% of the depths required from their tables, up to a maximum of 1.0m. Upfill in excess of 1.0 m is only permitted if the NHBC can be satisfied that the correct fill is being used and it is being compacted properly. This generally requires the builder to have the works supervised by qualified engineers who may carry out various tests to confirm the adequacy of the compaction.

There are situations where pseudo rafts can be placed at a lower level and the poorer fills removed can be replaced on top of the raft. This type of construction is often used on sites where a band of peat is present below firm soils and the peat is underlain by soft alluvium or soft silty clays. The advantages of this form of construction are that no materials require to be taken off site, and that the amount of imported stone is reduced. Where the formation is very soft it is good practice to place a layer of geotextile fabric down prior to placing the stone cushion (Fig. 2.17).

Rafts are also used in areas where old shallow mine workings are present and require grouting up. Where the old workings have rock cover of less than six times the seam thickness and have been grouted up there is still a risk of residual subsidence; a raft is better able to withstand localized crown hole subsidence.

Example 2.1   Semi-rigid raft design

Consider a typical semi-detached dwelling of brick and block construction to be built of a raft foundation placed on to weak clayey sands which have a high water table. The ground floor plan is shown in Fig. 2.18.

DESIGN INFORMATION

Loading: to BS 6399
Reinforced concrete design in accordance with BS 8110
Imposed loads on roof: 0.75 kN/m^2 for 25° pitch
Imposed loads on floors: 1.50 kN/m^2
Concrete: 35.0 N/mm^2 at 28 days
Reinforcement: high-yield bars and high-yield fabric.

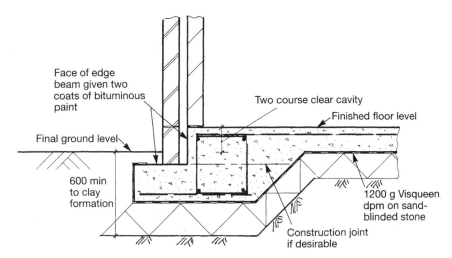

Fig. 2.16 Stepped-edge beam raft.

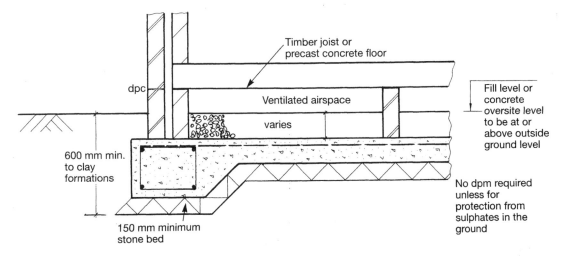

Fig. 2.17 Voided raft.

Foundation design

The site investigation reveals that the upper stratum of shallow fills overlies weak clayey sands. In view of the high water table it is considered prudent to adopt a pseudo or semi-rigid raft foundation with edge beams designed to span 3 m and cantilever 1.5 m.

Total unfactored floor loads = 2.40 kN/m²

Total factored floor loads = 3.66 kN/m²

LOADINGS

Roof

	Service loads (kN/m²)	Factored loads (kN/m²)
Tiles	0.55	
Battens and felt	0.05	
Trusses	0.23	
Ceiling board	0.15	
Insulation	0.02	
	1.00 × 1.40 =	1.40

Imposed loads

Snow 25° pitch	0.75	
Storage	0.25	
	1.0 × 1.60 =	1.60

Total unfactored roof load = 2.0 kN/m²

Total factored roof load = 3.0 kN/m²

First floor

	(kN/m²)	Factored loads (kN/m²)
Boarding (22 mm)	0.10	
Joists 225 × 50	0.15	
Ceiling board	0.15	
Partitions (stud)	0.50	
	0.90 × 1.40 =	1.26
Imposed loads = 1.50 kN/m² × 1.60	=	2.40

External walls

	(kN/m²)	(kN/m²)
Brick	2.25	
Block (100 mm)	1.25	
Plaster	0.25	
	3.75 × 1.40 =	5.25

Spine walls (kitchen/lounge)

100 mm blockwork	1.25	
Plaster	0.50	
	1.75 × 1.40 =	2.45

Staircase wall

100 mm blockwork	1.25	
Plaster	0.50	
	1.75 × 1.40 =	2.45

Party wall

2 skins blockwork	2.50	
2 coats plaster	0.50	
	3.0 × 1.40 =	4.20

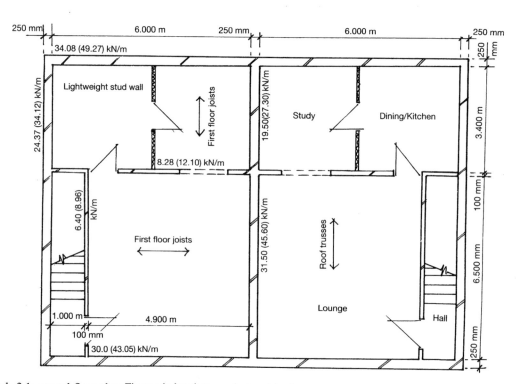

Fig. 2.18 Example 2.1: ground-floor plan. Figures in brackets are factored loads.

WALL LOADINGS

Ignore windows and doors.

Front walls

				Unfactored loads (kN/m)					Factored loads (kN/m)

Roof

$1.00 \times \dfrac{10.5}{2}$	$=$	5.25	$\times$	1.40	$=$	7.35
$1.0 \times \dfrac{10.5}{2}$	$=$	5.25	$\times$	1.60	$=$	8.40

Walls

3.75×5.20	$=$	19.50	$\times$	1.40	$=$	27.30
		30.00				43.05

Rear walls

(kN/m) (kN/m)

Roof

$1.00 \times \dfrac{10.5}{2}$	$=$	5.25	$\times$	1.40	$=$	7.35
$1.0 \times \dfrac{10.5}{2}$	$=$	5.25	$\times$	1.60	$=$	8.40

First floor

$0.90 \times \dfrac{3.40}{2}$	$=$	1.53	$\times$	1.40	$=$	2.14
$1.50 \times \dfrac{3.40}{2}$	$=$	2.55	$\times$	1.60	$=$	4.08

Walls

3.75×5.20	$=$	19.50	$\times$	1.40	$=$	27.30
		34.08				49.27

Party walls

(kN/m) (kN/m)

Average height $6.50 \times 3 =$	19.50	$\times$	1.40	$=$	27.30
First floor $2 \times 0.90 \times 2.5 =$	4.50	$\times$	1.40	$=$	6.30
First floor $2 \times 1.50 \times 2.5 =$	7.50	$\times$	1.60	$=$	12.00
	31.50				45.60

Spine wall

(kN/m) (kN/m)

First floor $0.90 \times \dfrac{3.4}{2}$	$=$	1.53	$\times$	1.40	$=$	2.14
$1.50 \times \dfrac{3.4}{2}$	$=$	2.55	$\times$	1.60	$=$	4.08
100 mm block $1.25 \times 2.4 =$		3.00	$\times$	1.40	$=$	4.20
Plaster 0.50×2.4	$=$	1.20	$\times$	1.40	$=$	1.68
		8.28				12.10

Staircase wall

First floor $0.90 \times \dfrac{4.90}{2}$	$=$	2.20	$\times$	1.40	$=$	3.08
100 mm block $1.25 \times 2.4 =$		3.00	$\times$	1.40	$=$	4.20
Plaster 0.50×2.4	$=$	1.20	$\times$	1.40	$=$	1.68
		6.40				8.96

Gable wall

	(kN/m)			(kN/m)
3.75×6.5	$= 24.37$	$\times$	1.40	$= 34.12$

Total load (unfactored)

	(kN)
Front wall: 30.00×12.75	$= 383$
Rear wall: 34.08×12.75	$= 435$
Gable wall: $24.37 \times 2 \times 10$	$= 487$
Spine wall: $8.28 \times 2 \times 6$	$= 100$
Stair wall: $6.40 \times 2 \times 6.50$	$= 83$
Party wall: 31.50×6.5	$= 205$
Party wall: 19.50×3.50	$= 68$
Total	$= 1761$

Self-weight of raft $= 0.2 \times 24$	$= 4.8$ kN/m^2
Edge thickenings	$= 1.6$ kN/m^2
	6.40 kN/m^2

Area of raft	$= 10.90 \times 13.150 =$	143 m^2
Weight of raft	$= 143 \times 6.40 =$	915 kN

Ground pressure $= \dfrac{1761 + 915}{143} = 18.71$ kN/m^2

Plus imposed load	$= 1.50$
	20.21 kN/m^2

Maximum line load $= 34.08$ kN/m

Consider edge beam and raft slab as composite with slab acting as a tie. If overall width is taken as 900 mm:

Line load	$=$	34.08 kN/m
Edge beam	$=$	6.85 kN/m
		40.93 kN/m

Pressure under edge strip $= \dfrac{40.93}{0.9} = 46$ kN/m^2 unfactored

This is less than the allowable ground bearing pressure of 50 kN/m^2.

With the pseudo raft the centroid of loading on the walls at foundation level generally coincides with the centroid of the edge strip. The raft slab is therefore not subjected to any torsional moments. In practice, the ground pressure under the edge strip is more likely to approach a triangular distribution with a higher edge pressure. The slab therefore needs to have sufficient reinforcement in the top section to cater for this rotating force. Its main function is to act as a structural tie while at the same time enhancing the edge beam stiffness. Where filled ground occurs it is prudent to provide a layer of bottom reinforcement to enable the raft slab to span over any soft spots.

To produce a theoretical edge pressure twice the allowable would result in an equivalent eccentricity of 150 mm. The slab will therefore have to cater for an ultimate moment of $0.15 \times 40.93 \times 1.5 = 9.20$ kN/m.

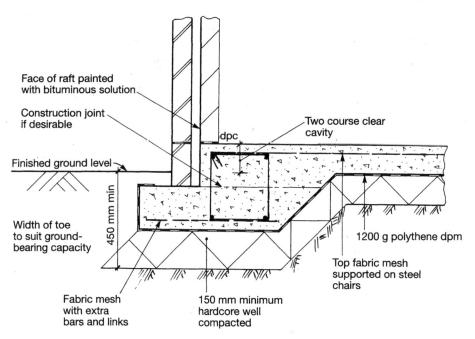

Fig. 2.19 Example 2.1: raft edge beam details.

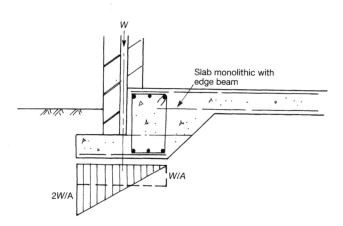

Fig. 2.20 Example 2.1: pressure under edge beam.

$F_{cu} = 35$ N/mm^2 and $F_y = 460$ N/mm^2

Therefore

$$\frac{M}{bd^2 f_{cu}} = \frac{9.20 \times 10^6}{10^3 \times 155^2 \times 35} = 0.11$$

Therefore l_a factor $= 0.95$

Therefore

$$A_s = \frac{9.20 \times 10^6}{0.87 \times 460 \times 0.95 \times 155} = 156 \text{ mm}^2$$

Minimum percentage $= \dfrac{0.13 \times 200 \times 1000}{100} = 260 \text{ mm}^2$

Provide a top and bottom layer of A252 fabric mesh in the raft slab.

PERIMETER GROUND BEAMS

Self weight of beam section per metre run:
$23.6\,[(0.45 \times 0.45) + (0.45 \times 0.15)] = 6.37$ kN/m.

Edge beams (sagging mode)

Use $wl^2/10$ to take account of some fixity.

Maximum line load $= (6.37 \times 1.4) + 49.27 = 58.18$ kN/m

Ultimate design moment $= \dfrac{58.18 \times 3.0^2}{10} = 52.36$ kN m

From Design Chart No.1:

$$\frac{M}{bd^2 f_{cu}} = \frac{52.36 \times 10^6}{450 \times 400^2 \times 35} = 0.021$$

l_a factor $= 0.95$.

Therefore $A_s = \dfrac{53.36 \times 10^6}{0.87 \times 460 \times 400 \times 0.95}$ $= 344$ mm^2

A252 Fabric 500 mm wide $= \dfrac{252 \times 500}{1000} = 126$ mm^2

Two T16 bars $= 402$ mm^2

528 mm^2

Provide A252 Fabric plus two T16 bars in bottom of edge beam.

Corner cantilever

Ultimate design moment $= \dfrac{58.180 \times 1.50^2}{2.0} = 65.45\ \text{kN m}$

$$\frac{M}{bd^2 f_{cu}} = \frac{65.45 \times 10^6}{450 \times 400^2 \times 35} = 0.025$$

l_a factor $= 0.95$.

Therefore $A_s = \dfrac{65.45 \times 10^6}{0.87 \times 460 \times 400 \times 0.95} = 430\ \text{mm}^2$

A252 Fabric 500 mm wide $= \dfrac{252 \times 500}{1000} = 126\ \text{mm}^2$

Two T16 bars in top $\qquad\qquad = 402$

$\qquad\qquad\qquad\qquad\qquad\qquad\qquad \overline{528\,\text{mm}^2}$

Provide A252 Fabric plus two T16 bars in top and bottom of edge beams at corners.
Place raft on a 1200 gauge polythene dpm laid on sand-blinded stone fill.

Toe design

Design as a cantilevered slab, to carry shear loading from outer leaf of cavity wall. Take height of wall as 6.8 m using a 150 mm thick toe reinforced with A252 fabric mesh in bottom.

Consider 1.0 m length.

Therefore $b_v = 1000\ \text{mm}$

$d = 150 - (40 + 3.5) = 106.50\ \text{mm}$

Loading / metre run = 110 mm brickwork= $2.30 \times 6.8 = 15.64\ \text{kN/m}$
Toe self-weight = $23.6 \times 0.15 \times 0.45 \qquad\quad = 1.60\ \text{kN/m}$

Therefore g_x $\qquad\qquad\qquad\qquad\qquad\qquad = \overline{17.24\ \text{kN/m}}$

Therefore design shear $V = 1.40 \times 17.24 = 24.136\ \text{kN/m}$.

Shear stress

$$v = \frac{V}{b_v d} = \frac{24.136 \times 10^3}{1000 \times 106.50} = 0.23\ \text{N/mm}^2$$

Reference BS 8110 clause 3.5.5 (shear resistance of solid slabs) and clause 3.4.5.2 (shear stress in beams).

Permissible concrete shear stress

Reference BS 8110 Table 3.9:

$\dfrac{100 / A_s}{b_v d} = \dfrac{100 \times 252}{1000 \times 106.5} = 0.236 \qquad v_c = 0.50$

v is less than v_c, therefore from BS 8110 Table 3.17 no shear steel is required.

Check toe for bending

Upward design load from GBP $= 1.50 \times 46 \times 0.45 = 31.05\ \text{kN/m}$

Downward design load (wall) $= 1.40 \times 15.64 = \quad 21.89\ \text{kN/m}$
(toe self-weight) $= 1.40 \times 1.60 \qquad\qquad\qquad = \quad 2.24$

$\qquad\qquad\qquad\qquad\qquad\qquad\qquad\qquad\quad \overline{24.13}$

Bending

$\dfrac{31.05 \times 0.45}{2.0} - 21.89 \times 0.10 - \dfrac{2.24 \times 0.45}{2.0}$

$= 6.98 \qquad\quad - 2.819 \qquad\quad - 0.504 \qquad = \quad 3.65\ \text{kN m}$

Reinforcement

Using design formulae method (BS 8110 Cl. 3.4.4.4):
By inspection k is greater than 0.95, therefore $z = 0.95\,d$.

Therefore

$$A_s = \frac{M}{0.87 f_y z} = \frac{3.65 \times 10^6}{0.87 \times 460 \times 0.95 \times 106.5} = 90\ \text{mm}^2$$

Minimum percentage $= 0.13\% = 195\ \text{mm}^2$, therefore use A252 Fabric in bottom of toe.

Internal ground beam (party wall)

Factored wall line load $\qquad\qquad\qquad\qquad\qquad = \quad 45.60$
Self-weight of beam $23.6 \times 0.45 \times 0.80 \times 1.40 \quad = \quad 11.90$

$\qquad\qquad\qquad\qquad\qquad\qquad\qquad\qquad \overline{57.50\ \text{kN/m}}$

Ground beam considered to act as a partially fixed element:

Ultimate design moment $= \dfrac{57.50 \times 3.0^2}{10} = 51.75\ \text{kN m}$

Use two layers of A252 in slab with one layer of A252 in thickening.
Consider A252 in bottom plus two T16. Therefore:

Area of fabric $= \dfrac{700}{1000} \quad = \quad 176\ \text{mm}^2$

Two T16 bars $\qquad\qquad = \quad 402\ \text{mm}^2$

$\qquad\qquad$ Total $\quad = \quad \overline{578\ \text{mm}^2}$

$d = 450 - (40 + 12) = 398.0\ \text{mm}$

$b = 800\ \text{mm}$, $z = 0.95\,d$. Therefore: $A_s = \dfrac{M}{0.87 f_y z}$

Therefore: Design moment of resistance $= A_s\, 0.87 f_y\, z$

$M = 578 \times 0.87 \times 460 \times 0.95 \times 398.0 \times 10^{-6} = 87.46\ \text{kN m}$.
OK > 51.75
Consider both layers of A252 in slab when checking cantilever mode.

Ultimate design moment $= 57.50 \times \dfrac{1.50^2}{2.0} = 64.68\ \text{kN m}$

Moment of resistance of two layers of mesh

$= 176 \times 0.87 \times 460 \times 0.95 \times 415 \times 10^{-6} = 27.76\ \text{kN m}$
$= 176 \times 0.87 \times 460 \times 0.93 \times 300 \times 10^{-6} = 19.65\ \text{kN m}$

Moment of resistance of two T16 bars $= 402 \times 0.87 \times 460 \times 0.95 \times 405 \times 10^{-6} = 61.89\ \text{kN m}$

Total moment of resistance $= 109.30\ \text{kN m}$

Use one layer of A252 in bottom of raft thickening with two layers in slab with two T16 in top and bottom of beams (Fig. 2.21).

Foundation design

Table 2.3. Summary of ground beam loadings

Wall location	Ground beam type	Maximum factored wall line load (kN/m)
External front wall	900 mm wide edge beam with two T16 bars in top and bottom with A252 fabric mesh in toe beam. T10 links at 250 mm centres	43.05
External rear wall	900 mm wide edge beam with two T16 bars in top and bottom with A252 fabric mesh in toe beams. T10 links at 250 mm centres	42.27
Party wall	800 mm wide slab thickening with A252 in base and two layers of A252 in 200 mm slab with two T16 top and bottom	45.60
Gable wall	750 mm wide edge beam with two T16 bars in top and bottom with A252 fabric mesh in toe beam T10 links at 250 mm centres	34.12
Spine wall	600 mm wide slab thickening with A252 in base and two layers of A252 in 200 mm slab	12.10
Staircase wall	As spine wall	8.96

Example 2.2   Structural calculations for three-storey flats

DESIGN INFORMATION

Design codes

- BS 6399 Part 1 *Loading*
- BS 648 1964 *Schedules of weights of building materials*
- BS 5628 *Structural use of masonry*
- BS 8110 *Structural use of concrete*

Foundation concrete: F_{cu} = 35 N/mm^2
Reinforcement: f_y = 460 N/mm^2
τ_f = 1.50 for dead plus imposed loads.

The site investigation has revealed that the upper firm clays are underlain at about 2 m below ground by a 100 mm band of peat. In addition, the site is in an area known to be affected by subsidence arising from solution features in gypsum strata at depth. In view of this, a stiff edge beam raft will be used.

LOADINGS

Roof

	(kN/m^2)
Concrete tiles	0.55
Battens and felt	0.05
Self-weight of trusses at 600 mm centres	0.23
Insulation	0.02
Plasterboard	0.15
Dead load	1.00
Imposed load: Snow	0.75
Ceiling	0.25
	1.0

Therefore total roof load taken as 1.00 + 1.0   =   2.00 kN/m^2

Floors

	(kN/m^2)
150 mm deep beam and block floor	2.25
50 mm concrete screed	1.20
Plasterboard	0.15
Stud partitions	0.50
Dead load	4.10
Imposed load	1.50

Therefore total floor load taken as 4.10 + 1.50 =   5.60 kN/m^2

External walls

	(kN/m^2)
100 mm blockwork	2.0
102 mm brickwork	2.0
12.50 mm plaster	0.25
	4.25

Internal walls

	(kN/m^2)
100 mm blockwork	2.0
Two coats of plaster	0.50
	2.50

Staircases

	(kN/m^2)
175 mm in-situ concrete	4.20
Finishes	0.30
	4.50
Imposed load	3.00
Total	7.50 kN/m^2

Loadings to walls

External wall A

		(kN/m)
Wall	4.25×9.0	38.25
Three floors	$\dfrac{3 \times 5.60 \times 4.0}{2.0}$	33.60
	Total	71.85

External wall B

Wall	4.25×7.80	33.15
Roof	$\dfrac{8.13 \times 2.0}{2.0}$	8.13
	Total	41.28

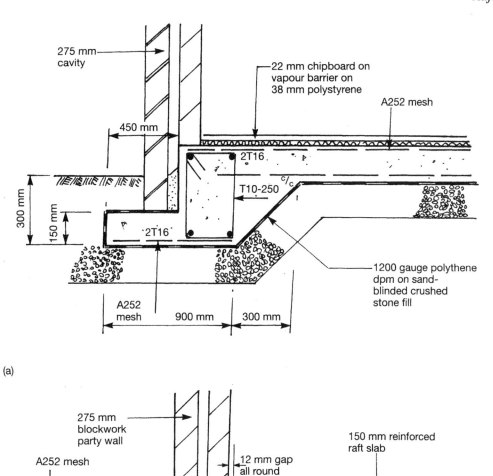

(a)

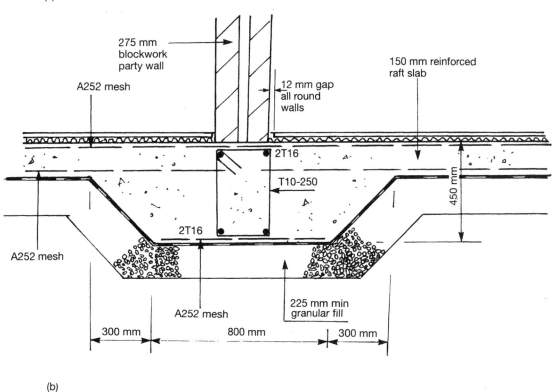

(b)

Fig. 2.21 Example 2.1: reinforcement details to raft foundation. (a) External edge beam; (b) party wall thickening.

Spine wall C

Wall	2.50×8.0	20.0
Three floors	$\dfrac{3 \times 5.60 \times 7.0}{2.0}$	58.80
	Total	78.80

Staircase wall

Wall	4.25×9.0 (average height)	38.25
Three floors	$\dfrac{3 \times 5.60 \times 3}{2.0}$	25.20

Three landings	$\dfrac{3 \times 7.500 \times 2}{2.0}$	22.50
	Total	85.25

GROUND FLOOR BLOCKWORK DESIGN

Maximum line load on spine wall = 78.80 kN/m. Using 10 N crushing strength blocks with mortar designation (iii), $f_k = 8.2$. Design vertical resistance of wall / unit length:

$$n_w = \frac{\beta t f_k}{\tau_m}$$

Foundation design

where β is a capacity reduction factor based on slenderness ratio (Table 7 BS 5628 (BSI, 1978)); f_k is the characteristic strength of the masonry obtained from Clause 23 to be taken as 8.20 N/mm²; τ_m is the partial safety factor for blockwork obtained from Clause 27, taken as 3.10; t = wall thickness in millimetres. Eccentricity at top of wall taken as 0.05 t.

$$\frac{h_{ef}}{t_{ef}} = \frac{2400}{100} \doteq 24$$

Therefore

$$\beta = 0.53$$

Therefore

$$\text{Vertical load resistance} = \frac{0.53 \times 100 \times 8.20}{3.10} = 140 \text{ N/mm}$$
$$= 140 \text{ kN/m}$$

Design load = 78.80 × 1.50 = 118 kN/m

Therefore for ground floor walls use 10 N crushing strength blocks.

Check first to second floor

Applied load $= 78.80 - \dfrac{58.80}{3.0} - (2.50 \times 2.70) = 52.55$ kN/m

Using 7.0 N strength blockwork, $f_k = 6.40$

Therefore vertical load resistance $= \dfrac{0.53 \times 100 \times 6.40}{3.10} = 109$ N/mm
$$= 109 \text{ kN/m}$$

Design load = 52.55 × 1.50 = 78.825 kN/m: OK

Therefore use 7.0 N crushing strength blocks from first floor to second floor.

Check second floor to roof level

Applied load $= 52.55 - \dfrac{58.80}{3.0} - (2.50 \times 2.70) = 26.20$ kN/m

Using 3.50 N strength blockwork, $f_k = 3.50$. Therefore

Vertical load resistance $= \dfrac{0.53 \times 100 \times 3.50}{3.10} = 59.00$ N/mm
$$= 59.00 \text{ kN/m}$$

Design load = 26.20 × 1.50 = 39.30 kN/m

Therefore use 3.50 N crushing strength blocks from second floor to roof level.

Check inner leaf of external walls

		(kN/m)
Wall A Floor load	=	33.60
Blockwork 2.50 × 9.0	=	22.50
		56.10

Using 7.0 N strength blockwork, $f_k = 6.40$. Therefore

Vertical load resistance $= \dfrac{0.53 \times 100 \times 6.40}{3.10} = 109$ N/mm
$$= 109 \text{kN/m}$$

Design load = 56.10 × 1.50 = 84.15 kN/m

Therefore use 7.0 N crushing strength blocks for all the inner leaf to the external walls from ground floor up to the second floor level.

RAFT DESIGN

Design edge beams to span a 4 m diameter solution feature and cantilever 2 m at the corners. Use $wl^2/10$ to take account of some

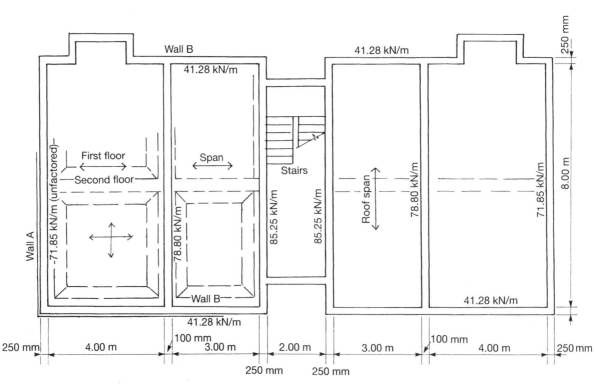

Fig. 2.22 Example 2.2: ground-floor plan of three-storey flats.

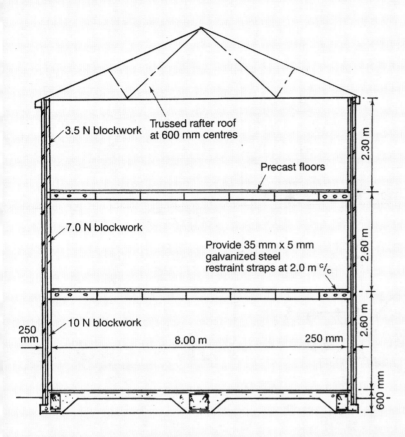

Fig. 2.23 Example 2.2: typical cross-section of three-storey flats.

fixity. Maximum line load = 85.25 kN/m. With load factor of 1.50 for dead plus imposed loads the ultimate line load equals 1.50 × 85.25 = 127.875 kN m. Therefore

$$\text{Ultimate moment (sagging)} = \frac{127.875 \times 4^2}{10} = 204.60 \text{ kN m}$$

$$\text{Ultimate moment (hogging)} = \frac{127.875 \times 2^2}{2} = 255.75 \text{ kN m}$$

Try 600 mm × 600 mm edge beam, $d = 600 - 40 - 10 = 550$ mm. From Fig. 2.11:

$$k = \frac{M}{bd^2 f_{cu}} = \frac{255.75 \times 10^6}{600 \times 550^2 \times 35} = 0.04$$

Therefore lever arm factor = 0.945, and therefore $z = 0.945 \times 550 = 519$ mm.

$$A_s = \frac{M}{0.87 fz} = \frac{255.75 \times 10^6}{0.87 \times 460 \times 519} = 1231 \text{ mm}^2$$

Therefore use four T20 high-yield bars top and bottom (1256 mm²).

Shear

$$V = 85.25 \times 2.0 \times 1.50 = 255.7 \text{ kN}$$

$$v = \frac{V}{bd} = \frac{255.75 \times 10^3}{550 \times 600} = 0.775 \text{ N/mm}^2 \quad \frac{100 A_s}{600 \times 550} = 0.38$$

$v_c = 0.45$ N/mm²

Therefore minimum links required
$s = 250$ mm

$$A_{sv} > \frac{0.40 bs}{0.87 f_y} = \frac{0.40 \times 600 \times 250}{0.87 \times 460} = 150 \text{ mm}^2$$

Therefore minimum two leg T10 links at 250 mm centres. A_{sv} = 157 mm²
Maximum spacing of links = 0.75 d = 0.75 × 550 = 412 mm.

Slab design

Total load on raft	(kN)
8 × 71.85 × 1.0 =	575
8 × 78.80 × 1.0 =	630
8 × 85.25 × 1.0 =	682
7.50 × 41.28 × 2.0 =	620
	2507

Load spread = 8.73 × 8.315 = 72.60 m². Therefore average bearing pressure below slab = 2507/72.60 = 34.50 kN/m². Assume slab spans simply supported. Maximum span = 4.0 m.

			(kN/m²)
Ultimate design load	= 1.50 × 34.50	=	51.75
Less weight of slab	= 1.40 × 0.20 × 24	=	6.72
			45.03

Therefore net uplift pressure = 45.03 kN/m². Try two-way spanning slab, simply supported at edges.
Provide spine beam across centre of raft.
$l_x = 4.0$ m, $l_y = 4.0$
$m_{sx} = \alpha_{sx} n l_x^2$
$m_{sy} = \alpha_{sy} n l_x^2$

From Table 3.14:

$\frac{l_y}{l_x} = \frac{4.0}{4.0} = 1.0 \qquad \alpha_{sx} = 0.062$

$\qquad \alpha_{sy} = 0.062$

$m_{sx} = 0.062 \times 45.03 \times 4^2 = 44.66$ kN m

$m_{sy} = 0.062 \times 45.03 \times 4^2 = 44.66$ kN m

$\frac{l_y}{l_x} = \frac{4.0}{3.0} = 1.33 \qquad \alpha_{sx} = 0.093$

$\qquad \alpha_{sy} = 0.055$

$m_{sx} = 0.093 \times 45.03 \times 3^2 = 37.69$ kN m

$m_{sy} = 0.055 \times 45.03 \times 3^2 = 22.28$ kN m

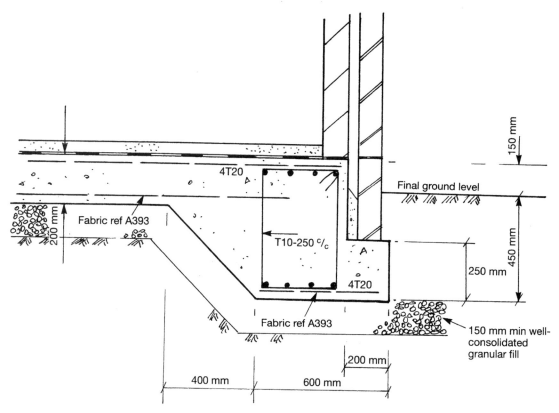

Fig. 2.24 Example 2.2: raft edge detail.

For $f_{cu} = 35$ N/mm^2 and $f_y = 460$ N/mm^2

$$k = \frac{M}{bd^2 f_{cu}} = \frac{44.66 \times 10^6}{1000 \times 155 \times 155 \times 35} = 0.053$$

From Fig. 2.11, $z = 0.93$. Therefore:

$$A_s = \frac{44.66 \times 10^6}{0.87 \times 155 \times 0.93 \times 460} = 775 \ \text{mm}^2/\text{m}$$

Use fabric A393 supplemented with T10 at 200 mm centres in top of slab both ways with A393 in bottom of slab.
For the 4.0 m × 3.0 m bay, moment = 37.69 kN m. Therefore:

$$\frac{M}{bd^2 f_{cu}} = \frac{37.69 \times 10^6}{10^3 \times 155 \times 155 \times 0.935} = 0.044$$

Therefore $z = 0.935$

Therefore $A_s = \dfrac{37.69 \times 10^6}{0.87 \times 460 \times 155 \times 0.935} = 650 \ \text{mm}^2$

Use fabric A393 supplemented with T10 at 300 mm centres in top in both directions with A393 in bottom of slab.

2.6 PAD AND PIER FOUNDATION

This type of foundation (Fig. 2.25) can be used in situations where piling is being considered. If only one or two dwellings are affected the cost of piling can be prohibitive because of the initial cost of getting the piling rig to the site. It is used in situations where very soft clays, peat and fill materials overlie firm or stiff strata at depths up to 3–4 m. On sites where foundations are on rock strata and a deep face is

encountered because of past quarrying or geological faulting, depths of up to 6 m are still more economic than piling and construction delays can be reduced.

The piers can often be constructed using manhole ring sections placed on a concrete pad foundation and filled with mass concrete. The top section of the pier can be reinforced to form a connection for the reinforced concrete ring beams placed at or close to ground level.

This method of construction can also be used where existing drainage is too close to a proposed wall foundation. In all cases the pad foundations must be wholly on similar bearing strata.

2.6.1 Disused wells

Quite often old disused wells are encountered on housing sites, usually during excavation for the foundations. When such wells are found it is wise to make the well safe without altering the water source which supplies the well. Filling a well with mass concrete is not a recommended solution: it could be very expensive and changes in the groundwater regime could occur. The most suitable method if the well is deeper than 2 is to fill the well with 150 mm single-size stone and, if in a garden area, provide a reinforced concrete cap twice the diameter of the well. If the well is under or very close to a foundation then, after filling, a beam system will be required to span over the well. The beams should extend a sufficient distance beyond the well; this minimum distance is generally taken as the well diameter each side.

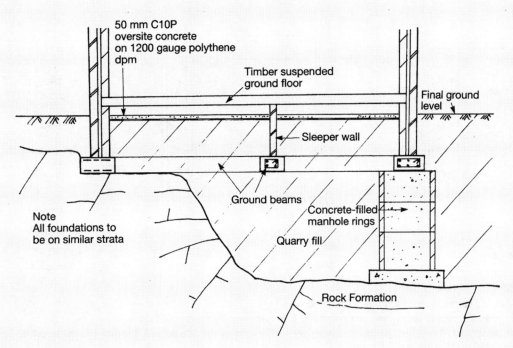

Fig. 2.25 Pad and pier foundations.

Example 2.3 Disused well

During excavation for a house footing a well was discovered following removal of a large sandstone cover (Fig. 2.26). The well was found to be 6 m deep and water was within 1.0 m of the ground

level. The well was approximately 2.0 m in diameter and the top 2.0 m was brick-lined. The well was positioned below the junction of the rear wall and party wall. It was not possible to reposition the dwelling so a beam system in the form of a tee configuration was adopted.

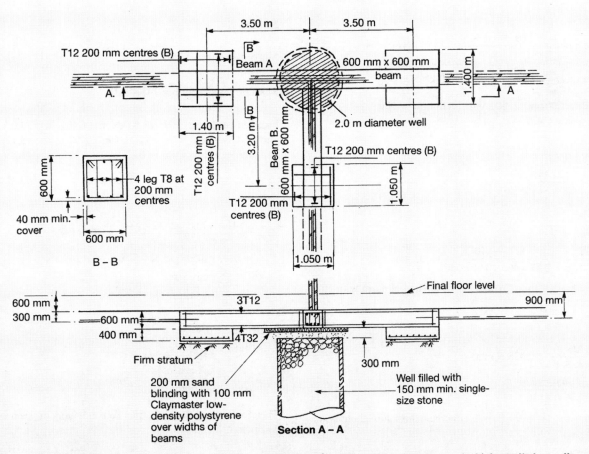

Fig. 2.26 Example 2.3: disused well – foundations. Concrete mix 30 N/mm² at 28 days; reinforcement to be high-tensile bars; allowable ground bearing pressure 80 kN/m² minimum.

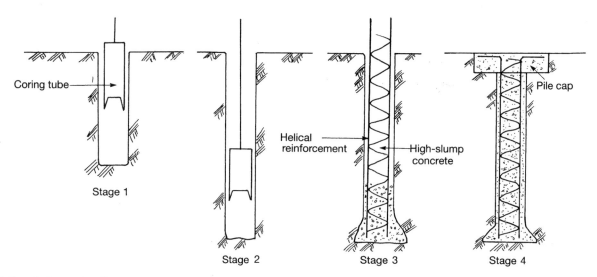

Fig. 2.27 Bored pile construction sequence.

Stage 1: Coring tube with cutting edge to suit ground conditions.

Stage 2: Shaft fully formed. A temporary casing may be required in water-bearing gravels and sands.

Stage 3: Helical reinforcement placed and spaced off sides of shaft. High-slump concrete poured. Temporary casing may be required while maintaining head of concrete in shaft.

Stage 4: Top of pile cut down and pile cap cast.

PARTY WALL: BEAM B

Factored dead load	=	$36\,\text{kN/m} \times 1.4$	=	50.40
Factored live load	=	$12\,\text{kN/m} \times 1.6$	=	19.20
Self weight of beam	=	$8\,\text{kN/m} \times 1.4$	=	11.20
		Total load	=	80.80 kN/m

Beam spans 3.50 m on to a pad foundation.

$$\text{Ultimate moment} = \frac{80.80 \times 3.50^2}{8} = 123.75 \text{ kN m}$$

$$\text{Ultimate shear} = \frac{80.80 \times 3.50}{2.0} = 141.40 \text{ kN}$$

Make beam 600 mm × 600 mm section. $d = 600 - 40 - 16 = 540$ mm. Using Fig. 2.11:

$b = 600$ mm

$f_{cu} = 30$ N/mm²

$$\frac{M}{bd^2 f_{cu}} = \frac{123.75 \times 10^6}{600 \times 540^2 \times 30} = 0.024$$

$z = 0.95\,d$

Therefore:

$$A_s = \frac{M}{0.87 f_y z} = \frac{123.75 \times 10^6}{0.87 \times 460 \times 0.95 \times 540} = 602 \text{ mm}^2$$

Therefore provide four T20 mm bars in bottom of beam with three T12 in top.

Shear

$$v = \frac{141.40 \times 10^3}{600 \times 540} = 0.438 \text{ N/mm}^2$$

$$\frac{100 A_s}{bd} = \frac{100 \times 4 \times 314}{600 \times 540} = 0.39\%$$

$v_c = 0.46$. As $v < v_c + 0.4$, provide minimum links throughout beam. Therefore:

$$A_{sv} = \frac{0.40 \times 300 \times 600}{0.87 \times 460} = 180 \text{ mm}^2/\text{m}$$

Therefore use four leg T8 links at 300 mm centres to satisfy clause 3.4.5.5, which stipulates that no longitudinal tension bar should be further than 150 mm from a vertical leg of a shear link.

BEAM A

Make span of beam 7.0 m

Reaction from beam B = 141.40 kN.

Factored dead load	=	$23.60\,\text{kN/m} \times 1.40$	=	33.0
Factored live load	=	$3.20\,\text{kN/m} \times 1.60$	=	5.12
Self-weight of beam	=	$8\,\text{kN/m} \times 1.40$	=	11.20
		Total load	=	49.32 kN/m

$$M_{ult} = \frac{141.4 \times 7.0}{4.0} + \frac{49.32 \times 7.0^2}{8.0} = 550 \text{ kN m}$$

$$V_{ult} = 141.40 \times 0.50 + 49.32 \times 0.50 \times 7.0 = 244 \text{ kN}$$

$$\frac{M}{bd^2 f_{cu}} = \frac{550 \times 10^6}{600 \times 540^2 \times 30} = 0.105$$

Therefore $z = 0.86\,d$
Therefore:

$$A_s = \frac{550 \times 10^6}{0.87 \times 460 \times 540 \times 0.86} = 2959 \text{ mm}^2$$

Therefore provide four T32 mm bars in bottom of beam with three T12 in top.

$$v = \frac{244 \times 10^3}{540 \times 600} = 0.76$$

Therefore

$$\frac{100 A_s}{bd} = \frac{100 \times 3220}{600 \times 540} = 0.99$$

Therefore $v_c = 0.63$, and hence A_{sv} is less than $v_c + 0.4$ and minimum links are required.

Hence

$$A_{sv} = \frac{0.4 \times 600 \times 300}{0.87 \times 460} = 180 \text{ mm}^2$$

Provide four leg T8 at 300 mm centres.

PAD FOUNDATIONS

Maximum unfactored line load on party wall = 48 + 8 = 56 kN/m. With a 600 mm wide footing, bearing pressure = 56/0.60 = 93 kN/m².
Maximum unfactored line load on external wall = 26.80 + 8 = 34.80 kN/m.
With a 450 mm wide footing, bearing pressure $= \frac{34.80}{0.45} = 77$ kN/m.

In order that differential settlements between the pad foundation and the strip footings can be kept within acceptable limits, a maximum allowable bearing pressure of 100 kN/m² should be adopted under the pad foundations even though the natural stiff clays are good for about 200 kN/m². In addition some section of the ground beam between the bearing pads and the well will be supported on natural ground, reducing ground bearing pressures even further.

Party wall pad

Unfactored reaction $= 56 \times \dfrac{3.50}{2} = 98$ kN

Self-weight of base $= 0.4 \times 24 = 9.6$ kN/m²

Area required $= \dfrac{98.0}{100 - 9.60} = 1.084$ m²;

say 1.05 m × 1.05 m × 400 mm thick pad foundation.

External wall pad

Unfactored reaction $= 34.80 \times \dfrac{7.0}{2} + \dfrac{98.0}{2} = 170.80$ kN

Area required $= \dfrac{170.80}{100 - 9.60} = 1.88$ m² $= 1.40$ m × 1.40 m ×

400 mm thick pad foundation

Base reinforcement

$d = 400 - 40 - 10 = 350$ mm
Maximum ultimate moment

$$= (100 - 9.6) \times 1.5 \times \frac{0.40^2}{2.0} = 10.848 \text{ kN m}$$

$$\frac{M}{bd^2 f_{cu}} = \frac{10.848 \times 10^6}{10^3 \times 350^2 \times 30} = 0.003$$

Therefore $l_a = 0.95d = 0.95 \times 350 = 332$ mm

Therefore $A_s = \dfrac{10.848 \times 10^6}{0.87 \times 460 \times 332} = 82 \text{ mm}^2$

Minimum percentage $= \dfrac{0.13}{100} \times 10^3 \times 350 = 455 \text{ mm}^2$

Therefore use T12 at 200 mm centres both directions in bottom.

Party wall pad

Maximum ultimate moment

$$= (100 - 9.6) \times 1.5 \times \frac{0.225^2}{2} = 3.43 \text{ kN m}$$

By inspection, nominal steel will be required.

Minimum percentage $= \dfrac{0.13}{100} \times 10^3 \times 350 = 455 \text{ mm}^2$

Therefore use 12 mm at 200 mm centres each way.

2.7 PILED FOUNDATIONS

Where a lot of dwellings require special foundations and the fills or weak ground are not suitable for ground improvement techniques, piling can be carried out at very little extra cost provided the design details are carefully considered. In most situations driven piles are used, driven to a predetermined set and subjected to a random load test of 1.50 times the working load.

Driven piles should be avoided in situations where the bedrock profile can vary over a short distance, as in infilled railway cuttings and backfilled stone quarries, for example. In these situations driven piles are prone to drifting out of plumb during the driving stage and often end up being damaged because of the eccentric loads applied.

If piling is being used then the ground beams should be kept as high as possible. It is often more economic to design a scheme based on large piles, especially if the piles are taken down on to rock or hard clays. One of the problems in using small mini-type piles in yielding strata, such as clays and sands, is that the end bearing component of the working load required is usually of a low magnitude and the piles need to be driven deeper to pick up sufficient skin friction. What may seem to be an economic piling scheme based on estimated driven lengths may, on final remeasure, end up being very costly.

Where piles are driven or bored through filled ground still settling under its own weight, due allowance must be made for the additional load arising from negative skin friction. Also, on sites where highly compressible strata such as peat are likely to be loaded, due to the site levels being raised, additional loads will be transferred to the pile shaft.

2.7.1 Bored piles

These are generally formed using a simple tripod rig. Where there are groundwater problems, or very soft clays which may cause necking, then temporary casings or permanent steel sleeves should be used. Great care must be exercised when withdrawing temporary casings, and a sufficient head of concrete should always be maintained in the pile shaft to prevent necking or concrete loss.

Bored piles usually rely on end bearing and skin friction to support the pile loads. They are best suited to clay sites where no groundwater problems exist and the upper strata are strong enough to maintain an open bore. Once steel casings have to be considered, the bored pile can be uneconomic compared with other faster systems.

Their main advantage is that they can be installed quietly with minimum vibration; ideal when piling close to existing buildings.

Where end bearing is required it is important to ensure that the stratum below the pile toe remains competent for a distance of at least 3 m. If no site investigation is available this can be checked by overboring several piles.

Figure 2.27 shows the sequence of operations for installing a bored pile. Different cutting tools are required for the various soil conditions.

Safe loads for bored piles in clay soils are generally calculated using the Skempton formulae which combine both end bearing and frictional properties of the pile.

For varying soil strengths the skin friction is considered for the separate elements of the pile shaft with negative skin friction being considered when passing through filled ground.

2.7.2 Design of a bored pile

Adopt a factor of safety of 3 for end bearing and a factor of safety of 2 for skin friction. Q_u = the ultimate resistance of the pile = $Q_s + Q_b$. Q_s is the ultimate value for skin friction = shaft area $\times 0.45c = \pi\,dh \times 0.45c$. Q_b is the ultimate value for end bearing = base area $\times 9c = \frac{1}{4}\pi\,d^2 \times 9c$ where c is the cohesion value of the clays determined by laboratory testing or by the use of field tests: i.e. penetrometer or shear vane tests. Thus

$$\text{Allowable working load for a pile } = \frac{Q_s}{2} + \frac{Q_b}{3}$$

$$= \frac{1}{2}\left(\pi dh \times 0.45c\right) + \frac{1}{3}\frac{\left(\pi d^2 \times 9c\right)}{4}$$

$$= \frac{\pi dc(2.7h + 9d)}{12}$$

Where the clay cohesions vary, the pile shaft is split into vertical elements using the appropriate values down the pile length.

2.7.3 Design of bored and driven piles

Estimation of approximate working loads is as follows.

The ultimate load capacity Q_u of a pile is

$$Q_u = Q_b + Q_s$$

where Q_s = ultimate shaft resistance, Q_b = ultimate base area resistance, q = ultimate unit end bearing resistance, A_b = effective cross-sectional area of the base of the pile, f = ultimate unit shaft resistance on sides of pile, A_s = effective surface area of pile shaft considered as loaded.

Base resistance: clay strata

The ultimate base resistance of a bored or driven pile in cohesive strata is given by

$$q = N_c\,C_b$$

where N_c = bearing capacity factor, C_b = undrained shear strength of the cohesive strata at the pile base. Values of N_c for cohesive strata can be variable and are dependent on the angle of internal friction, ϕ. For estimating purposes a value of 9 is generally used for pile diameters up to 450 mm diameter.

Base resistance: granular strata

The base resistance in a granular stratum is given by

$$q = \tau D N_q$$

where N_q is the bearing capacity factor obtained from the Berezantsev graph (Berezantsev, 1961) based on the angle of shearing resistance ϕ for the stratum. The value of ϕ is usually determined from the standard penetration test results carried out in the field. For bored piles an N_q value appropriate for loose soil conditions is recommended as the boring operation loosens the strata, and ϕ values of 28–30° can be used. τ = average effective unit weight of soil surrounding the pile, and D = depth to the base of the pile. The value of q should not exceed 11 MN/m².

Shaft resistance in clay soils

The shaft resistance f is given by

$$f = \alpha C_s$$

where α = 0.45 for bored piles, C_s = undrained shear strength. α is taken as 1.0 for driven piles in contact with strata with $C_s < 50$ kN/m².

For strata with $C_s > 50$ kN/m² the value of α lies between 0.25 and 1.0 and is dependent on the depth of penetration into the clay strata and the prevailing ground conditions.

The shaft resistance f in granular soils is given by

$$f = \frac{1}{2}\tau(D + d)K_s\tan\delta$$

where D = depth to base of pile or base of the granular strata, whichever is the lesser; d = depth to the top of the granular strata; δ = angle of friction between the granular strata and the pile shaft; K_s = earth pressure coefficient dependent on the relative density of the soil.

Broms (1966) related the values of K_s and δ to the effective angle of shearing resistance of granular soils ϕ for various types of piles and relative densities. For driven piles:

Pile type	δ	K_s	
		Low relative density	High relative density
Steel	20°	0.5	1.0
Concrete	$\frac{3}{4}\phi$	1.0	2.0

ϕ is generally taken to be the value of ϕ as obtained from the SPT tests. For bored piles, values of δ = 22° and K_s = 1.0 should be used to cater for the loosening effect when boring out.

The expression $\frac{1}{2}\tau(D+d)K_s\tan\delta$ can only be used for penetration depths up to 10–20 times the pile diameter. Between 10 and 20 times the pile diameter a peak value of unit skin friction is reached and this value is not exceeded at greater depths of penetration. It is prudent to adopt a peak value of 100 kN/m² for straight-sided piles.

Tomlinson (1980) suggested that the following approximate value can be adopted for f (kN/m²):

Relative density		f (kN/m²)
<0.35	Loose	10
0.35–0.65	Medium dense	10–25
0.65–0.85	Dense	25–70
>0.85	Very dense	70 but < 110

A factor of safety of 2.5 should be adopted to these ultimate values to obtain the allowable or safe working load on the pile.

Example 2.4   Bored piles

A site investigation has revealed that loose colliery waste fills about 4 m thick overlie firm-to-stiff clays which are underlain by weathered mudstones. The maximum pile loading is 300 kN but an additional load resulting from negative skin friction has to be catered for as the fills have only been in place for one year and are still consolidating under their own weight. The strata are described in Figs 2.28 and 2.29.

An existing culvert passes under the proposed dwelling and its condition is not known. It has been decided therefore to provide bored piles down to the strong mudstones at about 6.50 m below ground level.

Based on information from borehole no. 1 (Fig. 2.28), no test results are available for the firm/stiff brown clays and weathered mudstones. Assume a value for c of 50 kN/m² which is very conservative (Fig. 2.30).

t = unit negative skin friction
p_o = effective overburden pressures
ϕ_e = the effective angle of shearing resistance
τ_{act} = $k\,p_o\tan\phi_e$

From Bjerrum, $\tau_{act}=0.20\,p_o$ for clays of low plasticity. The negative skin friction factor $\alpha=0.20$

Skin friction ultimate $=\dfrac{\gamma_h\alpha A_s}{2}$

Required pile capacity = 300 kN.

NEGATIVE SKIN FRICTION

This is approximately $=\frac{1}{2}\tau_{act}\,p_o A_s$

$$=\frac{0.20(18\times3.80)}{2}\times(\pi\times0.45\times3.80)=38\text{ kN}$$

Therefore total pile capacity required = 300 + 38 = 338 kN.

END BEARING

For N values of 49 the unconfined compressive strength of the mudstone equates to $13.30\times49=652$ kN/m² (soft rock).

The shear strength $=\dfrac{652}{2}=326$ kN/m²

Allowable end bearing pressure

$$=\frac{N_c\times c}{\text{Factor of safety}}=\frac{9\times326}{3.0}=978\text{ kN/m}^2$$

End bearing capacity $=978\times\dfrac{\pi\times0.6^2}{4}=277$ kN

SKIN FRICTION

α = adhesion factor = 0.40.

$$U_s=\frac{c\times A_s}{\text{Factor of safety}}=\frac{0.40\times50\times(\pi\times0.45\times2.45)}{2.0}$$
$$=34\text{ kN}$$

Therefore pile capacity = 277 + 34 = 311 kN.

With a 1.0 m penetration into the mudstone this value is considered adequate for borehole no. 1.

Based on soils strata in borehole no. 2 (Figs 2.29, 2.31):

Required pile capacity	= 300 kN
Negative skin friction =	

$$\frac{0.2(18\times7.2)}{2}\times(\pi\times0.45\times7.2) \quad = 132\text{ kN}$$
$$+\,0.1(18\times7.2+0.65\times20)(\pi\times0.45\times1.3) \quad = 26\text{ kN}$$

Therefore pile capacity required = 458 kN

For $N=152$, unconfined compressive strength
$=13.3\times152$ = 2022 kN/m²

Shear strength $=\dfrac{2022}{2}$ = 1011 kN/m²

Allowable end bearing capacity

$$=\frac{9\times1011\times(\pi\times0.45^2)/4}{3} \quad = 482\text{ kN}$$

This is greater than the 458 kN required. The piles must therefore penetrate at least 1.0 m into the shaley mudstones, ideally below the weathered zones. The minimum cement content in piles should be 370 kg/m³ sulphate-resisting cement with a free water ratio not exceeding 0.45 to cater for Class 2 sulphates in soils.

PILE REINFORCEMENT

$N=1.60\times458=732$ kN
$f_{cu}=40$ N/mm²

With 75 mm tolerance on pile position the bending moment on the pile shaft equates to

$$\frac{75}{10^3}\times(1.60\times458)=54.96\text{ kN (ult)}$$

$$\frac{N}{d^2 f}=\frac{732\times10^3}{450\times450\times40}=0.090$$

$$\frac{M}{d^3 f_{cu}}=\frac{54.96\times106}{450^3\times40}=0.015.$$

Therefore use nominal steel only. Provide seven R12 vertical bars with nominal R6 helical binders at 150 mm pitch.

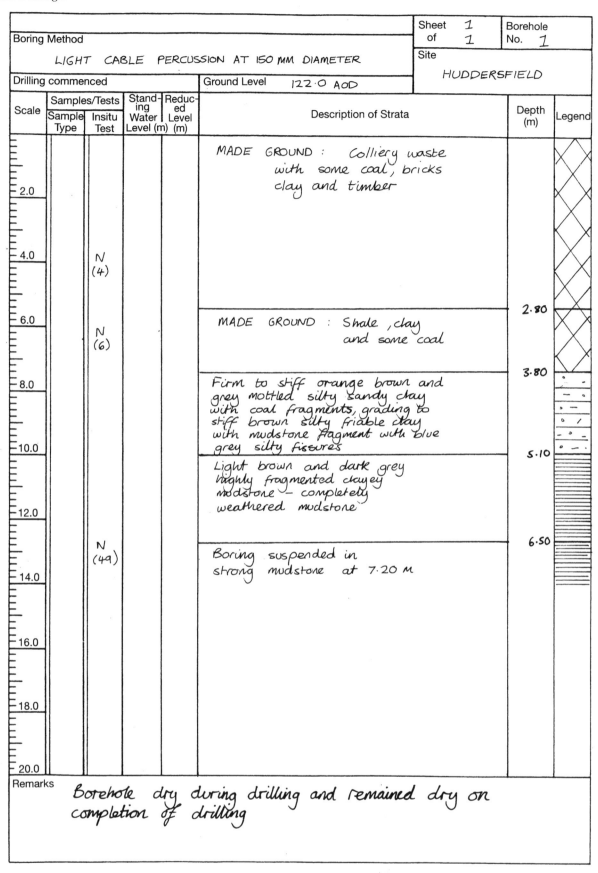

Boring Method					Sheet 1 of 1		Borehole No. 1
LIGHT CABLE PERCUSSION AT 150 MM DIAMETER					Site HUDDERSFIELD		
Drilling commenced				Ground Level 122·0 AOD			

Scale	Samples/Tests		Stand-ing Water Level (m)	Reduc-ed Level (m)	Description of Strata	Depth (m)	Legend
	Sample Type	Insitu Test					
2.0					MADE GROUND : Colliery waste with some coal, bricks clay and timber		
4.0		N (4)					
6.0		N (6)				2·80	
					MADE GROUND : Shale, clay and some coal		
8.0					Firm to stiff orange brown and grey mottled silty sandy clay with coal fragments, grading to stiff brown silty friable clay with mudstone fragment with blue grey silty fissures	3·80	
10.0							
12.0					Light brown and dark grey highly fragmented clayey mudstone - completely weathered mudstone	5·10	
14.0		N (49)			Boring suspended in strong mudstone at 7·20 M	6·50	
16.0							
18.0							
20.0							

Remarks
Borehole dry during drilling and remained dry on completion of drilling

Fig. 2.28 Example 2.4: borehole 1.

2.7.4 Driven piles

In the housing field these usually consist of steel tubes filled with concrete, precast concrete segmental piles, concrete segmental shell piles or other similar types.

The main advantage of these types of pile is that they can be placed through weak or water-bearing strata without any changes occurring in their cross-section. They are generally end bearing piles but additional loads resulting from skin friction (adhesion) can be developed when driven through

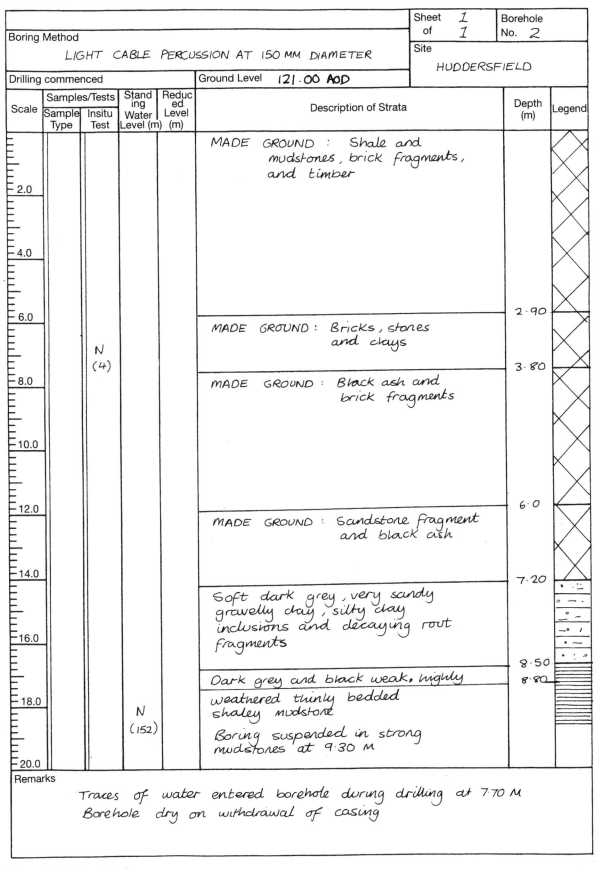

Boring Method					Sheet 1 of 1		Borehole No. 2	
LIGHT CABLE PERCUSSION AT 150 MM DIAMETER					Site HUDDERSFIELD			
Drilling commenced			Ground Level 121.00 AOD					

Scale	Samples/Tests		Standing Water Level (m)	Reduced Level (m)	Description of Strata	Depth (m)	Legend
	Sample Type	Insitu Test					
		N (4)			MADE GROUND : Shale and mudstones, brick fragments, and timber	2.90	
					MADE GROUND : Bricks, stones and clays	3.80	
					MADE GROUND : Black ash and brick fragments		
					MADE GROUND : Sandstone fragment and black ash	6.0	
					Soft dark grey, very sandy gravelly clay, silty clay inclusions and decaying root fragments	7.20	
		N (152)			Dark grey and black weak, highly weathered thinly bedded shaley mudstone	8.50 8.80	
					Boring suspended in strong mudstones at 9.30 M		

Remarks

Traces of water entered borehole during drilling at 7.70 M
Borehole dry on withdrawal of casing

Fig. 2.29 Example 2.4: borehole 2.

the clays or gravels overlying the bearing strata.

These piles are usually driven to a predetermined set based on the **Hiley formula** which is a dynamic criterion related to the weight of the driving hammer and height of the drop.

When using this concept it is essential that a random pile be load tested using kentledge or by jacking against tension piles to confirm that the pile driving assumptions and established set criteria are valid.

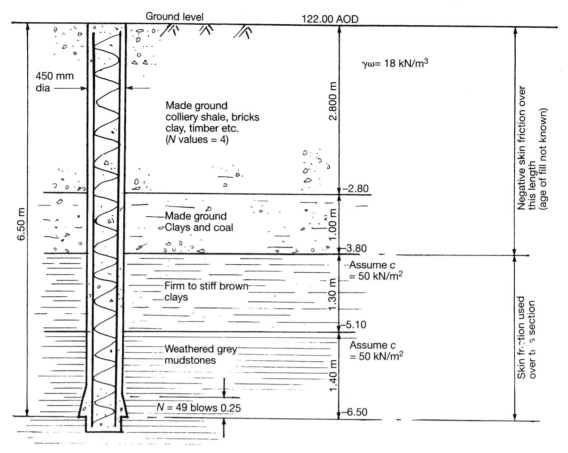

Fig. 2.30 Example 2.4: ground conditions at borehole 1.

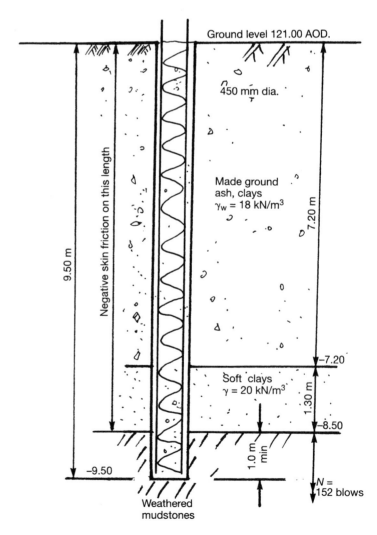

Fig. 2.31 Example 2.4: ground conditions at borehole 2.

The Hiley piling equation is

$$R_\mathrm{u} = \frac{Whn}{s+c/2}$$

Note: This equation should never be used when driving into silts or soft clays.

where R_u = ultimate resistance to be overcome by driving; h = effective drop height of piling hammer; w = weight of the piling hammer; n = hammer blow efficiency, which can vary for different arrangements of pile, helmet and driving dolly; s = set or penetration per blow; c = total temporary compression of pile, soil, dolly and driving cap.

When driving piles a good check on the driving formula is to re-strike the piles. Re-strike is best carried out the day following driving to allow a period of rest, especially in ground with pore water pressures.

If a smaller set is achieved on re-strike this indicates the validity of the predetermined set. If the re-strike set exceeds the previous set obtained before rest then the piles should be driven further or load tested. It may at this stage be necessary to examine the soils data and review the piling design. Provided the weight of the drop hammer is adequate to drive the pile through the upper strata at an acceptable rate without damaging the pile head, a satisfactory set is generally considered to be 10 blows giving a penetration of 25 mm, i.e. 2.5 mm/blow, but these sets can vary for different piling systems. Some piling contractors and firms specializing in remedial mini-piling use vibratory hammers; these should only be used when the operating characteristics of the vibratory hammers, penetration and soils data have been proven on similar soils.

2.7.5 Driving precast piles

These are usually driven using the modified Hiley formula:

$$R_\mathrm{u} = \frac{E}{s+c/2}$$

where R_u = ultimate load = total working load × factor of safety of 2; E = transfer energy at pile top = 0.7×10^4 kN mm; c = temporary compression of pile and ground per blow, taken as 12 mm; s = set blow count. Therefore

$$s = \frac{E}{R_\mathrm{u}} - \frac{c}{2} = \frac{7000}{267\times2} - \frac{12}{2} = 7.11 \text{ mm/blow}$$

Using a 3 tonne hydraulic hammer with a 400 mm drop this should result in a set for 10 blows of 70 mm or less.

General notes

(a) The Hiley formula should only be applied to piles which obtain their support in cohesionless soils such as sands and gravels, or in very stiff or hard clays or rock strata.

   The formula is not applicable to pure 'friction piles' penetrating into and obtaining their support in soft clays or silts throughout most of the pile length.

(b) If a reduction of resistance is shown following a re-drive check, the Hiley formula should not be used, and R_u should be determined by means of a load test.

(c) When driving piles in backfilled stone quarries or infilled railway cuttings a rock shoe should be provided to enable the pile toe to obtain good penetration. Driven piles can be unsuitable in such situations because the rock profile is not visible and may be steeply sloping. This could give rise to instability at the toe of the pile. Extreme care should be exercised during the driving operation for any signs of the pile losing its verticality.

(d) The recommended tolerances for pile positions are:
   75 mm out of position on plan;
   maximum 1 in 75 out of plumb.
   Where piles are outside the 75 mm allowed, various alterations can be made to the ground beams to accommodate the eccentric moments developed. Where piles are placed out of plumb it may be sufficient just to downrate the pile load capacity. However, if the out of plumb is as a result of sloping rock at the pile toe then a new pile should be installed using a rock socket.

2.7.6 Test loading

Two types of test loading are in general use. In the first, the **maintained load test**, the kentledge load or adjacent tension piles are used to provide a reaction for the test load. This load is applied by hydraulic jacks placed over the pile being tested and the load applied is measured by a proving ring or similar load-measuring device which should be calibrated before and after each series of tests. The load is increased in fixed incremental steps and is held at each stage of loading until all settlement has either ceased or does not exceed a specified amount in a given period of time. Figure 2.32 shows a typical pile-testing arrangement using tension piles and Figs 2.33 and 2.34 show the load / settlement recorded.

The second type of test is the **constant rate of penetration** method. The test arrangement is the same as for the maintained load test but the load is increased continuously at a rate such that settlement of the pile occurs at a constant rate per minute. Generally the applied 'proof' load is 1.50 times the working load of the pile. In proof-loading tests it is usually required that the pile shall not have failed at the specified proof load, and that the net or gross settlement of the pile head shall not exceed a stated figure.

Example 2.5   Detached house: pile and ground beam design

The site investigation has revealed made ground resulting from a filled-in clay pit close to a brickworks. The fills consist of soft clays and mixed rubble which were of loose consistency. No organic materials were evident.

The maximum depth of the clay pit was 4.50 m and the base of the pit revealed stiff brown clays containing small boulders underlain by sandstone rock at a depth of 8.0 m. Owing to the condition of the fill it was decided that 150 mm steel tube driven piles should be used.

DESIGN PHILOSOPHY

Where continuous beams are approximately equal spans it is reasonable to design them as simply supported and provide top steel over the supports equivalent to the bottom steel. This will satisfy the serviceability requirements of BS 8110 in preventing any

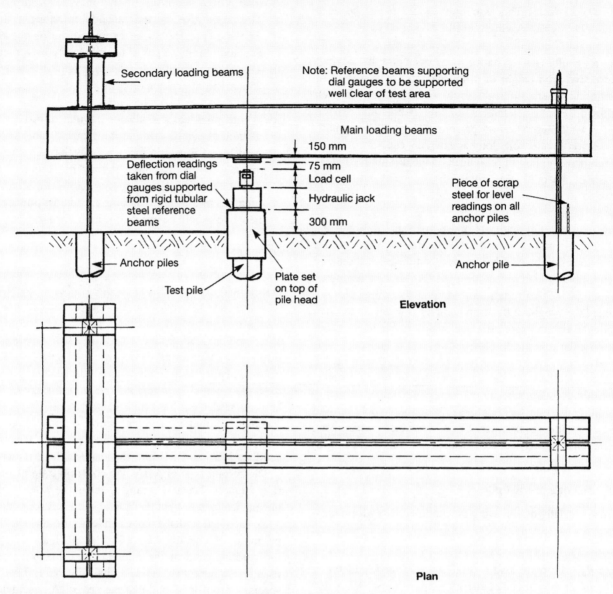

Fig. 2.32 Static load pile-testing arrangement using tension piles.

cracking over the supports. Ground floor construction to be fully suspended using full-span timber joists at 400 mm centres. Provide a 400 mm wide by 600 mm deep reinforced concrete ground beam with tie beam across the lounge. Piles to be 6.5 mm thick, 150 mm diameter steel tubes driven to a predetermined set and tested to 1.50 times the safe working load.

ASSESSMENT OF WALL LOADINGS

BS 6399 (1984) Part 1 *Code of practice for dead and imposed loads*
BS 648 (1964) *Schedule of weights of building materials.*
If we consider a traditional two-storey brick and blockwork construction with timber ground and first floors and a trussed rafter roof with concrete tiles the following loads are applicable for calculating the line loads to each wall.

- For residential self-contained housing units imposed loads are 1.50 kN/m^2 for floors and for flat roofs up to 10° pitch to which access is possible.
- For flat roofs and sloping roofs up to 10° pitch where no access is provided the imposed load (including snow load) is to be 0.75 kN/m^2.
- For sloping roofs with a pitch in excess of 10° where no access is provided the imposed load is to be 0.75 kN/m^2 up to 30°.
- For a roof slope greater than 75° pitch the imposed load is zero.
- For roofs of pitch between 30° and 75° the imposed load is determined by linear interpolation between 0.75 kN/m^2 for a 30° slope and zero for a 75° slope.
- For residential balconies and staircases imposed loads are taken as 1.50 kN/m^2 for two-storey dwellings which have single occupancy.
- For apartment dwellings exceeding two storeys, corridors, staircases, landings etc, the imposed loads are to be 3.0 kN/m^2.
- Balconies are the same imposed load as the rooms to which they give access but with a minimum of 3.0 kN/m^2.

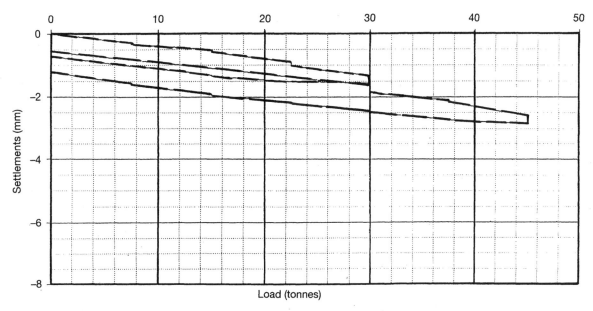

Fig. 2.33 Load–settlement plot.

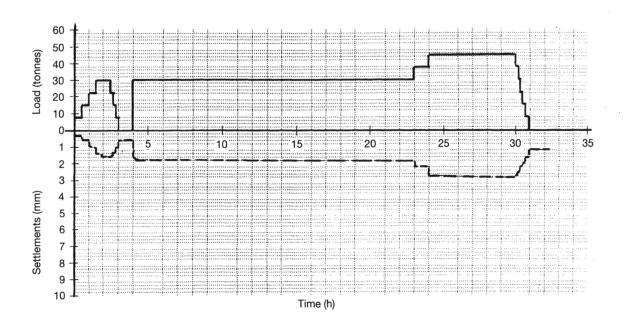

Fig. 2.34 Graph of load and settlement versus time.

Roof

Dead loads

	(kN/m²)	Factored
Tiles	0.55	
Battens and felt	0.05	
Trusses	0.23	
Plasterboard ceiling	0.15	
Insulation	0.02	
Total =	1.00	× 1.40 = 1.40 kN/m²

Imposed loads

	(kN/m²)
25° pitch snow load	0.75
Ceiling access loads	0.25
	1.0 × 1.60 = 1.60 kN/m²

Total unfactored dead load + imposed load = 2.00 kN/m²

Total factored dead load + imposed load   = 3.00 kN/m²

55

Foundation design

Ground floor and first floor

Dead loads

	(kN/m²)	Factored
22 mm boarding	0.10	
225 × 50 mm joists	0.15	
Plasterboard and skim	0.15	
Stud partitions	0.50	
	0.90 × 1.40 = 1.26 kN/m²	

Imposed loads

1.50 × 1.60 = 2.40 kN/m²

Total unfactored dead load + imposed load = 2.40 kN/m²

Total factored dead load + imposed load = 3.66 kN/m²

External walls

	(kN/m²)	Factored
Brick skin 102.5 mm	2.25	
Blockwork skin 100 mm	1.25	
Plaster	0.25	
	3.75 × 1.40 = 5.25 kN/m²	

Internal block walls

	(kN/m²)	Factored
Blockwork 100 mm	1.25	
Two coats plaster	0.50	
	1.75 × 1.40 = 2.45 kN/m²	

Party walls

	(kN/m²)	Factored
Two skins of blockwork	2.50	
Two coats plaster	0.50	
	3.0 × 1.40 = 4.20 kN/m²	

Wall line loads

Consider typical detached house with plan dimensions as shown in Fig. 2.35.

Wall A

		Unfactored (kN/m)	Factored (kN/m)
Roof: nominal 0.5 m width	= 2.00 × 0.5 =	1.00	1.50
First floor	= 2.40 × $\frac{4.0}{2.0}$ =	4.80	7.32
Ground floor	= 2.40 × $\frac{4.0}{2.0}$ =	4.80	7.32
Walls: average height	= 6.50 m × 3.75 =	24.375	34.125
Wall (A) total	=	34.975	50.265

Wall B1 (staircase section)

		Unfactored (kN/m)	Factored (kN/m)
Roof (6.0 + 0.6 m overhang)=	$\frac{6.6 \times 2.0}{2.0}$ =	6.66	9.90
First floor	= $\frac{2.4 \times 2.0}{2.0}$ =	2.40	3.66
Ground floor	= $\frac{2.4 \times 2.0}{2.0}$ =	2.40	3.66
Walls	= 3.75 × 5.5 =	20.62	28.86
Wall (B1) total =		32.02	46.08

Wall B2

32.02 − (2.40 × 2.0)		= 27.22	38.70

Wall C1

		Unfactored (kN/m)	Factored (kN/m)
Roof	= $\frac{2.0 \times 6.6}{2.0}$ =	6.60	9.90

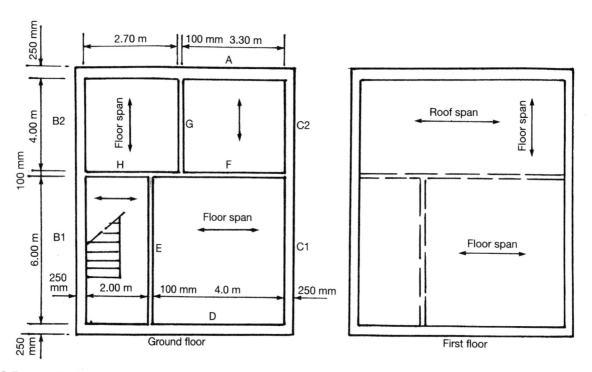

Fig. 2.35 Example 2.5: floor plans.

First floor	$= \dfrac{2.4 \times 4.0}{2.0}$	$= 4.80$	$= 7.32$
Ground floor	$= \dfrac{2.4 \times 4.0}{2.0}$	$= 4.80$	$= 7.32$
Walls	$= 3.75 \times 5.5$	$= 20.62$	$= 28.86$
Wall (C1) total		$= 36.82$	53.40

Wall C2

$$36.82 - (2 \times 4.80) = 27.22 = 38.70$$

Wall D

		Unfactored (kN/m)	Factored (kN/m)
Roof: nominal 0.5 m width	$= 0.5 \times 2.0$	$= 1.00 =$	1.50
Walls	$= 3.75 \times 6.5$	$= 24.375 =$	34.125
Wall (D) total		25.375	35.625

Wall E

		Unfactored (kN/m)	Factored (kN/m)
First floor	$= \dfrac{2.4 \times 2.0}{2.0}$	$= 2.40 =$	3.66
First floor	$= \dfrac{2.4 \times 4.0}{2.0}$	$= 4.80 =$	7.32
Ground floor	$= \dfrac{2.4 \times 2.0}{2.0}$	$= 2.40 =$	3.66
Ground floor	$= \dfrac{2.4 \times 4.0}{2.0}$	$= 4.80 =$	7.32
Wall: 100 mm blockwork	$= 1.75 \times 2.6$	$= 4.55 =$	6.37
Wall (E) total		$= 18.95 =$	28.33

Wall F

		Unfactored (kN/m)	Factored (kN/m)
First floor	$= \dfrac{2.4 \times 4.0}{2.0}$	$= 4.80 =$	7.32

Ground floor	$= \dfrac{2.4 \times 4.0}{2.0}$	$= 4.80$	$= 7.32$
Walls	$= 1.75 \times 2.6$	$= 4.55$	$= 6.37$
Wall (F) total		$= 14.15$	$= 21.01$

Wall G

$$100 \text{ mm blockwork} = 1.75 \times 2.6 \quad = 4.55 \times 1.4 = 6.37$$

Wall H (as wall F)

$$= 14.95 \times 1.4 = 21.01$$

The calculated wall line loads are summarized in Fig. 2.36.
The pile layout is shown in Fig. 2.37.
Ground beams: 600 mm × 400 mm wide
Self-weight 5.76 kN/m
Tie beams: 300 mm × 300 mm wide
Self-weight 2.16 kN/m

PILE LOADINGS

Pile 1

	Unfactored (kN)		Factored (kN)
$\dfrac{34.975 \times 2.875}{2.0}$	$= 50.27$	$\dfrac{50.625 \times 2.875}{2.0}$	$= 72.25$
$\dfrac{27.22 \times 4.175}{2.0}$	$= 56.82$	$\dfrac{38.70 \times 4.175}{2.0}$	$= 80.78$
$\dfrac{5.76 \times (4.175 + 2.875)}{2.0}$	$= 20.30$	8×3.525	$= 28.20$
Total	$= 127.39$		$= 181.23$

Pile 2

	Unfactored (kN)		Factored (kN)
$34.975 \times \dfrac{2.875}{2.0}$	$= 50.27$	$50.265 \times \dfrac{2.875}{2.0}$	$= 72.25$

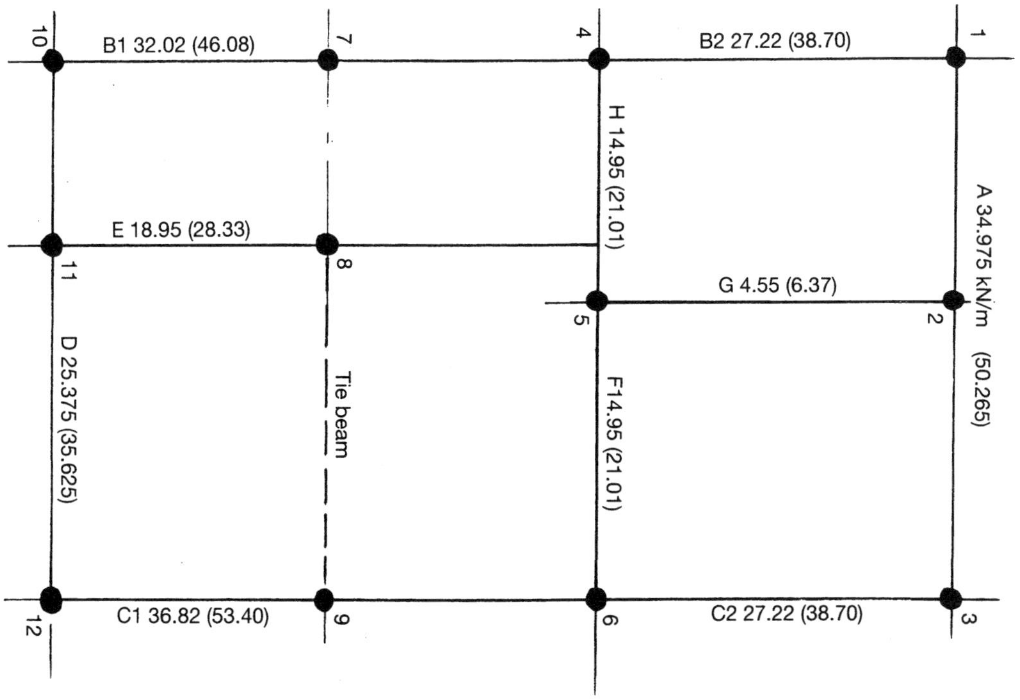

Fig. 2.36 Example 2.5: wall line loads (kN/m). Factored line loads shown in brackets.

Foundation design

$$34.975 \times \frac{3.475}{2.0} = 60.76 \qquad 50.265 \times \frac{3.475}{2.0} = 87.33$$

$$4.55 \times \frac{4.175}{2.0} = 9.50 \qquad 6.37 \times \frac{4.175}{2.0} = 13.29$$

$$5.76 \times \frac{(2.875 + 3.475 + 4.175)}{2.0} = 30.31 \qquad 8.0 \times 5.26 = 42.08$$

$$\text{Total} = 150.84 \qquad\qquad 214.95$$

$$14.15 \times \frac{2.875}{2.0} = 20.34 \qquad 21.01 \times \frac{2.875}{2.0} = 30.20$$

$$39.22 \times \frac{0.70}{2.875} = 9.54 \qquad 57.67 \times \frac{0.70}{2.875} = 14.04$$

$$\text{Total} = 166.97 \qquad\qquad = 239.07$$

Pile 5

	Unfactored (kN)		Factored (kN)

$$4.55 \times \frac{4.175}{2.0} = 9.50 \qquad 6.37 \times \frac{4.175}{2.0} = 13.29$$

$$5.76 \times \frac{4.175}{2.0} = 12.020 \qquad 8 \times \frac{4.175}{2.0} = 16.70$$

$$14.15 \times \frac{2.875}{2.0} = 20.34 \qquad 21.01 \times \frac{2.875}{2.0} = 30.20$$

$$14.15 \times \frac{3.475}{2.0} = 24.58 \qquad 21.01 \times \frac{3.475}{2.0} = 36.50$$

$$5.76 \times \frac{(2.875 + 3.475)}{2.0} = 18.280 \qquad 8 \times 3.175 = 25.40$$

$$39.22 \times \frac{2.175}{2.875} = 29.67 \qquad 57.67 \times \frac{2.175}{2.875} = 43.62$$

$$\text{Total} = 114.39 \qquad\qquad = 165.71$$

Pile 3

	Unfactored (kN)		Factored (kN)

$$27.22 \times \frac{4.175}{2.0} = 56.82 \qquad 38.70 \times \frac{4.175}{2.0} = 80.78$$

$$34.975 \times \frac{3.475}{2.0} = 60.76 \qquad 50.265 \times \frac{3.475}{2.0} = 87.33$$

$$5.76 \times \frac{(4.175 + 3.475)}{2.0} = 22.00 \qquad 8.0 \times 3.825 = 30.60$$

$$\text{Total} = 139.58 \qquad\qquad = 198.71$$

Pile 6

	Unfactored (kN)	Factored (kN)

$$27.22 \times \frac{4.175}{2.0} = 56.82 \qquad 38.70 \times \frac{4.175}{2.0} = 80.78$$

$$36.82 \times \frac{3.175}{2.0} = 58.45 \qquad 53.40 \times \frac{3.1p75}{2.0} = 84.77$$

Pile 4

Reaction from pile 8 to pile 4–5 = 39.22 kN

	Unfactored (kN)	Factored (kN)

$$27.22 \times \frac{4.175}{2.0} = 56.82 \qquad 38.70 \times \frac{4.175}{2.0} = 80.78$$

$$32.02 \times \frac{3.175}{2.0} = 50.83 \qquad 46.08 \times \frac{3.175}{2.0} = 73.15$$

$$5.76 \times \frac{(3.175 + 4.175)}{2.0} = 21.16 \qquad 8 \times 3.675 = 29.40$$

$$5.76 \times \frac{2.875}{2.0} = 8.28 \qquad 8 \times \frac{2.875}{2.0} = 11.50$$

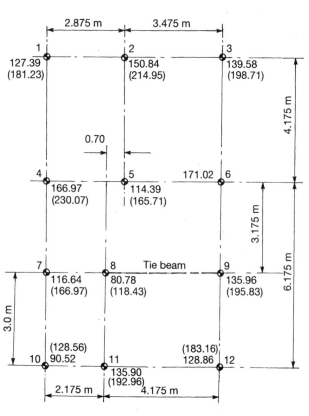

Fig. 2.37 Example 2.5: pile layout. Ultimate loads (kN) shown in brackets.

58

$$14.15 \times \frac{3.475}{2.0} = 24.58 \qquad 21.01 \times \frac{3.475}{2.0} = 36.50$$

$$5.76 \times \frac{(4.175 + 3.175 + 3.475)}{2.0} = 31.17 \qquad 8 \times 5.41 = 43.28$$

Total $= 171.02 \qquad\qquad = 245.33$

Pile 7

	Unfactored (kN)		Factored (kN)
$32.02 \times \frac{3.175}{2.0} =$	50.83	$46.08 \times \frac{3.175}{2.0} =$	73.15
$32.02 \times \frac{3.0}{2.0} =$	48.03	$46.08 \times \frac{3.0}{2.0} =$	69.12
$5.76 \times \frac{(3 + 3.175)}{2.0} =$	17.78	$8 \times 3.0875 =$	24.70
Total $=$	116.64		$= 166.97$

Pile 8

	Unfactored (kN)		Factored (kN)
$18.95 \times \frac{3.175}{2.0} =$	30.08	$28.33 \times \frac{3.175}{2.0} =$	44.97
$18.95 \times \frac{3.0}{2.0} =$	28.42	$28.33 \times \frac{3.0}{2.0} =$	42.50
$5.76 \times \frac{6.175}{2.0} =$	17.78	$8 \times \frac{6.175}{2.0} =$	24.70
$2.16 \times \frac{4.175}{2.0} =$	4.50	$3 \times \frac{4.175}{2.0} =$	6.26
Total $=$	80.78		118.43

Pile 9

	Unfactored (kN)		Factored (kN)
$36.82 \times \frac{6.175}{2.0} =$	113.68	$53.40 \times \frac{6.175}{2.0} =$	164.87
$5.76 \times \frac{6.175}{2.0} =$	17.78	$8 \times \frac{6.175}{2.0} =$	24.70
$2.16 \times \frac{4.175}{2.0} =$	4.50	$3 \times \frac{4.175}{2.0} =$	6.26
Total $=$	135.96		$= 195.83$

Pile 10

	Unfactored (kN)		Factored (kN)
$32.02 \times \frac{3.0}{2.0} =$	48.03	$46.08 \times \frac{3.0}{2.0} =$	69.12
$25.375 \times \frac{2.175}{2.0} =$	27.59	$35.625 \times \frac{2.175}{2.0} =$	38.74
$5.76 \times \frac{(3 + 2.175)}{2.0} =$	14.90	$8 \times 2.58 =$	20.70
Total $=$	90.52		$= 128.56$

Pile 11

	Unfactored (kN)		Factored (kN)
$25.375 \times \frac{(2.175 + 4.175)}{2.0} =$	80.56	$35.625 \times 3.175 =$	113.10
$18.95 \times \frac{3.0}{2.0} =$	28.425	$28.33 \times \frac{3.0}{2.0} =$	42.50
$5.76 \times \frac{(2.175 + 4.175 + 3.0)}{2.0} =$	26.92	$8.0 \times 4.67 =$	37.36
Total $=$	135.90		192.96

Pile 12

	Unfactored (kN)		Factored (kN)
$25.375 \times \frac{4.175}{2.0} =$	52.97	$35.625 \times \frac{4.175}{2.0} =$	74.36
$36.82 \times \frac{3.0}{2.0} =$	55.23	$53.40 \times \frac{3.0}{2.0} =$	80.10
$5.76 \times \frac{(4.175 + 3)}{2.0} =$	20.66	$8 \times 3.58 =$	28.70
Total $=$	128.86		$= 183.16$

The calculated pile loadings are summarized in Table 2.4.

Maximum unfactored pile loads have not taken into account the additional load due to elastic shears. If these are to be considered then it is appropriate to multiply the working loads by 1.25:

Maximum working load based on pile 6 $= 171.02 \times 1.25$
$= 213.77 \text{ kN}$

Use 165 mm diameter steel tube piles driven to a predetermined set to give a maximum working load of 225 kN.

All piles to be subjected to a re-strike and one random pile to be load-tested using kentledge. The test load applied to be $1.50 \times 225 = 337.50 \text{ kN}$.

GROUND BEAM ANALYSIS

This analysis assumes the ground beams are continuous and designed to cater for bending moments top and bottom of $wl^2/10$.

Some engineers use a simple supported design philosophy with the use of anti-crack reinforcement over the pile supports. This is very conservative for the bottom steel but problems of serviceability cracking over the supports may result which could affect the durability of the concrete beams.

On sites where clay heave is unlikely the ground beams can be cast against the earth face using 75 mm cover to the links or the beams can be poured in shutter moulds (BS 8110 Clause 3.3.1.4)

All design in accordance with BS 8110 Part 1 (1985).

C 30 P mix $f_{cu} = 30 \text{ N/mm}^2$

High-yield bars $f_y = 460 \text{ N/mm}^2$

$b = 400 \text{ mm}, h = 600 \text{ mm}$

Figure 2.38 shows a typical beam section.

From BS 8110 Table 6.1, Minimum cement content to be 370 kg/m^3 with a water/cement ratio of 0.45. This gives Class 2 sulphate protection (concrete exposed to sulphate attack).

59

Foundation design

A_s = area of steel in mm^2
d = effective depth of beam = $600 - 40 - 8 - \phi/2 = 552 - 12 = 540$ mm

z is limited to 0.95d

Design ultimate moment, $M = A_s \times 0.87 \times f_y \times z$
$$= 380.20 \, A_s \, d \times 10^6 \text{ kN m}$$

$$k = \frac{M}{bd^2 f_{cu}} = \frac{M}{400d^2 \times 30} = \frac{M \times 10^6}{12\,000 d^2}$$

$$z = d\left[0.5 + \sqrt{\left(0.25 - \frac{k}{0.9}\right)}\right]$$

z is less than 0.95d

From BS 8110 Clause 3.4.4.4 :

$$k = \frac{M}{bd^2 f_{cu}} \qquad A_s = \frac{M}{0.87 f_y z}$$

Therefore

$$z = \frac{M}{0.87 f_y A_s}$$

$$z = d\left[0.5 + \sqrt{\left(0.25 - \frac{k}{0.9}\right)}\right]$$

$$\left(\frac{z}{d} - 0.5\right)^2 = 0.25 - \frac{\kappa}{0.9}$$

$$\frac{z^2}{d^2} - \frac{z}{d} = -\frac{k}{0.9}$$

Table 2.4. Calculated pile loadings for Example 2.5

Pile number	Service load (kN)	Ultimate load (kN)
1	127.39	181.23
2	150.84	214.95
3	139.58	198.71
4	166.97	230.07
5	114.39	165.71
6	171.02	245.33
7	116.64	166.97
8	80.78	118.43
9	135.96	195.83
10	90.52	128.56
11	135.90	192.96
12	128.86	183.16

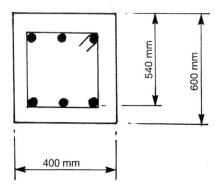

Fig. 2.38 Example 2.5: typical beam section.

Substituting $\dfrac{M}{0.87 f_y A_s}$ for z and $\dfrac{M}{bd^2 f_{cu}}$ for k :

$$\frac{M^2}{\left(0.87 f_y A_s d\right)^2} - \frac{M}{0.87 f_y A_s d} = \frac{-M}{0.9 bd^2 f_{cu}}$$

Therefore

$$\frac{M}{\left(0.87 f_y A_s d\right)^2} - \frac{1}{0.87 f_y A_s d} = -\frac{1}{0.9 bd^2 f_{cu}}$$

Therefore

$$M = 0.87 f_y A_s d - \frac{\left(0.87 f_y A_s\right)}{0.9 b f_{cu}}$$

When $f_y = 460$, $f_{cu} = 30$ and $b = 400$, then

$$M = 0.87 \times 460 \times A_s d - \frac{\left(0.87 \times 460 A_s\right)^2}{400 \times 30 \times 0.9}$$

Therefore $M = \dfrac{400 A_s d - 14.83 A_s^2}{10^6}$ kN m

This formula only applies when z is less than $0.95d$.

Maximum span

Deflection criteria are as follows.

Clause 3.4.6.3 Table 3.10:

$$\frac{\text{Span}}{\text{Effective depth}} = 20 \text{ simply supported}$$

Clause 3.4.6.5 Table 3.11: Modification factor for tension steel is equal to

$$0.55 + \frac{477 - 5 f_y/8}{120\left(0.9 + M/bd^2\right)}$$

but no greater than 2.0.

Therefore

$$\text{Maximum span} = \left(0.55 + \frac{1.57}{0.9 + M/400d^2}\right) 20d$$

Shear reinforcement

Design concrete shear stress, $v_c = 0.79 \dfrac{\left(100 A_s/bvd\right)^{1/3}\left(400/d\right)^{1/4}}{\gamma_m}$

where $\gamma_m = 1.25$.

$$v_c = 0.63\left(\frac{0.25 A_s}{d}\right)^{1/3}\left(\frac{400}{d}\right)^{1/4} \text{ N/mm}^2$$

Clause 3.4.5.2 Table 3.8:
Minimum shear steel using T8 links

$$A_{sv} = \frac{0.4 b_v s_v}{0.87 f_{yv}}$$

where $b_v = 400$ mm, $f_{yv} = 460$ N/mm^2, $A_{sv} = 100.5$ mm^2. Therefore

$$s_v = \frac{100.50 \times 0.87 \times 460}{0.4 \times 400} = 250 \text{ mm}$$

Minimum links T8 at 250 mm centres.
With minimum links $v = v_c + 0.4$ with $v = V/b_v d$. Therefore

$$V = v b_v d = \frac{\left(v_c + 0.4\right) 400 d}{10^3}$$

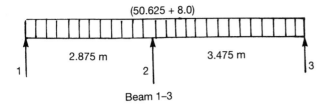

(50.625 + 8.0)

2.875 m   3.475 m

1   2   3

Beam 1–3

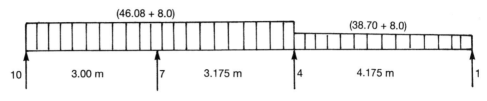

(46.08 + 8.0)

(38.70 + 8.0)

10   3.00 m   7   3.175 m   4   4.175 m   1

Beam 1–4–7–10

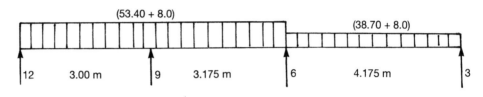

(53.40 + 8.0)

(38.70 + 8.0)

12   3.00 m   9   3.175 m   6   4.175 m   3

Beam 3–6–9–12

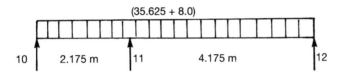

(35.625 + 8.0)

10   2.175 m   11   4.175 m   12

Beam 10–11–12

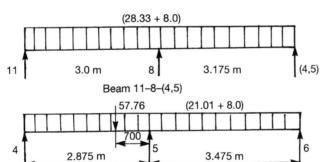

(28.33 + 8.0)

11   3.0 m   8   3.175 m   (4,5)

Beam 11–8–(4,5)

57.76   (21.01 + 8.0)

700

4   2.875 m   5   3.475 m   6

Beam 4–5–6

Fig. 2.39 Example 2.5: beam-loading diagrams.

Shear steel T8 links

$$A_{sv} = \frac{b_v s_v (v - v_c)}{0.87 f_{yv}} \qquad v = \frac{A_{sv} \times 0.87 f_{yv} + v_c}{b_v s_v}$$

Therefore

$$V = \frac{A_{sv} \times 0.87 f_{yv} d}{s_v \times 10^3} + \frac{v_c b_v d}{10^3} \text{ kN}$$

s_v	V
100	$0.4d + 0.4v_c d$
125	$0.4d + 0.4v_c d$
150	$0.4d + 0.4v_c d$
175	$0.23d + 0.4v_c d$
200	$0.20d + 0.4v_c d$
225	$0.18d + 0.4v_c d$

Beam design

Beam 1–3 (Fig. 2.39(a))

$$\text{Maximum ultimate moment} = \frac{58.265 \times 3.475^2}{10} = 70.358 \text{ kN m}$$

Beam 1–4–7–10 (Fig. 2.39(b))

$$\text{Maximum ultimate moment} = \frac{46.7 \times 4.175^2}{10} = 81.40 \text{ kN m}$$

Beam 3–6–9–12 (Fig. 2.39(c))

$$\text{Maximum ultimate moment} = \frac{46.7 \times 4.175^2}{10} = 81.40 \text{ kN m}$$

61

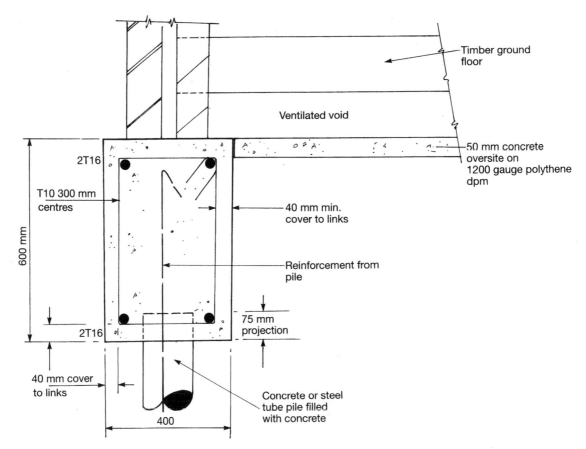

Fig. 2.40 Example 2.5: ground beam detail.

Beam 10–11–12 (Fig. 2.39(d))

Maximum ultimate moment $= \dfrac{43.625 \times 4.175^2}{10} = 76.04 \text{ kN m}$

Beam 11–8–(4, 5) (Fig. 2.39(e))

Maximum ultimate moment $= \dfrac{36.33 \times 3.175^2}{10} = 36.62 \text{ kN m}$

Beam 4–5–6 (Fig. 2.39(f))

$R_a = 29.01 \times \dfrac{2.875}{2} = 41.70 \text{ kN}$

$57.67 \times \dfrac{0.7}{2.875} = 14.04 \text{ kN}$

$R_a = \overline{55.74} \text{ kN}$

$R_b = 41.70 + \left(57.67 \times \dfrac{2.175}{2.875} \right) = 85.32 \text{ kN}$

Moment $4 - 5 = 55.74 \times 1.92 - 29.01 \times \dfrac{1.92^2}{2} = 53.55 \text{ kN m}$

Maximum shear $= 101.23 \text{ kN}$

Using C30 N/mm^2 concrete with 40 mm cover and Design Chart No. 1 (Fig. 2.11):

$k = \dfrac{M}{bd^2 f_{cu}} = \dfrac{81.40 \times 10^6}{400 \times 540^2 \times 30} = 0.023$

Therefore lever arm factor $= 0.95$

$A_s = \dfrac{M}{0.87 f_y z} = \dfrac{81.40 \times 10^6}{0.87 \times 460 \times 0.95 \times 540} = 397 \text{ mm}^2$

Use two T16 mm bars top and bottom in all beams.

Maximum shear stress $= \dfrac{101.23 \times 10^3}{400 \times 540} = 0.468 \text{ N/mm}^2$

$v_c = 0.36$ N/mm^2; provide minimum links throughout beams.

$A_{sv} = \dfrac{0.40 \times 400 \times 300}{0.87 \times 460} = 119 \text{ mm}^2$

Use T10 links at 300 mm centres throughout.

Example 2.6

Poor ground conditions on part of a housing site require several plots to be built off piled foundations. Maximum pile loading is approximately 450 kN. Piles to be 200 mm square precast concrete driven to a predetermined set and this set to be checked by re-striking the piles. Minimum factor of safety on piles to be 2.25. The piles will be driven with a 4 tonne hydraulic hammer (Banut type) with a 400 mm drop (Table 2.5).

Using the modified Hiley formula :

$R_u = \dfrac{E}{S + C/2.0}$

Where R_u = total load (450 kN) × factor of safety (2.25) = 1012.50 kN;
E = transfer energy at top of pile = 0.85×10^4 kN mm;
C = temporary compression of pile and ground per blow (assume 10 mm);
S = set per blow.
Therefore:

$S = \dfrac{E}{R_u} - \dfrac{C}{2.0} = \dfrac{8500}{1012.50} - \dfrac{10}{2} = 3.39 \text{ mm/blow}$

Therefore set for 10 blows = 34 mm or less.

Table 2.5. Hammer transfer energy table. Rig type: hydraulic hammer (Banut type)

Hammer weight (tonnes)	Transfer energy (tonne m)				
	Hammer drop (mm)				
	300	400	500	600	700
1.50	0.25	0.35	0.45		
3.0	0.55	0.70	0.90		
4.0		0.85	1.10		
5.0		1.05	1.40		

LOADINGS

Floor plan is shown in Fig. 2.41.

Roof

30° pitch

	Unfactored dead loads (kN/m²)	Unfactored imposed loads (kN/m²)
Concrete tiles	0.55	
Battens and felt	0.05	
Trusses	0.23	
Insulation	0.02	0.75
Ceiling: plasterboard	0.15	0.25
Total	1.00	1.00

First floor

Tongued and grooved boarding	0.10	
Plasterboard and skim	0.15	
Stud partitions	0.50	1.5
Total	0.75	

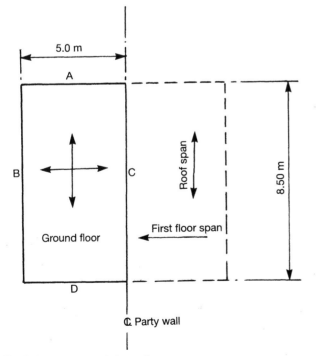

5.0 m

A

B — C

Roof span

8.50 m

Ground floor

First floor span ←

D

℄ Party wall

Fig. 2.41 Example 2.6: floor plans.

Suspended *in-situ* concrete ground floor

200 mm slab	4.70	
Stud partitions	0.50	1.50
Total	5.20	

External walls

	Unfactored dead loads (kN/m²)
102.5 mm brickwork	2.25
100 mm blockwork	1.25
Plaster	0.25
Total	3.75

Party walls

215 mm brickwork	4.50
Plaster two sides	0.50
Total	5.00

Walls A and D

	Dead load (kN/m)		Imposed load (kN/m)	
Roof and ceiling	$\dfrac{1.00 \times 8.50}{2.0}$	= 4.25	$\dfrac{1.0 \times 8.50}{2.0}$	= 4.25
First floor nominal	0.75×1.0	= 0.75	1.50×1.0	= 1.50
Ground floor	$5.20 \times 5.0 \times 0.33$	= 8.58	$1.50 \times 5 \times 0.33$	= 2.50
Wall	3.75×5.0	= 18.75		
Underbuild and ground beam		= 10.00		
Total		42.33		8.25

Wall B

Roof and ceiling	1.00×0.60	= 0.63	$1.0 \times 0.6p0$	= 0.60
First floor	$0.75 \times \dfrac{5.0}{2.0}$	= 1.50	$1.50 \times \dfrac{5.0}{2.0}$	= 3.75
Ground floor	$5.20 \times \dfrac{5.0}{2.0}$	= 13.00	$1.50 \times \dfrac{5.0}{2.0}$	= 3.75
Wall	3.75×6.0	= 22.50		
Underbuild and ground beam		= 10.00		
Total		47.97		8.10

Wall C (party wall)

Roof and ceiling	$1.00 \times 0.6 \times 2$	= 1.20	$1.0 \times 0.6 \times 2$	= 1.20
First floor	$0.75 \times \dfrac{5.0}{2n.0} \times 2$	= 3.75	$1.50 \times \dfrac{5.0}{2.0} \times 2$	= 7.50

Foundation design

Ground floor $5.20 \times \dfrac{5.0}{2.0} \times 2 = 24.50$ $1.50 \times \dfrac{5.0}{2.0} \times 2 = 7.50$

Wall $5.00 \times 6.0 = 30.00$

Underbuild and ground beam $= 10.00$

Total $\overline{70.95}$ $\overline{16.20}$

PILE LOADINGS

Pile 1

$(42.33 + 8.25) \times \dfrac{5.0}{2.0} + (47.97 + 8.25) \times \dfrac{4.25}{2.0}$ $= 245.60$ kN

Pile 2

$(42.33 + 8.25) \times \dfrac{5.0 \times 2.0}{2.0} + (70.95 + 16.20) \times \dfrac{4.25}{2.0} = 438.00$ kN

Pile 3

$(47.97 + 8.10) \times 4.25$ $= 238.30$ kN

Pile 4

$(70.95 + 16.20) \times 4.25$ $= 370.38$ kN

Pile 5

As Pile 1 $= 245.60$ kN

Pile 6

As Pile 2 $= 438.00$ kN

BEAMS

Beam moments

Beams A and D

Ultimate moment $=$ $42.33 \times 1.40 \times \dfrac{5.0^2}{10}$ $=$ 148.15   (kN m)

$8.25 \times 1.60 \times \dfrac{5.0^2}{10}$ $=$ 33.00

Total $=$ 181.15

Beam B

Ultimate moment $=$ $47.97 \times 1.40 \times \dfrac{4.25^2}{10}$ $=$ 121.30   (kN m)

$8.10 \times 1.60 \times \dfrac{4.25^2}{10}$ $=$ 23.41

Total $=$ 144.71

Beam C

Ultimate moment $=$ $70.95 \times 1.40 \times \dfrac{4.25^2}{10}$ $=$ 179.41   (kN m)

$16.20 \times 1.60 \times \dfrac{4.25^2}{10}$ $=$ 46.818

Total $=$ 226.23

Beam shears

Beams A and D

Ultimate shear $=$ $42.33 \times 1.40 \times \dfrac{5}{2}$ $=$ 148.155   (kN)

$8.25 \times 1.60 \times \dfrac{5}{2}$ $=$ 33.00

Total $=$ 181.15

Beam B

Ultimate shear $=$ $47.97 \times 1.40 \times \dfrac{4.25}{2}$ $=$ 142.71   (kN)

$8.10 \times 1.60 \times \dfrac{4.25}{2}$ $=$ 27.54

Total $=$ 170.25

Beam C

Ultimate shear $= (70.95 \times 1.40 + 16.20 \times 1.60) \times \dfrac{4.25}{2} = 266.15$ kN

Beams A and D

Ultimate moment = 181.15 kN m   Ultimate shear = 181.15 kN

$k = \dfrac{M}{bd^2 f_{cu}} = \dfrac{181.15 \times 10^6}{400 \times 390^2 \times 25} \, 0.119$

Therefore lever arm factor = 0.83

$A_s = \dfrac{181.15 \times 10^6}{0.87 \times 460 \times 390 \times 0.83} = 1398 \text{ mm}^2$

Use five T20 mm bars (1571 mm²)

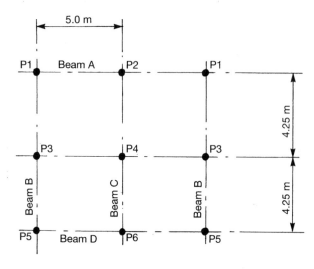

Fig. 2.42 Example 2.6: pile layout.

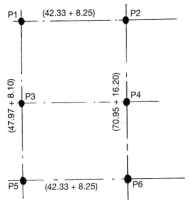

Fig. 2.43 Example 2.6: line loads (kN/m).

Shear stress $v = \dfrac{181.15 \times 10^3}{400 \times 390} = 1.16 \text{ N/mm}^2$

$\dfrac{100 A_s}{b_v d} = \dfrac{100 \times 1571}{400 \times 390} = 1.0$

Therefore $v_c = 0.64$

Because $(v_c + 0.4)$ is less than v provide nominal links to beam.
Using T8 links (Table 2.6) :

$s_v = \dfrac{100.50 \times 0.87 \times 460}{0.40 \times 400} = 251 \text{ mm}$

Provide T8 links at 250 mm centres for full length of beam.

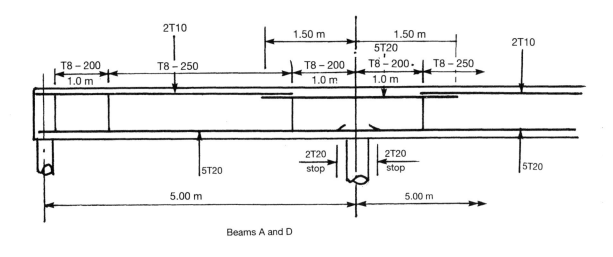

Beams A and D

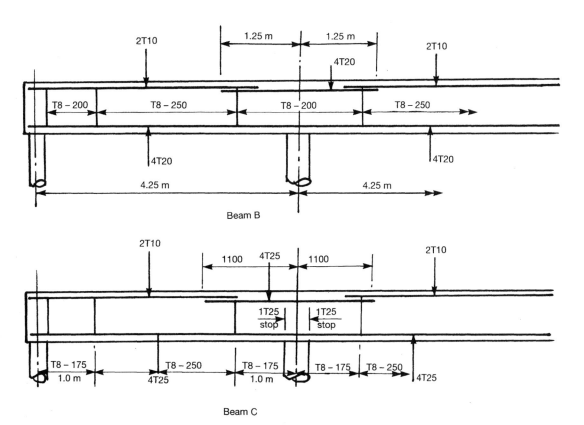

Fig. 2.44 Example 2.6: reinforcement details.

Table 2.6. Shear stress v using T8 links

Tension steel	A_s	d	v_c	v (kN) Spacings (mm)						
				250	225	200	175	150	125	100
3T12	339	396	0.38	122	130	138	152	165	185	220
4T12	452	396	0.40	128	135	145	155	170	190	225
3T16	603	395	0.46	135	143	150	162	177	200	230
4T16	804	395	0.50	140	150	158	170	185	205	236
3T20	943	390	0.52	145	154	160	173	189	210	240
4T20	1260	390	0.57	153	160	170	180	195	217	250
3T25	1470	388	0.62	157	165	175	185	200	220	251
4T25	1960	388	0.68	168	175	182	195	210	230	260

Beam B

Ultimate moment = 144.71 kN m   Ultimate shear = 170.25 kN

$$\frac{M}{bd^2 f_{cu}} = \frac{144.71 \times 10^6}{400 \times 390^2 \times 25} = 0.095$$

$k = 0.86$

Therefore

$$A_s = \frac{144.71 \times 10^6}{0.87 \times 460 \times 390 \times 0.86} = 1078 \text{ mm}^2$$

Use four T20 top and bottom.

$$v = \frac{170.25 \times 10^3}{400 \times 390} = 1.09 \text{ N/mm}^2$$

$$\frac{100 A_s}{b_v d} = \frac{100 \times 1256}{400 \times 390} = 0.805$$

Therefore $v_c = 0.60$

As $(v_c + 0.4)$ is less than v use nominal links throughout beam's length. Provide T8 links at 250 mm centres.

Beam C

Ultimate moment = 226.23 kN m   Ultimate shear = 266.15 kN

$$\frac{M}{bd^2 f_{cu}} = \frac{226.23 \times 10^6}{400 \times 390^2 \times 25} = 0.148$$

$k = 0.76$

$$A_s = \frac{226.23 \times 10^6}{0.87 \times 460 \times 0.76 \times 390} = 1907 \text{ mm}^2$$

Use four T25 mm bars top and bottom (1960 mm²)

$$\frac{100 A_s}{b_v d} = \frac{100 \times 1960}{400 \times 390} = 1.25 \qquad v_c = 0.68 \text{ N/mm}^2$$

$$v = \frac{266.15 \times 10^3}{400 \times 390} = 1.70 > 0.40 + v_c,$$

therefore design links required.

Using T8 links:

$$s_v = \frac{100.50 \times 0.87 \times 460}{400 \times (1.25 - 0.68)} = 176 \text{ mm}$$

Therefore use two leg T8 at 175 mm centres for 1000 mm from support face and then two leg T8 at 250 mm for remaining length of beam.

BIBLIOGRAPHY

Berezantsev, V.G. (1961) Load bearing capacity and deformation of piled foundations. *Proc. Fifth International Conference on Soil Mechanics*, Paris, Vol. 2, pp. 11–12.

BRE (1991) BRE Digest 363: *Concrete in sulphate-bearing soils and groundwater*, Building Research Establishment.

Broms, B (1966) Methods of calculating the ultimate bearing capacity of piles: a summary. *Sols (Soils)*, **5** (18/19), 21–31.

BSI (1964) BS 648: *Schedule of weights of building materials*, British Standards Institution.

BSI (1978) BS 5628: *Structural use of masonry*. Part 1: Unreinforced masonry.

BSI (1984) BS 6399: *Design loading for buildings*. Part 1: Code of practice for design and imposed loads, British Standards Institution.

BSI (1985) BS 5628: *Structural use of masonry*. Part 2: Structural use of reinforced and prestressed masonry, British Standards Institution.

BSI (1985) BS 8110: *Structural use of concrete*. Part 1: Code of practice for design and construction. Part 2: Design charts for singly reinforced beams.

NHBC (1977) *NHBC Foundation Manual: Preventing Foundation Failures in New Buildings*, National House-Building Council, London (now rewritten as *NHBC Standards* Chapter 4.1).

Tomlinson, M.J. (1980) *Foundation Design and Construction*, 4th edn, Pitman.

Vesic, A. (1966) Tests on instrumented piles. Ogecnee River Site. *Journal of Soil Mechanics and Foundation Division, American Society of Civil Engineers*, **96**, SM2.

Chapter 3
Foundations in cohesive soils

3.1 INTRODUCTION

Foundations transmit the total load from a building on to the ground by direct contact pressure. The foundation must distribute the building loads in such a way that it ensures that the bearing stratum is not overstressed and that total settlements are within acceptable limits. The foundation designer therefore needs to have some knowledge of the type of strata present below the foundations.

Site investigations seldom reveal an allowable bearing capacity in simple terms. This is because the various strata are composed of many different soil types, each having different properties.

Cohesive soils are subjected to plastic deformation when they are loaded. If the pressures applied to such soils are just sufficient to cause shear failure, this pressure is described as the **ultimate bearing capacity** of the soil. By applying a factor of safety to this value, we have a reduced bearing pressure which is referred to as the **safe bearing capacity**.

The **allowable bearing pressure** is the maximum net loading that a soil can sustain, taking into account the **safe bearing pressure** and the magnitude of settlement that the building can accommodate safely.

The **net bearing pressure** is the difference between the actual pressure below the foundation and the pressure from the removed overburden. This principle is often adopted when designing buoyant foundations.

3.2 SETTLEMENTS IN COHESIVE SOILS

Soil conditions can change considerably from before to during and following construction of foundations. Most cases of excessive settlement arise because of unforeseen soil conditions which suddenly arise. It is therefore useful to examine the types of ground movement mechanism which are potential causes of settlement in cohesive soils.

(a) **Consolidation settlements**. In cohesive soils which are saturated, the effect of loading the soils is to squeeze out some of the porewater. This is called **consolidation**. A change of loading is required for this consolidation to take place and it may take several years before it finishes settling. The most susceptible strata are the normally consolidated clays and silts, and organic clays such as soft alluvium and clayey peats.

(b) **Moisture movements**. Some types of clay show a marked volumetric change as their moisture content is changed. Clays which fall into this category are referred to as **shrinkable** or **expansive clays**. They are usually found in southern and eastern counties of the UK, but they can occur in other areas in localized pockets.

(c) **Effects of trees and vegetation.** A major factor which can affect cohesive soils with medium-to-high plasticity is the effects of trees and vegetation. The tree-root system abstracts water from the clays, resulting in surface subsidence. If trees and vegetation are removed the clays are allowed to rehydrate with the result that swelling takes place. This subject is dealt with in more detail in Chapter 6.

(d) **Groundwater lowering**. Clays containing a high water table can be affected if this water table level is drawn down, by pumping for example. First, the resulting reduction in moisture will cause the clay to shrink and settle and, second, the weight of the overburden will increase as the soils are no longer buoyant. With soft clays and peat strata this could result in further consolidation stresses occurring because of the increase in effective stresses.

(e) **Temperature changes**. Frost can cause severe ground heave in sustained low-temperature conditions. Most silts, fine sands and chalks are frost-susceptible. Great care must be taken when designing foundations for cold-storage buildings.

(f) **Lateral displacement**. This is often caused by deep trenches being excavated parallel and too close to existing foundations. In effect the clays are subjected to a shear-slip type of failure similar to that experienced on sloping sites.

(g) **Mining subsidence**. Settlement can occur at the surface as a result of longwall coal extraction or as a result of crown hole collapses in pillar and stall workings. This subject is covered in more detail in Chapter 5.

3.3 CONSOLIDATION SETTLEMENT

There are two types of settlement:

- **Immediate settlements**. These occur within seven days as a result of elastic and plastic deformation of the clays. Because they occur rapidly and are relatively small, they seldom pose a problem.
- **Consolidation settlements**. These are time-dependent and can take from several months to several years to develop (usually 1–5 years).

Consolidation settlement analysis is applied to all saturated or partially saturated fine-grained soils. To analyse foundation settlements it is necessary to obtain the **coefficient of compressibility** m_v from consolidation test results. The values of m_v vary as the pressure ranges vary.

From the applied vertical stresses, the overburden pressures and thicknesses of the various strata, the total amount of settlement can be 'predicted' using the m_v values. Despite the many advances in soil mechanics the prediction of settlements is still not an exact science. There are no hard-and-fast rules for settlement computation.

When designing foundations on cohesive soils there are two criteria which must be considered and satisfied separately:

- There must be an adequate factor of safety against a bearing capacity failure of the clay.
- The total settlements and, more particularly, the differential settlements must be kept within reasonable limits.

For foundations on clay soils either of the above criteria may govern the foundation design.

3.3.1 Bearing capacity of cohesive soils

In cohesive soils the ultimate bearing capacity is calculated by using total stress parameters. As settlements in such soils take longer to develop, this gives the final construction case, which is the most onerous, and allows the foundation design to be based on undrained shear strength tests which can be easily obtained at no great expense.

Consider a simple strip footing of width B where the bearing stratum is a rigid plastic material and the shear strengths are derived from Coulomb's equations:

$$\tau = c' + \sigma \tan \phi'$$

where τ = shear strength, c' = cohesion (kPa), σ = normal effective stress component (kPa), and ϕ' = angle of internal friction.

Base shear often dictates the recommended ground bearing capacity.

Allowable bearing capacity $q_a = \dfrac{q_u}{\text{Factor of safety}}$

Using the Terzaghi bearing capacity equations for a soil with $\phi = 0$:

$$q_{ult} = c N_c + p \quad \text{and net ult.} = c N_c$$

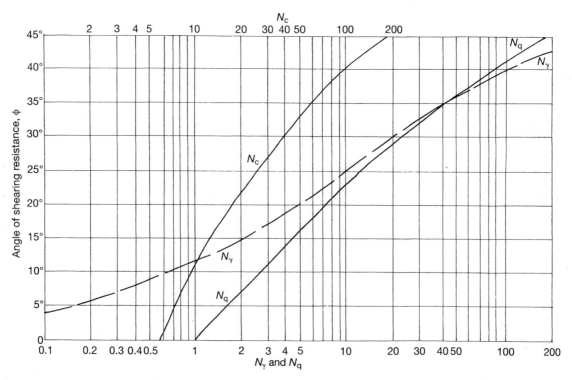

Fig. 3.1 Bearing capacity factors N_c, N_q and N_γ (after Terzaghi). For sands, $c = 0$, therefore $cN_c = 0$. N_q and N_γ based on SPT values for ϕ. For clays, $\phi = 0$, giving $N_\gamma = 0$, $N_q = 1.0$, $N_c = 2 + \pi = 5.14$.

where c = undrained cohesion,
N_c = bearing capacity factor,
p = total overburden pressures at foundation level.
When $\phi = 0$, $N_c = 2 + \pi = 5.14$.

In 1943 Terzaghi produced an equation for q_u which allowed for the effects of cohesion and friction under the foundation base; this was applicable to shallow foundations, i.e. where z/B is less than 1.0. For a strip footing Terzaghi's equation is

$$q_u = c\,N_c + \gamma\,z\,N_q + 0.50\,\gamma\,B\,N_\gamma$$

The values of N_c, N_q and N_γ for various values of ϕ can be obtained from Fig. 3.1. The value of N_c increases to 5.70 for a surface foundation due to the frictional allowance. Therefore

$$q_{ult} = 5.70\,c + \gamma\,z$$

The coefficient N_q allows for the surcharge effects arising from the overburden, and N_γ allows for the size of the foundation, B. When $\phi = 0$, $N_\gamma = 0$, $N_c = 5.70$ and $N_q = 1.0$. Skempton showed that N_c increases with foundation depth increase for $\phi = 0$ soils and these values can be obtained from Table 3.1.

Table 3.1. N_c values for depth factor in soils with $\phi = 0$ (after Skempton)

Foundation type	Depth foundation width ratio, Z/B				
	0	0.50	1.0	2.0	4.0
Circle or square	6.2	7.1	7.7	8.40	9.0
Strip footing	5.1	5.9	6.40	7.0	7.50

Using these coefficients, the ultimate net bearing capacity of a strip footing is given by

$$q_{nu} = c\,N_c + p_o\,(N_q - 1) + 0.5\,\gamma\,B\,N_\gamma$$

For a square or circular foundation

$$q_{nu} = 1.2\,c\,N_c + p_o\,(N - 1) + 0.4\,\gamma\,B\,N_\gamma$$

where γ = the bulk density of the soil below the foundation,
c = the undrained shear strength of the soil,
p_o = the effective overburden pressure at foundation level,
and B = the foundation width (or diameter).
The ultimate bearing capacity P_u is given by

$$P_u = P_{nu} + p$$

where p = the total overburden pressure at the foundation level. If the water table is at or above the foundation level then the value for the density must be the submerged density. When calculating p_o, a similar density must be used when the water table is at or above the foundation level.

Meyerhof (1952) modified the Terzaghi equations to make allowance for the foundation shape, depth and roughness of the base.

These values for N_c, N_q and N_γ are shown in Fig. 3.2. When using Meyerhof values a shape factor λ must be applied; this can be obtained from Fig. 3.3. As for the

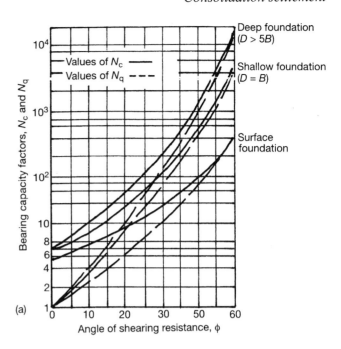

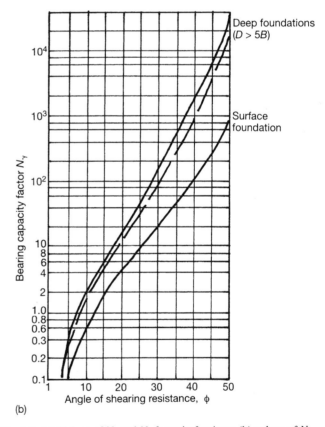

Fig. 3.2 (a) Values of N_c and N_q for strip footings; (b) values of N_γ for strip footings (after Meyerhof, 1952).

Terzaghi formula, the submerged density must be used if the water table is at or above the foundation level. For pure cohesive soils, i.e. $\phi = 0$, $N_q = 0$ and $N_\gamma = 0$. Therefore values of N_c can be taken from Fig. 3.4.

Using the Skempton formula, the allowable bearing capacity q_a is given by

$$q_a = \frac{N_c\,C_u}{F} + p_o$$

Foundations in cohesive soils

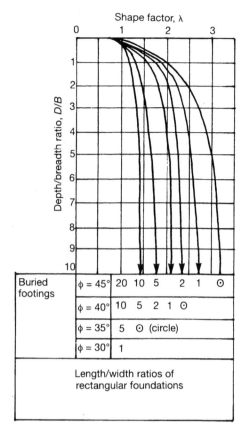

Fig. 3.3 Values of shape factor λ for strip footings (after Meyerhof, 1952).

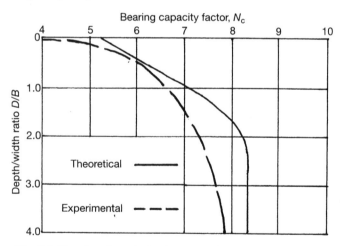

Fig. 3.4 N_c values for strip footings on soils with ø = 0 (Meyerhof, 1952).

where N_c = bearing capacity factor, C_u = average undrained shear strength of soils below the foundation, p_o = vertical pressure applied to the soil at footing level, F = factor of safety (usually a minimum of 3.0). Therefore

$$q_{net} = \frac{N_c C_u}{F}$$

For a strip footing, N_c = 6.50 for z/B = 0.75/0.60 = 1.25; therefore

$$q_{ult} = \frac{6.50\, C_u}{3.0} = 2.15\, C_u$$

Since the bearing capacity factor N_c is never less than 5–6 then a reasonable net allowable bearing pressure for shallow

strip footings can be obtained by using the undrained shear strength $C_u \times 2.0$.

3.3.2 Vertical stress distribution

When a foundation load is applied to a soil a pressure bulb is generated. The stress on the ground decreases with depth and by using graphs the values of vertical pressure can be obtained from Table 3.2.

Table 3.2. Vertical pressure factors

B/Z	Factor for shear stress		Factor for vertical pressure
0	0		0
0.1	0.032		0.065
0.2	0.063		0.127
0.5	0.15	strip	0.30
0.6		footings	0.358
0.8	0.22		0.46
1.0	0.25		0.55
1.50	0.30		0.71
2.0	0.32		0.82
2.2	0.31		0.88
3.0	0.29		0.92
3.50	0.27		0.94
4.0	0.25		0.96
4.50	0.23		0.97
5.0	0.22		0.978
5.50	0.20		0.985
6.0	0.19		0.988
6.50	0.18		0.99
7.0	0.17		0.991
8.0	0.15		0.994
9.0	0.13		0.996
10.0	0.12		0.997
100.0	0.01		1.0

q = contract pressure; B = foundation width; Z = depth of soil element below foundation base.
Shear stress = $q \times$ shear factor
Vertical pressure = $q \times$ vertical pressure factor

3.3.3 Construction problems on clay sites

Some clay soils are very variable. They often contain water-bearing lenses of sands, gravels and silts as a result of past glaciation. When these are encountered in an excavation, many building inspectors ask the groundworks foreman to excavate deeper in the hope of finding clays at a lower level. Quite often, excavating deeper can lead to a costly foundation. If the clays are not encountered within a reasonable distance and the sands are water-bearing, or contain perched water, the sides of the trench will collapse, and a large, soft hole will result. The only solution left is to pump out the excavation and fill it with mass concrete to within 900 mm of the ground level.

It will be then be possible to compact granular fill in discrete layers over the mass concrete and provide a raft foundation or wide reinforced stiff ground beam.

70

To avoid this situation occurring on site, always excavate a trial hole about 3 m away from the foundation trench to determine whether good clays are present at shallow depth below the sand. If they are not present at a depth of 1.2 m then consideration should be given to using either a raft foundation or a wide reinforced stiff ground beam.

When excavating for deep trench-fill foundations, precautions should be taken to prevent collapse of the trench sides due to heavy ingress of water from wet sandy lenses. Generally, such flows can be controlled by pumping from a lower sump at one end of the excavations.

When constructing foundations in clays with medium to high plasticities the base of the excavation should be concreted immediately following excavation to reduce the risk of swelling from seepage or rainwater. Failure to protect the base of the excavation can result in the clay's bearing capacity being reduced as the water makes it more compressible.

Trench excavations which are left unconcreted in dry weather for long periods can be subject to swelling as the wet concrete restores the moisture levels. If the soil's moisture content increases, the soil will heave and upward movement of the concrete will occur. It is therefore prudent on clay sites to install as much as possible of the main drainage so that in wet weather the working conditions are improved.

Clay sites can give rise to problems of slope stability, especially if major cut-and-fill operations are taking place on the site. The placing of fills on to clay slopes should be avoided as the natural slope drainage is blocked, and an unstable slope could result.

3.3.4 Foundation designs on clay soils

Where firm or stiff clays overlie soft clays, soft silty alluvium which reduce in strength with depth, then it is recommended that a detailed trial pit or borehole investigation be carried out and shear strengths obtained for the various strata at depths likely to be affected by the proposed foundation loads. In trial pits the use of a hand shear vane tester with extended rods is recommended, as trial pits in excess of 1.20 m should not be entered, especially if soft ground or peats are suspected.

When using the shear vane it is most important to take a range of readings at each level so that an average value can be obtained. This is particularly important in the soft to very soft clays. In such clays account may need to be taken of the adhesion of the clay on the barrel of the vane tester and this value can be determined by using the dummy vane and deducting the amount obtained.

Once these *in-situ* undrained shear values are obtained the allowable bearing capacities can be determined from Terzaghi's equations. These can then be compared with the calculated stresses at foundation level and in the stressed levels below the foundation base within the pressure bulb (Table 3.4).

A correction factor μ can be used depending on the plasticity index of the clays. These fall within the range shown in Table 3.3. For a strip footing $q_a = 1.90S\mu$, neglecting overburden pressures.

Table 3.3. Correction factor, μ

Clay category	Undrained shear strength (kN/m²)	μ
Very stiff	> 150	1.0
Stiff	100–150	0.90
Firm to stiff	75–100	0.75
Firm	50–75	0.65
Soft to firm	40–50	0.60
Soft	20–40	0.55
Very soft	< 20	0.50

Table 3.4. Proposed allowable bearing values for clays (after Terzaghi and Peck (1968))

Description of clay	N	c	q_m	
			Square	Strip
Very soft	< 2	< 13.50	< 32	< 24
Soft	2–4	13.50–27	32–64	24–48
Medium	4–8	27–54	64–128	48–96
Stiff	8–15	54–107	128–260	96–190
Very stiff	15–30	107–215	260–515	190–385
Hard	> 30	> 215	> 515	> 385

N = number of blows per foot in standard penetration test;
c = cohesion (kN/m²);
q_m = proposed normal allowable bearing value (kN/m²);
F = factor of safety with respect to base failure

Example 3.1   Strip footing on clay soil

A strip footing 1.00 m wide is placed at a depth below ground level of 1.00 m. Vane test readings down to 3.0 m show that the clays are firm down to 2.0 m with shear strengths of 60 kN/m² changing to 30 kN/m² from 2.0 m down. Determine the allowable bearing capacity of the soils at foundation level, and check that the width used is acceptable to carry a line load of 60 kN/m along its length.

Using Meyerhof values for N_c, N_q for ø = 0:

$N_c = 5.14, N_q = 1.0, N_\gamma = 0$

Therefore net ultimate bearing capacity = $5.14 C_u = 5.14 \times 60 = 308$ kN/m²

Using a factor of safety of 3.0:

Allowable bearing pressure $= \dfrac{308}{3.0} = 103$ kN/m²

Actual pressure $= \dfrac{60}{1.0} = 60$ kN/m² < 103 OK.

Check at 2.0 m depth:

Allowable bearing pressure $= \dfrac{5.14 \times 30}{3.0} = 51.50$ kN/m²

From Table 3.2, vertical pressure factor for $z/B = 1.0/1.0 = 1.0$ equals 0.55.

Actual pressure = $60 \times 0.55 = 33$ kN/m², < 51.50.

The safe bearing capacity = $C_u N_c/3.0$ + overburden pressures γz. Therefore at 2.0 m, safe bearing capacity = $51.50 + 18 \times 2 = 87.50$ kN/m². Should groundwater levels rise these pressures should be halved. The width of the foundations is therefore satisfactory.

3.3.5 Settlements in clay soils

The distribution of vertical stress below a foundation is proportional to its width. When the foundation is loaded, a pressure bulb is formed below the formation, and the pressure distribution depends on the shape of the foundation. Strip footings stress the ground to a much greater depth than square pad foundations.

In the housebuilding industry there is often a reluctance to carry out a site investigation because of the cost. On larger projects the cost of borehole investigations is generally acceptable, especially if piling or basements are involved. For clay soils with bearing capacities in excess of $50\,kN/m^2$ the total settlements on a traditional two-storey masonry construction would be well within acceptable limits. Table 3.5 indicates the range of allowable bearing pressures which are based on a minimum foundation width of 1.0 m with a maximum total settlement of 25 mm.

Table 3.5. Allowable bearing pressures

Category of strata	Type of stratum	Allowable bearing value (kN/m²)	Comments
Non-cohesive soils	Dense gravel, dense sand and gravel	> 600	Width of the foundation not less than 1.0 m
	Medium dense gravel or medium dense sands and gravel	< 200–600	Groundwater level assumed to be at a depth not less than 1.0 m
	Loose gravel or loose sand and gravel	< 200	below the base of the foundation
	Compact sands	> 300	
	Medium dense sands	100–300	
	Loose sands	< 100	
Cohesive soils	Very stiff boulder clays and hard clays	300–600	Soft clays and silts are subject to long-term consolidation settlement
	Stiff clays	150–300	
	Soft clays and silts	75–150	
	Very soft clays and silts	Not suitable	

However, where the soil strengths are less than $50\,kN/m^2$ and are also reducing in strength with increased depth, total settlements must be considered. In such soils it is most important to obtain the correct soils data in order that settlement predictions are reasonably accurate. Estimation of such settlements should always be conservative.

(a) Basic settlement calculations

The total amount of settlement on a uniform thickness of compressible stratum of thickness H (Fig. 3.5) is

$$P_{oed} = H\, m_v\, \sigma_z$$

where P_{oed} = oedometer settlement (mm); σ_z = imposed pressure on the layer; m_v = coefficient of compressibility. The value of σ_z is usually determined from elastic analysis using Boussinesq theory; there are many standard tables and charts available derived by Jurgenson, Newmark and Schmertmann.

The relationship between m_v and E, the Young's modulus for the clays, can be considered as

$$\frac{1}{m_v} = \frac{(1-v)E}{(1+v)(1-2v)}$$

where v = Poisson's ratio (normally 0.5 in a saturated soil). Though m_v is not strictly an elastic constant, as displacements measured in the oedometer tests are not elastic, the relationship between m_v and E can be considered as relevant to the stress–strain relationship for a single cycle of loading.

If there are several layers of varying clay strata below a foundation then the average value of σ_z is calculated for each individual layer and the total settlement is the sum of the calculated values for all the layers affected:

$$P_{oed} = H_1\, m_{v1}\, \sigma_{z1} + H_2\, m_{v2}\, \sigma_{z2} + H_3\, m_{v3}\, \sigma_{z3} + \dots$$

The value of the vertical stress will change considerably over a thick compressible layer so it is advisable to split such layers into a series of thinner layers using an average value of σ_z for each layer and summating the sum total of the individual layers to obtain the total settlement.

On certain clays, the settlements calculated from consolidation test results can result in an overestimation. A correction factor μ must be applied to the calculated value if desired such that

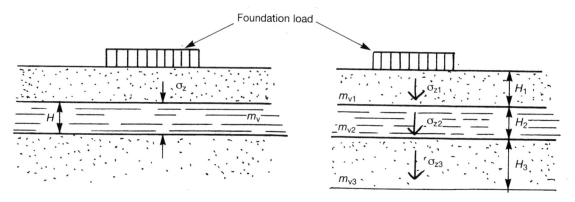

Fig. 3.5 Total settlement.

$$P = \mu\, P_{\text{oed}}$$

Values of μ are given in Table 3.3 for the different types of clay.

Because of the variations which can occur in soils, settlement calculations can only be considered an approximate guide.

Differential settlements are generally taken as one half of the total settlements calculated. Though this is only a rule of thumb, it is adequate for simple structures on fairly uniform strata.

For sites where the strata consist of soft uniform clays to depths in excess of twice the foundation width, there are quick approximate methods available for calculating the total consolidation settlements. Consider a simple strip footing loaded to give a contact pressure of p kN/m^2 at the base of the footing. The vertical normal stress beneath the centre of the footing can be considered to be in the form of a triangular dispersal as illustrated in Fig. 3.6.

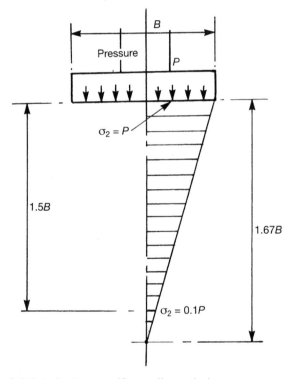

Fig. 3.6 Strip footing on uniform soils: vertical pressure distribution.

Directly below the footing the contact pressure will be p, which equals σ_z. At a depth of 1.50 B below the footing, σ_z is approximately equal to $0.1\, p$. Therefore the maximum depth of stressed soil is equal to

$$\frac{1.50\, B}{0.90} = 1.67\, B \text{ m}$$

For an average pressure σ_z of $0.50\, p$ the maximum total settlement is given by

$$P_{\text{oed}} = 1.67\, B \times 0.50\, p \times m_v = 0.835\, m_v\, B \text{ mm}$$

This is an approximation. On a soil of infinite thickness with m_v decreasing with the total normal vertical pressures, it would not be realistic to consider settlements deeper than $2B$ below the footings.

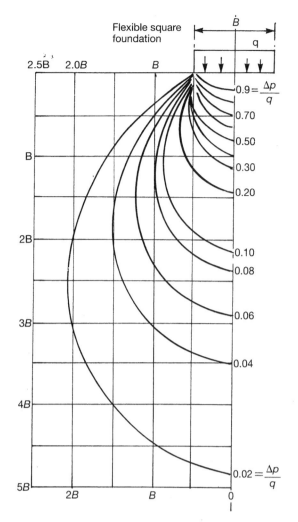

Fig. 3.7 Vertical pressure under a uniformly loaded square foundation.

(b) Influence line factors method

Where a footing is on an infinitely thick layer of soft compressible stratum which has a constant strength it is possible to calculate the total settlements using depth influence factors.

As m_v decreases with depth (as the vertical pressures decrease) the settlements between z_1 and z_2 (variable depths) can be obtained by using the coefficients obtained from Fig. 3.8:

$$P_{\text{oed}} = m_v\, B\, p\, (I_2 - I_1)$$

where I_1 and I_2 are the influence factors for depths z_1 and z_2 respectively.

Example 3.2   Settlements on clay soil

A house foundation 600 mm wide supports a three-storey gable wall. The applied line load equals 65 kN/m run. Vane test results taken at various depths in the soils directly below the footing indicate that there is a stiff desiccated clay crust for 1.0 m below ground level underlain by a soft silty clay alluvium. The values of the vane shear strengths are 75 kN/m^2 down to 1.0 m depth and 30 kN/m^2 down to a depth of 3.0 m below ground level. Determine the total settlements under the gable foundation.

73

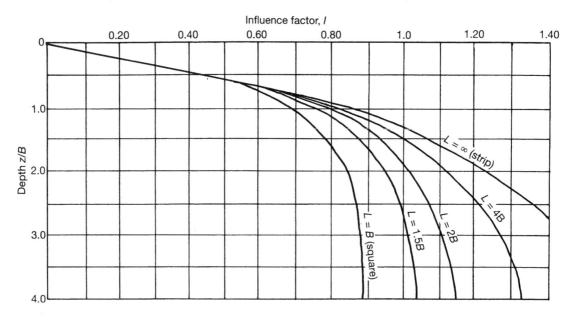

Fig. 3.8 Influence line factors I at the centre of a foundation.

Stress applied at footing level:

$$\frac{65 \times 10^3}{600} = 108 \text{ kN/m}^2$$

Consider the loading to be on an infinite footing length. The vertical stress distribution graph is shown in Fig. 3.10. The values of p_v are obtained as follows:

$$\frac{z}{B} = \frac{0.25}{0.60} = 0.41$$

Therefore

$$\frac{p_v}{p} = 0.75$$

$$p_v = 0.75p = 0.75 \times 108 = 81 \text{ kN/m}^2$$

$$\frac{z}{B} = \frac{0.75}{0.60} = 1.25 \quad p_v = 0.50p = 0.50 \times 108 = 54 \text{ kN/m}^2$$

$$\frac{z}{B} = \frac{1.25}{0.60} = 2.08 \quad p_v = 0.30p = 0.30 \times 108 = 33 \text{ kN/m}^2$$

$$\frac{z}{B} = \frac{1.75}{0.60} = 2.91 \quad p_v = 0.20p = 0.20 \times 108 = 21 \text{ kN/m}^2$$

$$\frac{z}{B} = \frac{2.25}{0.60} = 3.75 \quad p_v = 0.14p = 0.14 \times 108 = 15 \text{ kN/m}^2$$

The stiff clay has an undrained shear strength of 75 kN/m². Therefore an approximate value of E_v^1 (the modulus of compressibility, $= 1/m_v$) is given by

$$E_v^1 = 130\,C_u = 130 \times 75 = 9750 \text{ kN/m}^2$$

The soft clay has an undrained shear strength of 30 kN/m². Therefore

$$E_v^1 = 130\,C_u = 130 \times 30 = 3900 \text{ kN/m}^2$$

The total settlement is equal to the area of the pressure diagram in Fig. 3.10. For an approximation take a triangular distribution. Then

$$\sigma_1 = \frac{108+81}{2} \times \frac{0.25 \times 10^3}{9750} \qquad = 2.42 \text{ mm}$$

$$\sigma_2 = \frac{81+54}{2} \times \frac{0.50 \times 10^3}{3900} \qquad = 8.65 \text{ mm}$$

$$\sigma_3 = \frac{54+33}{2} \times \frac{0.50 \times 10^3}{3900} \qquad = 5.57 \text{ mm}$$

$$\sigma_4 = \frac{33+21}{2} \times \frac{0.50 \times 10^3}{3900} \qquad = 3.46 \text{ mm}$$

$$\sigma_5 = \frac{21+15}{2} \times \frac{0.50 \times 10^3}{3900} \qquad = 2.30 \text{ mm}$$

$$\text{Total settlement} \qquad = 22.40 \text{ mm}$$

This is less than 25 mm and is considered acceptable. A correction factor μ could be applied where $\mu = 0.55$–0.6 which would result in a settlement of about 14 mm.

Using the quick approximate method:

$$P_{oed} = 0.835\,Bm_v\,p = \frac{0.835 \times 0.60 \times 1 \times 108 \times 10^3}{3900} = 13.87 \text{ mm}$$

3.4 MOISTURE MOVEMENTS

One of the commonest problems with cohesive soils is the effect of the soil's drying out as a result of extreme dry weather or from moisture abstraction by roots of large trees.

The slow volume changes which occur when moisture evaporates from a clay soil can be predicted by assuming the lower limit of the soil's moisture content to be the shrinkage limit. Desiccation beyond this value will not bring about any further reduction in volume.

Many clay soils in the UK, especially in the southeast of England, possess a large potential for slow volumetric change. However, the mild damp climate which generally prevails means that any significant deficits in soil moisture content

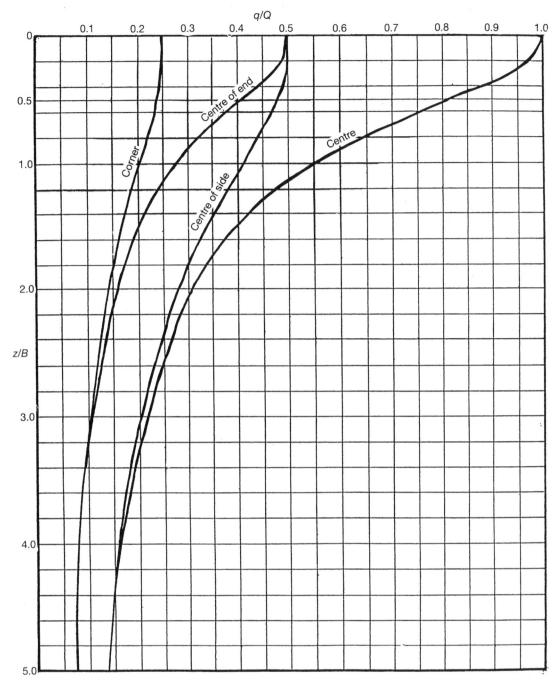

Fig. 3.9 Distribution of vertical stress beneath a long strip footing.

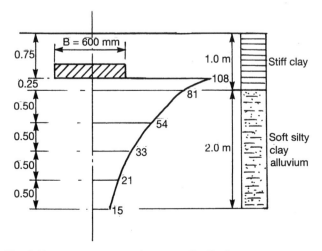

Fig. 3.10 Example 3.2: vertical stress distribution.

which occur in the summer months are generally limited to the top 1.0–1.50 m below ground level, and these soils generally recover over the winter period. However, it is recognized that deeper permanent deficiencies can be caused by large high water demand trees.

Driscoll states that the moisture content of a clay at values of suction e_F equal to 2 and 3 are 0.50 liquid limit and 0.40 liquid limit. These provide a crude estimate of the moisture content at the beginning of the desiccation process and when it becomes significant.

In addition to shrinking, these types of clay are also prone to swelling when the clays rehydrate. Poor or inadequate drainage can introduce excess water into the soils and weaken them. Expansive clays can be identified from their plasticity characteristics. One of the soil properties most widely used

Foundations in cohesive soils

to predict swelling potential is the **activity** of the clay. This was researched by Skempton in 1953. A clay's activity is defined as

$$\text{Activity} = \frac{\text{Plasticity index}}{\text{Clay percentage}}$$

Clays with large activities are referred to as active clays; they show plastic properties over a wide range of moisture content values. Figure 3.11 (After Skempton) indicates the relationship between the plasticity index and clay percentage.

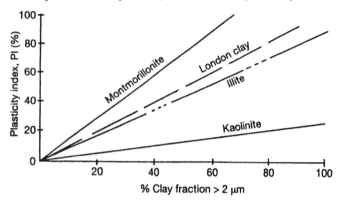

Fig. 3.11 Clay 'activity' graph (after Skempton).

To design foundations in cohesive soils, such that the effects of moisture depletion or excess moisture do not result in settlements or heave, it is essential to determine the Atterberg limits of the clays. These tests will enable the clays to be fully classified. They were designed by Atterberg in 1911. The tests determine the various values of moisture content at which changes in a soil's strength characteristics occur; as a

silt or clay dries out its strength increases and it becomes less compressible.

The moisture content at which the clays stop acting as a liquid and start acting as a plastic solid is known as the **liquid limit**. As further moisture is removed from the clays it becomes possible for the clays to resist large shearing stresses. Eventually the soil simply fractures with no plastic deformation taking place. The limit at which plastic failure changes to brittle fracture is referred to as the **plastic limit**.

The **plasticity index** is the range of moisture content within which a clay is plastic. The finer the grain particles in the soils the greater is its plasticity index.

Plasticity index = Liquid limit – Plastic limit

PI = LL – PL

3.4.1   Liquid limit test

BS 1377 specifies two methods for determining the liquid limit of a clay sample. Field samples for these tests should be a minimum of 2 kg or greater.

(a)   Cone penetrometer

The clay sample to be tested is first kiln-dried and thoroughly mixed. 200 g of the sample are then sieved through a 425 micron sieve and placed on a glass slide. The sample is then mixed with enough distilled water to form a paste. A standard metal mould, approximately 55 mm in diameter and 40 mm deep, is filled with the clay paste and levelled off at

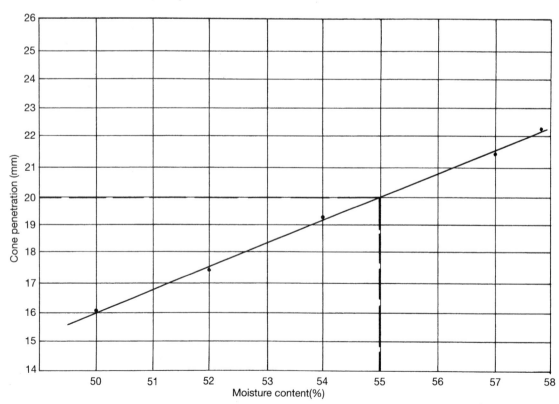

Fig. 3.12 Liquid limit graph using cone penetrometer test. The liquid limit is the water content corresponding to 20 mm penetration, i.e. 55%.

76

the surface. The cone penetrometer is placed at the centroid of the sample and level with it. The cone is then released so that it penetrates into the sample and the full penetration depth over a period of 5 s is measured.

This test is repeated by lifting the cone clear, filling in the first depression with more paste and allowing the cone to fall again. If the difference between the two measurements is less than 0.50 mm then the test is considered to be valid. The average penetration is noted and the moisture content of the sample is determined in the normal way.

The procedure is repeated at least four times with increasing moisture contents. The amount of water added to the samples should be just sufficient to produce depths of penetrations within the range of 15 to 25 mm.

The liquid limit is then found by plotting the variation of cone penetration on the vertical scale against the various moisture content values on the horizontal scale. A best-fit straight line should be drawn through the points on the graph. The liquid limit is taken to be the moisture content which corresponds to a cone penetration of 20 mm (Fig. 3.12).

(b) Casagrande apparatus

This method was superseded by the cone penetrometer, but it is still widely used. The process of drying the sample and making a paste is similar to that for the cone method. The paste is then placed in a special brass cup and a 2 mm wide groove is cut in the top of the sample using a special profiled grooving tool. The brass cup is then inserted into the apparatus and the handle is turned at a rate of 2 rev/s. This actuates the cam which causes the brass cup to lift 10 mm and then fall on to the base plate. The number of blows to close the 2 mm gap over 13 mm is recorded and the moisture content is determined in the usual way (Fig. 3.13).

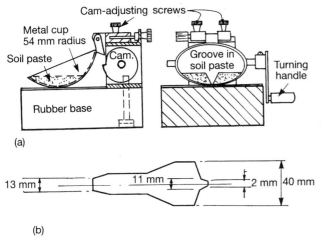

(a)

(b)

Fig. 3.13 Casagrande apparatus: (a) Casagrande liquid limit test apparatus; (b) grooving tool.

The test is repeated at least four times and the moisture contents are plotted on the vertical scale against the number of blows on the horizontal scale, using a log scale. The moisture content which corresponds to 25 blows is the liquid limit, expressed as a whole number (Fig. 3.14).

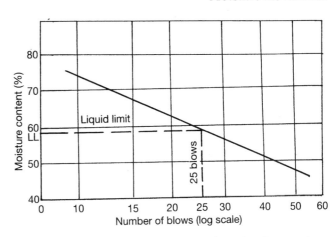

Fig. 3.14 Graph for liquid limit determination using Casagrande apparatus.

3.4.2 Plastic limit test

Take 20–25g of the clay sample after it has been kiln dried. The sample is then placed onto a glass plate and sufficient water is mixed with the sample to form a paste which can be rolled out between the palm of the hand and the glass plate. The sample is said to be at its plastic limit when it just begins to crumble at a thread diameter of 3 mm. The moisture content of the sample is determined and the test is repeated several times.

Once the soil plasticity characteristics have been found, the clays can be classified by using the Casagrande plasticity chart (BS 5930:1981) which will enable comparisons to be made.

(a) Casagrande plasticity chart

To use the plasticity chart (Fig. 3.15) the coordinates for plasticity index and the corresponding liquid limits are plotted. The sample can then be classified from its position on the chart relative to the A line: an empirical boundary between inorganic clays which come above the line and organic silts and clays which come below the line. The A line is drawn through the baseline where the PI is equal to zero and the liquid limit is 20%.

The main soil types are given specific designation letters and additional designatory lettering is used to denote the grading and plasticity (Table 3.6).

(b) The triaxial test (undrained compressive test)

This test, to determine the values of the total shear strength parameters of a soil, is carried out in the triaxial test apparatus but the sample is prevented from draining during shearing and is therefore sheared immediately after the application of the cell pressure. The test is quick and the results are expressed in terms of total stress.

The shear strength of a clay soil is made up of two components: cohesion and frictional resistance. Samples of the clay are subjected to quick undrained triaxial

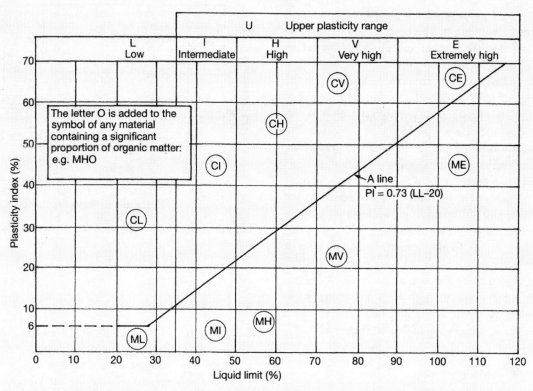

Fig. 3.15 Plasticity chart (British Soil Classification System, BS 5930). Silts (M soils) plot below the A line; clays (C) plot above the A line.

compressive tests, using three specimens 38 mm in diameter and 76 mm long taken from a U4 sample and tested under

lateral pressures of 70, 140 and 210 kN/m² at a constant rate of strain of 2%/min.

When the results of these tests are available, Mohr circles of stress can be drawn which will enable values of ϕ and c to be determined.

Example 3.3 Triaxial test

A clay sample tested using the quick undrained triaxial test gave the results listed in Table 3.7 when tested using lateral pressures of 70, 140 and 210 kN/m².

Additional axial stress required for failure = $\sigma_1 - \sigma_3$. Mohr circles are constructed from these values (Fig. 3.16). The intercept on the vertical axis is the apparent cohesion c, and the angle of frictional resistance, ϕ, is obtained from the slope of the best line drawn tangential to the curves; i.e. tan ϕ (Fig. 3.17).

Table 3.6. Casagrande classification

		Descriptive name	Letter
Coarse components	Main terms	**Gravel**	G
		Sand	S
	Qualifying terms	Well graded	W
		Poorly graded	P
		Uniform	Pu
		Gap graded	Pg
Fine components	Main terms	**Fine soil, Fines**	F
		Silt (M-soil) plots below A line of plasticity chart (of restricted plastic range)	M
		Clay plots above A line (fully plastic)	C
	Qualifying terms	Of low plasticity (ll<35%)	L
		Of intermediate plasticity (ll 35–70%)	I
		Of high plasticity (ll 50–<70%)	H
		Of very high plasticity (70:<ll<90%)	V
		Of extremely high plasticity (ll>90%)	E
		Of upper plasticity range incorporating groups I,H,V and E	U
Organic components	Main terms	**Peat**	Pt
	Qualifying term	Organic may be suffixed to any group	O

Table 3.7. Undrained triaxial test results for Example 3.3

Cell pressure, σ_3 (kN/m²)	Deviator stress, $\sigma_1 - \sigma_3$	Major principal stress, σ_1
Sample 5 at 0.50 m		
70	107	177
140	118	258
210	130	340
Sample 5 at 1.50 m		
70	53	123
140	72	212
210	86	296

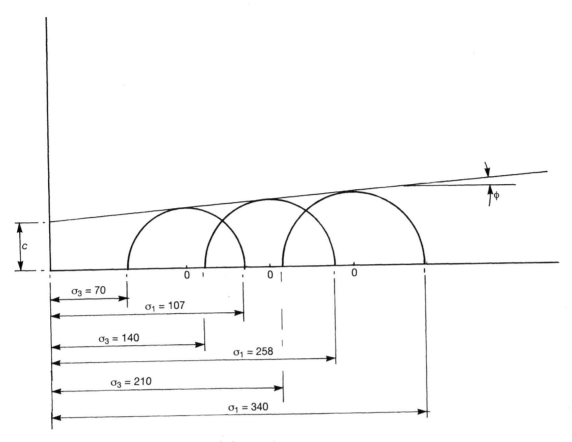

Fig. 3.16 Determination of c and ϕ from Mohr stress circles.

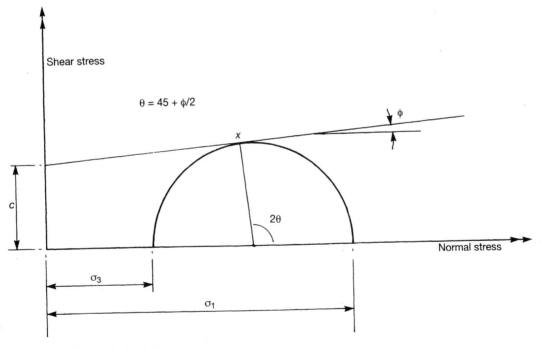

Fig. 3.17 Construction of Mohr's circle of stress.

BIBLIOGRAPHY

BSI (1999) BS 5930: *Code of practice for site investigations*, British Standards Institute.

BSI (1990) BS 1377: *Method of testing for civil engineering purpose*, British Standards Institute.

Carter, M. (1983) *Geotechnical Engineering Handbook*, Pentech Press, London.

Casagrande, A. (1947) Classification and identification of soils. *Proc. American Society of Civil Engineers*, No. 73.

Department of the Environment (1991) *Approved Documents*, HMSO, London.

Meyerhof, G.G. (1952) The ultimate bearing capacity of foundations. *Geotechnique*, **2** (4), 301–332.

Peck, R.B., Hanson, W.E. and Thornburn, T.H. (1974) *Foundation Engineering*, John Wiley, New York.

Powell, M.J.V. (ed.) (1979) *House-Builder's Reference Book*, Newnes-Butterworth, London.

Skempton, A.W. (1951) The bearing capacity of clays. *Building Research Conference*, Institution of Civil Engineers, Div. 1, 180.

Terzaghi, K. and Peck, R.B. (1968) *Soil Mechanics in Engineering Practice*, 2nd edn, John Wiley, New York.

Tomlinson, M.J. (1980) *Foundation Design and Construction*, 4th edn, Pitman.

Chapter 4
Foundations in sands and gravels

4.1 CLASSIFICATION OF SANDS AND GRAVELS

Sands and gravels are classified in the laboratory by carrying out a sieve test. In this test the soil samples are washed, kiln dried, and then run through a set of graded sieves. The soil mass retained on each sieve is recorded and the results are plotted on a particle size distribution chart as shown in Fig. 4.1. From these grading curves it is possible to determine for each soil sample the total percentage of a particular particle size and the percentage of particle sizes larger or smaller than any particular particle size.

- A sand or gravel is deemed to be **well graded** if the curve on the chart is not too steep and is constant over the full range of the soil's particle sizes with no excess or deficiencies of any particular size of particle.
- A sand or gravel is deemed to be **poorly graded** if the majority of the curve on the chart is too steep, the soils

have a limited particle size distribution and most of the particles tend to be about the same size.
- If the curve on the chart shows a large percentage of larger and smaller particles with only a small fraction of the intermediate sizes then the sample is deemed to be **gap-graded**.

Figure 4.1 illustrates the curves for three samples using sieve analysis:

1. a gap-graded sandy gravel;
2. a uniformly graded sand;
3 a fine silty clay.

4.1.1 Composite sands and gravels

Gravels laid down in the form of alluvial deposits are usually mixed with sands in various proportions. Table 4.1 lists the required description for such mixed soils based on their composition.

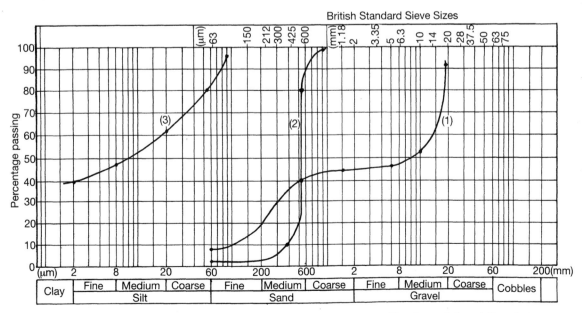

Fig. 4.1 Particle size distribution chart. Sample 1: gap-graded sandy gravel. Sample 2: uniformly graded sand. Sample 3: fine, silty sand.

Table 4.1. Required descriptions for composite sands and gravels

Description	Composition
Slightly sandy gravel (gravel with a little sand)	< 5% sand
Sandy gravel (gravel with some sand)	5–20% sand
Very sandy gravel	> 20% sand
Sand/gravel	50% sand 50% gravel
Very gravelly sand	> 20% gravel
Gravelly sand	5–20% gravel
Slightly gravelly sand	< 5% gravel

4.1.2 Dilatant sands

Fine sands and silts can exhibit dilatancy, and in soils with a high water table construction problems can arise as such sands are prone to 'boiling'. In addition, because of their composition, such soils can be difficult to dewater using conventional well-pointing systems.

In the field, soils can be checked for dilatancy by a simple test. Take a small amount of the sample and bounce it in the palm of the hand until water is evident on the surface of the sample. If the water is seen to recede when the sample is pressed with the thumb then it is either a fine sand or a silt. Difficulty can arise on site in deciding whether a soil is a fine sand or a silt, and Table 4.2 provides guidance on field examination.

Table 4.2. Field examination of sands and silts

Fine sand	Silt	Clay
Individual particles visible	Some particles visible	No particles visible
Exhibits dilatancy	Exhibits dilatancy	Not dilatant
Easily crumbles and drops off one's hands when dried	Some plasticity when moulded	Hard to crumble
Feels gritty	Feels grainy	Feels smooth
Non-plastic	—	Plasticity evident

4.1.3 Calcareous sands

These types of sand deposits were formed from the past evaporation of saline and lime-rich waters and when cemented together they have the appearance of a crusty coral-like stratum.

In their *in-situ* condition they exhibit good bearing qualities, but once disturbed and exposed to flowing water these materials rapidly break down. Once the material has deteriorated it is highly unstable. Bands of this calcareous sand occur in and around Kirton Lindsey in South Humberside, and in some cases the sands underlie peat bands. These soils can give rise to construction problems and, where possible, any foundations should be placed in compotent strata below such sands.

4.2 RELATIVE DENSITIES OF GRANULAR SOILS

When sands or gravels are encountered on a site there is often some difficulty in assessing the *in-situ* density or strength of the material. It is a stratum which is not liked by some engineers and building inspectors, particularly when water is present.

In cohesive soils a simple penetrometer or shear vane test will provide a quick indication of the soil's strength in the field. There are no similar types of quick field test for sands and gravels but the relative densities can be assessed by:

(a) visual examination of the trial pit or excavation;
(b) hand digging with spade, pick or bar;
(c) the use of a Mackintosh probe or a Perth penetrometer (forms of SPT more used for testing soil density in trial pits);
(d) standard penetration tests.

These field test methods are crude but they give a reasonable guide as to the soil's variability.

4.2.1 Field density assessment

- **Very loose.** Sides of trench collapse whilst excavating.
- **Loose.** Easily excavated with a shovel. A flat-ended 50 mm square wooden peg can be easily driven 150 mm into the ground. A 10 mm steel bar can be pushed approximately 500–600 mm into the stratum.
- **Medium dense.** Excavation is only possible with a pick. It is hard to drive a 50 mm square wooden peg more than 150 mm into the ground.
- **Dense.** It is not possible to drive a 50 mm square peg into the ground. The materials have the consistency of a soft rock.

The BS peg test is described in BS 8004 (BSI, 1986). A 50 mm square flat-bottomed wooden peg is driven by hand into the formation for approximately 150 mm and an assessment of the density can be made based on the difficulty experienced in the driving.

4.2.2 Visual observations

The visual observations made during the trial pit excavation can be very important in assessing the relative density of a granular soil. The following factors should be applied in the assessment.

1. Did the trench sides remain standing during and after excavation of the trial pit? If not, and the trial pit sides collapsed during the digging, this indicates that the soils are loose or very loose. In such soils the allowable bearing capacities will be of a low order and excessive settlements could result, especially in wet ground. Such soils can be improved by vibrocompaction types of ground improvement. If the loose conditions are not too deep, the soils can be excavated and replaced in discrete layers using a suitable vibrating trench roller.

2. Was any groundwater evident? The level of groundwater in granular soils can affect the allowable bearing capacity and settlement criteria by a factor of 2. The level of the groundwater lowers the effective stress parameters of the soils, thus reducing their ultimate bearing capacity.

3. Did the sands flow laterally because of water ingress? Fine sands in a wet loose condition are susceptible to lateral movement when excavated. Major problems can occur on sites where drainage excavations follow on after foundations are placed resulting in a loss of confinement of the sands below the foundation base with resulting subsidence.

4. Did the sands contain soft clay or soft silty lenses? Such soils, if encountered, will require the bearing capacity of the sands to be re-assessed and most likely reduced to avoid excessive settlements.

In general, therefore, well-graded dry sands or gravels of medium dense or dense composition have higher ultimate bearing capacities than most cohesive soils. In addition, the settlements under load take place quickly but these settlements can increase if the groundwater levels rise to within a distance equal to the foundation width.

4.2.3   Groundwater levels

It is most important to make sure that the full effect of a rising water table due to seasonal movements is allowed for in the foundation design. Trial pits excavated in a dry summer period may not be so dry during a wet winter period. If there is any doubt, it should be assumed that groundwater could rise and consequently the allowable bearing capacity should be halved.

Table 5 in BS 8004:1986 gives the density classification for granular soils based on standard penetration test results. These are shown in soil reports as N values.

4.2.4   The standard penetration test

This test is the most widely used method of determining the relative density of a granular soil. The test involves placing a split spoon sampler with a bottom steel driving shoe on to the drilling rods. When the borehole has been sufficiently advanced the split spoon sampler is lowered down the hole and driven into the soil by means of hammer blows on the top of the drilling rods. The hammer weighs 63.50 kg and is dropped a distance of 760 mm. The number of blows required to drive the sampler through three 150 mm intervals is recorded. The sum of the number of blows required to drive the last two 150 mm increments is recorded as the N value. The first 150 mm increment is a seating in allowance and is disregarded (Fig. 4.2).

SPT values are usually obtained at 1.50–2.0 m intervals. Though used predominantly for granular soils the test can be used in granular fills, mixed soils and clays, but the results obtained must be tempered with caution.

Table 4.3 illustrates the relative densities of sands and gravels based on SPT results.

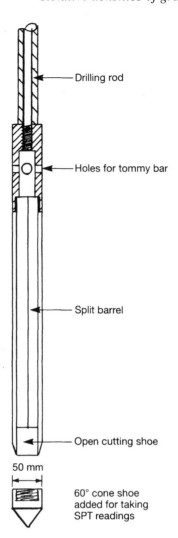

Fig. 4.2 SPT apparatus, showing drilling rod, split barrel sampler and SPT cone shoe.

Table 4.3. Relative densities of sands and gravels based on SPT results

Relative density	N, blow count/300 mm
Very loose	< 4
Loose	4–10
Medium dense	10–30
Dense	30–50
Very dense	> 50

The SPT is an empirical test, based on experience, and suitable precautions need to be taken when carrying out the test to ensure accurate results. The base of the hole must be carefully cleaned out, removing any disturbed soil. When drilling in sands below the water table, the positioning of the casing can be critical to obtaining accurate results. If the casing is not advanced far enough, the wet sands can surge into the borehole, which will result in low N values. If the casing is extended too far, the sands may be compacted and high N values will result.

Interpretation of the test results is based on experience, and many researchers in soil mechanics have produced correlations to be applied for various soil conditions. At

Foundations in sands and gravels

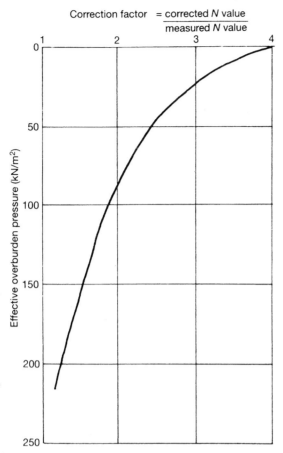

Fig. 4.3 Depth-correction values for SPT *N* values (Gibbs and Holtz, 1957).

shallow depths the recorded *N* values tend to be under-estimated, and correction factors were derived by Gibbs and Holtz (1957). These can be obtained from Fig. 4.3 to take account of the overburden pressures.

Silty sands and saturated silts usually produce an over-estimation of the relative density and a modified *N* value, N_m, can be obtained from the formula

$$N_m = 15 + \frac{1}{2}(N - 15)$$

where $N > 15$. If $N < 15$ then no correction is required.

4.2.5 Interpretation of SPT results

Terzaghi and Peck (1968) produced correlations for bearing capacity factors based on the relative density of the soils obtained from standard penetration results. Figure 4.4 illustrates the factors N_q and N_γ derived by Terzaghi, Peck and Hanson using the angle of shearing resistance ϕ. In addition, the SPT results can be used in Fig. 4.5 to determine the allowable bearing pressures for foundations in excess of 1.0 m width based on settlements not being greater than 25 mm as indicated in BS 8004 Table 1.

However, where foundations narrower than 1.0 m are being used in sands the allowable bearing capacity must be checked, as research has shown that these reduce rapidly and bearing capacity failure becomes the criterion with a suitable factor of safety applied.

A set of charts are indicated in Figs 4.6, 4.7 and 4.8 based on different ratios of depth to foundation width and based on

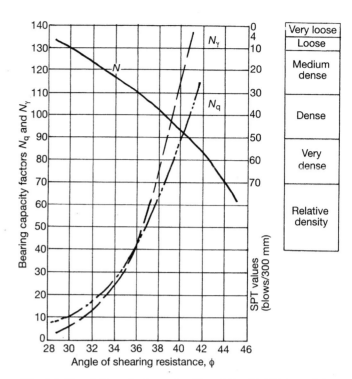

Fig. 4.4 Correlation of values of ø, N_q and N_γ with SPT tests (after Peck, Hanson and Thornburn).

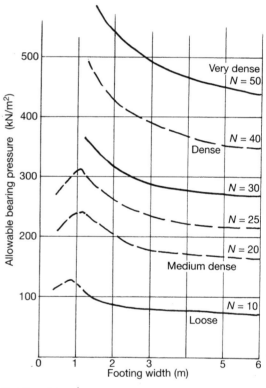

Fig. 4.5 Allowable bearing pressures on sands based on SPT *N* values (25 mm settlement criterion).

a straight-line relationship for simplicity. These values are for dry soils and based on a factor of safety of 2. Should the water levels rise to within a distance equal to the foundation width, then these values should be halved.

4.2.6 Ultimate bearing capacities

For granular soils where the dissipation of pore water pressures is usually fairly rapid, the effective shear strengths

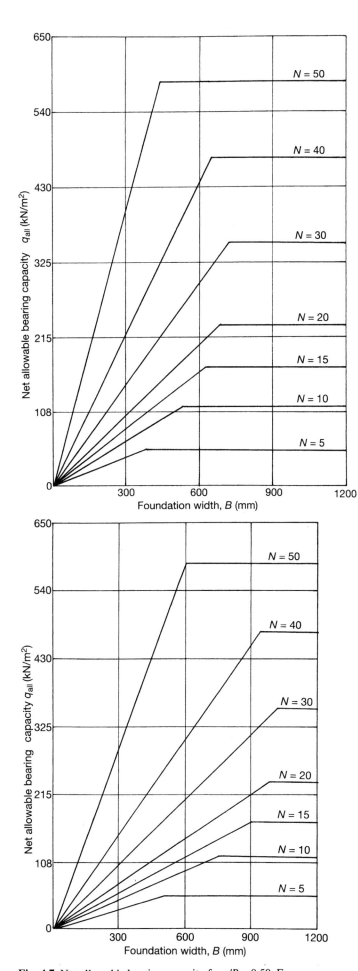

Fig. 4.6 Net allowable bearing capacity for $z/B = 1.0$. For foundation settlements not exceeding 25 mm. Factor of safety = 2.0.

Fig. 4.7 Net allowable bearing capacity for $z/B = 0.50$. For foundation settlements not exceeding 25 mm. Factor of safety = 2.0.

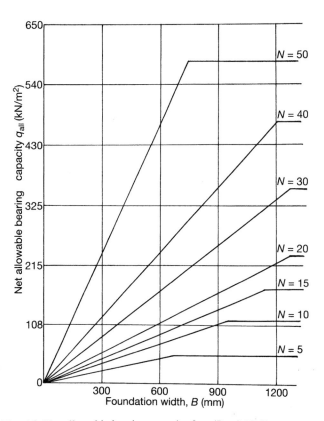

Fig. 4.8 Net allowable bearing capacity for $z/B = 0.25$. For foundation settlements not exceeding 25 mm. Factor at safety = 2.0.

85

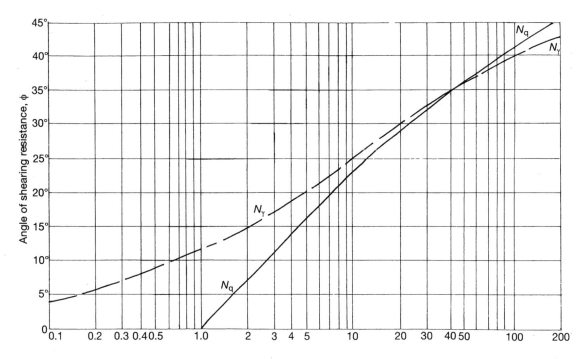

Fig. 4.9 Bearing capacity factors N_γ and N_q for sands and gravel strata. N_γ and N_q values based on SPT N values and $\emptyset$ relationship.

Table 4.4.

Relative density	Value of ø (degrees)				
	Round grain Uniform	Angular grain Well graded	Silty sands	Sandy gravels	Inorganic silts
Loose	27.50	33	27–33	35	27–30
Dense	34	45	30–34	50	30–35

are used when considering the allowable bearing capacities. Because of the difficulty in obtaining undisturbed samples in the field for laboratory testing, the ground strength parameters are usually obtained from SPT results using the various correlations.

Using Terzaghi equations, the ultimate net bearing capacity of a shallow foundation is given by:

Strip footings

$$P_{nu} = cN_c + p_o\left(N_q - 1\right) + \frac{\gamma B N_\gamma}{2}$$

Pad foundations

$$P_{nu} = 1.2\, c\, N_c + p_o\, (N_q - 1) + 0.4\, \gamma\, B\, N_\gamma$$

where c = the shear strength of the soil;
γ = the bulk density of the soil;
B = the foundation width;
p_o = the effective overburden pressure;
N_c, N_γ and N_q are Terzaghi bearing capacity factors.
When the soils are granular, and c = zero, $cN_c = 0$.

Figure 4.9 and Table 4.4 can be used to determine the N_q and N_γ based on ϕ values.

4.3 CONSTRUCTION PROBLEMS IN GRANULAR SOILS

Granular deposits make good founding strata if they are in a medium or dense state of compaction. In fact, such sands and gravels perform better than firm clays in that the foundation settlements occur immediately they are loaded.

The major construction problems are caused when granular deposits have water-bearing levels either in the form of wet lenses or as a standing water table. Any excavation below the water table will cause instability in the sides of the excavation. In addition the base of the excavation can 'boil' if the sands are very loose and start to flow.

If the water ingress into excavations is not too heavy it may be possible to control it by pumping from a lower sump. Consideration must be given to the effects of temporarily lowering the groundwater level especially if there are existing buildings close by. Any groundwater lowering could remove the fine particles from the surrounding locality and result in subsidence of existing buildings. It may be necessary to consider the use of chemical injection methods which are suitable for fine-to-coarse sands and gravels.

On open sites where deep drainage is to be installed through water-bearing granular soils, well pointing can be adopted if the soils have a fine-to-coarse grading. This system can also be used closer to existing building as the filtering system does not remove as much of the finer particles as occurs when pumping from open sumps. Where the soils have a grading less than 0.06 mm, such as a fine silt, well pointing is not a suitable method of groundwater lowering and it may be necessary to use electro-osmosis or ground-freezing methods.

If water levels on a site cannot be lowered owing to various circumstances it will be necessary to consider using a raft foundation or piles. If a raft is used the main house drainage should be installed prior to forming the raft formation. If piling is used, it is preferable and more economic to use a precast concrete driven pile or steel pile than a bored pile with a temporary casing. Alternatively a continuous flight auger pile can be used if wet granular bands are evident at depth.

4.4 FOUNDATION DESIGN IN GRANULAR SOILS

The allowable bearing capacity of granular soils is usually limited by settlement considerations. The allowable bearing pressures shown in Fig. 4.4 are based on a maximum settlement of 25 mm with no factors of safety included for. They are also based on the assumption that the water table is at a depth of at least B below the foundation base.

For foundations less than 1.0 m wide the ultimate bearing capacity should be checked, with a factor of safety of 2.50–3.0 being applied against bearing capacity failure. Bearing capacity failure of soil below a foundation is generally accompanied by high settlements and rotational movements. It is differential settlements that give rise to most structural failures, and to avoid such occurrences the foundation should be designed to keep total settlements within acceptable limits. In situations where sands are waterlogged narrow foundations could give rise to a bearing capacity failure.

For most practical uses it is sufficiently accurate to obtain the values of N_q and N_γ from the SPT results and obtain ϕ from Table 4.4 and Table 4.5 derived by Terzaghi and Peck (1968).

Table 4.5. Factor N_q (from Terzaghi and Peck)

SPT blow count, N	Angle of internal friction, ϕ (degrees)	N_q
10	30	18
20	33	26
30	36	37
40	39	55
50	41	72

Example 4.1   Strip footing on granular soil

A strip footing for a factory is to be 750 mm wide. The maximum line loading on the wall is 55 kN per metre run. Field tests have shown that the site is underlain by medium dense sands with an angle of internal friction of 35° with average N values of 15 at 1.20 m depth. Determine the net allowable bearing pressures and check the foundation width if the depth of the foundation is 1.20 m below ground. Bulk density of soil is 18.0 kN/m³.

From Fig. 4.4, $N_q = 30$ and $N_\gamma = 30$. Therefore

$$q(\text{net ult.}) = \left[\frac{N_\gamma}{2} + (N_q - 1.0)\frac{D_f}{B}\right]\gamma B$$

where N_q and N_γ are bearing capacity factors;
D_f = depth of foundation below ground level;
B = foundation width;
γ = bulk density of soil below foundations.
Therefore

$$q(\text{net ult.}) = \left[\frac{30}{2.0} + (30 - 1.0)\frac{0.60}{0.75}\right]18 \times 0.75$$

$$= (15 + 23.20) \times 13.50 = 515 \text{ kN/m}^2$$

With a factor of safety of 3.0 the allowable bearing capacity = 515/3.0 = 170 kN/m².
Actual bearing pressure = 55.0/0.75 = 73 kN/m².

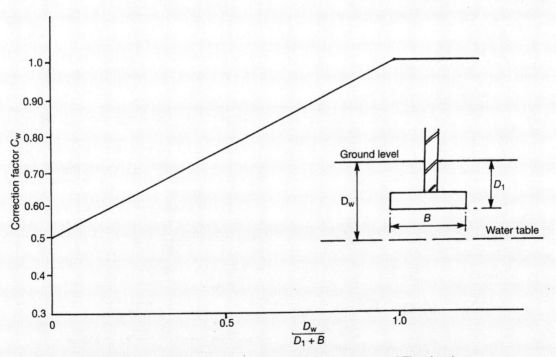

Fig. 4.10 Correction factor to SPT blow count for depth of water table (after Peck, Hanson and Thornburn).

As a rise in the groundwater level could result in the allowable bearing capacity being halved to 85 kN/m² this is a suitably sized foundation which will keep settlements within acceptable limits. Calculated settlements:

$$\text{settlement } \rho = \frac{P}{0.47N}$$

where P = applied bearing pressure; N = SPT blow count.

Overburden pressure = $18 \times 1.20 = 21.60$ kN/m²

From Fig. 4.3, correction factor = 3; therefore modified blow count N' is given by

$$N' = 3 \times 15 = 45$$

Assume groundwater can rise up to foundation level, and apply a correction factor of 0.50 from Fig. 4.10. Then

$$\rho = \frac{73.0}{0.47 \times 45 \times 0.50} = 6.90 \text{ mm}$$

When excavations on site result in variations in the formation bearing stratum with sands alternating with firm clays it is only necessary to provide a bottom layer of mesh reinforcement, say B283 as required in BS 8004.

Example 4.2 Pad foundation on sand

A pad foundation 3.0 × 3.0 m is founded at a depth of 1.0 m in a thick sand layer which has been assessed as being in the dense category. The value of shearing resistance $\phi = 35°$ and the sand's *in-situ* bulk density is 19 kN/m³. Determine the safe bearing capacity of the sands using a factor of safety of 3.0.

$$\text{Safe bearing capacity} = \frac{q_u \text{net}}{3.0} + \gamma Z$$

$$q_u \text{net} = \gamma Z (N_q - 1) S_q D_q + 0.5 \gamma B N_\gamma S_\gamma D_\gamma$$

where S_q = shape factor = $1 + (B/L) \tan \phi$ (from De Beer);
D_q = depth factor = $1 + 2 \tan \phi (1 - \sin \phi)^2 (z/B)$ for z/B = < 1.0;
S_γ = shape factor = $1 - 0.4 B/L$;
D_γ = depth factor = 1.0 for z/B = < 1.0.
Using Table 4.6 for $\phi = 35°$, $N_q = 33.30$ and $N_\gamma = 48.03$:

$$S_q = 1 + \frac{3.0}{3.0} \tan 35 = 1 + 1 \times 0.70 = 1.70$$

$$S_\gamma = 1 - 0.4 \times \frac{3.0}{3.0} = 0.60$$

$$D_q = 1 + 2 \times 0.70(1 - 0.573)^2 \times \frac{Z}{B}$$

$$= 1 + (0.427)^2 \times \frac{1}{3} = 1.086$$

$$q_{ult} = 19 \times 1.0(33.30 - 1)1.70 \times 1.086$$
$$+ 0.5 \times 19 \times 3.0 \times 48.03 \times 0.6 \times 1.0$$

$$= 1133 + 821.31 = 1954.31 \text{ kN/m}^2$$

$$\text{Safe bearing capacity} = \frac{1954.31}{3.0} + 19 \times 1.0$$

$$= 651 + 19 = 670 \text{ kN/m}^2$$

This value would be halved if the water table rose to within 3.0 m of the foundation base, i.e. 335 kN/m².

Table 4.6. Typical bearing capacity factors

ϕ (degrees)	N_c	N_q	N_γ
0	5.13	1.0	0
5	6.50	1.57	0.10
10	8.34	2.47	1.22
15	10.98	3.94	2.65
20	14.83	6.40	5.39
25	20.72	10.66	10.88
30	30.14	18.40	22.40
35	46.12	33.30	48.03
40	75.31	64.20	109.41
45	133.87	134.87	271.76
50	266.88	319.06	762.89

Example 4.3 Strip footing on sand, high water table

A continuous strip footing 1.0 m wide is founded at a depth of 1.0 m in a well-graded angular sand which has a bulk density of 18.50 kN/m³. The water table is known to fluctuate to within foundation level. Determine the ultimate bearing capacity of the sands if the soil strength parameters are based on SPT results of $N =$ 12 at 1.0 m depth and $N = 15$ at 1.20 m depth.

Consider that $\phi = 30°$ for $N = 12$ and $\phi = 31°$ for $N = 15$. For a continuous footing with $\phi = 30$, $N_c = 30.14$, $N_\gamma = 22.40$ and $N_q = 18.40$:

$$q_{ult} = \gamma Z (N_q - 1) + 0.50 \gamma B N_\gamma$$

$$= 18.50 \times 1.0 \times 17.40 + 0.5 \times 18.50 \times 1.0 \times 22.40$$

$$= 321.90 + 207 = 529 \text{ kN/m}^2$$

Applying a factor of safety of 3.0:

$$\text{Safe bearing capacity} = \frac{529}{3.0} + \gamma Z$$

$$= 176 + 18.50 \times 1.0 = 194.50 \text{ kN/m}^2$$

As the groundwater level can rise to the foundation base it is prudent to halve this value and use a figure of 93 kN/m².

Example 4.4 Bearing pressure of granular soil

A granular stratum was tested at depths of 2.0 m and 3.0 m and the N blow counts recorded were 18. Groundwater was measured at a depth of 1.30 m below ground level. The saturated sands had a bulk density of 19 kN/m³.
A strip footing is to be placed at a depth of 1.0 m and is required to be 1.20 m wide. Determine the allowable bearing pressure.
Corrected N value:

$$N_m = 15 + \tfrac{1}{2}(N - 15) = 15 + \tfrac{1}{2}(18 - 15)$$
$$= 16.50$$

Depth correction factor:

Overburden pressure = $2 \times 19 = 38$ kN/m²

From Fig. 4.2 the depth correction factor = 2.50. Therefore

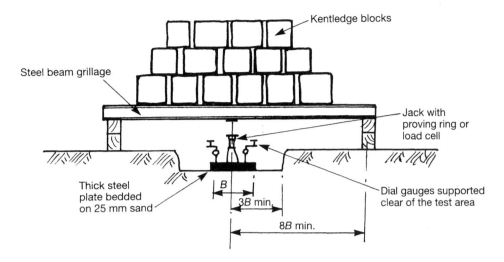

Fig. 4.11 Plate bearing test apparatus.

$N' = 16.50 \times 2.50 = 41$

For $N' = 41$ and $B = 1.20$ using Fig. 4.4:

Allowable bearing pressure = 490 kN/m²

This value is for a dry sand and should be halved to 245 kN/m² to allow for the groundwater levels rising.

4.5 PLATE BEARING TESTS

The allowable bearing pressure of a granular soil can be determined by carrying out a plate bearing test (Fig. 4.11). However, such tests must be carried out with full knowledge of the underlying strata as the plate will only stress a limited depth of ground below the foundation level and it is most important that the groundwater level is known.

The main drawback in attempting to determine settlements with this method is the effect of the small plate stressing a shallower zone of stratum than that stressed by a wider foundation. The plates therefore should be as large as possible and never less than 300 mm.

The test plate should be rigid enough to avoid bending and can be between 400 mm and 800 mm, square or circular. The kentledge should be placed in incremental stages, each increment of load being about one fifth of the proposed bearing pressure. The loading is then increased up to two or three times the proposed loading, and settlement readings should be recorded for each stage. Where there is no definitive failure point the ultimate bearing capacity is taken to be the pressure which causes a settlement equal to one fifth of the plate width. The results of settlement against load intensity should be plotted on a log scale to determine the failure point.

Terzaghi established that the settlement of a 300 m square plate at a given load can be related to the settlement of a foundation by using the formula

$$S_2 = S_1 \left(\frac{2B}{1+B} \right)^2$$

where S_1 = the settlement of a foundation of width B, where

B is taken as the ratio of foundation width to plate width; S_2 = the settlement of a test plate.

The ultimate bearing capacity of the foundation can be assessed from that of the plate by applying the following formula for granular soils:

$$\frac{Q_2}{Q_1} = \frac{B_2}{B_1}$$

where Q_2 and Q_1 are the ultimate bearing capacities of foundation and plate, respectively, and B_2 and B_1 are their respective widths.

Settlement predictions from plate bearing tests are often inaccurate but Peck, Hanson and Thornburn (1974) developed analysis based on field experience of small diameter plates and actual foundations and their conclusions were that, for foundations on granular strata:

$$\rho = \frac{P}{0.47N}$$

where P = applied bearing pressure in kN/m²; N = SPT blow count; ρ = settlement in mm. The SPT values should be the test results with correction factors for depth from Fig. 4.3 and for groundwater levels from Fig. 4.10.

4.6 PILING INTO SANDS AND GRAVEL STRATA

Piles in sands and gravel strata can be bored, *in-situ* cast in place driven type, precast or steel driven or continuous flight auger (CFA) piles. Generally the preferred system is the driven pile, which in driving increases the density of the granular strata. In addition the risk of having a poorly formed pile is ruled out when steel sections or precast concrete piles cast in a factory are used.

Bored piles and CFA piles can be very useful on sites where there are existing buildings and vibrations need to be kept to a minimum. CFA piles are generally used when water-bearing or very soft strata are encountered and a bored type of pile is needed.

4.6.1 Bored piles

These can be formed using a conventional three-leg tripod rig. For large-diameter piles a special rig design is generally required, especially if the piles are to be under-reamed, to enlarge the pile base.

Where the piles have to pass through very weak soils or water-bearing strata, the piling contractor may need to use temporary or permanent casings. If the casing is temporary and the pile is an *in-situ* concrete system in which the casing is extracted during the construction of the pile shaft, it is important to ensure that a sufficient head of concrete is maintained in the pile shaft to prevent 'necking' of the pile during withdrawal of the casing.

Great care must be exercised when withdrawing the pile casings to avoid lifting the pile reinforcement cage and surrounding concrete up with the casing. To prevent this from being a problem the concrete should be a rich mix and have a high workability.

In extreme situations where groundwater inflows are high it may be desirable to form the pile shaft using a tremie pipe operation. Concrete placed by tremie operation should be easily workable, have a slump between 100 and 175 mm, and should have a high cement content of at least 400 kg/m³.

In some situations the casing can be replaced by using a bentonite slurry. The use of a tremie pipe in these situations requires that an adequate head of concrete is kept in the tremie pipe to overcome the pressure of the bentonite mud which has to be displaced by the outflowing concrete.

4.6.2 Continuous flight auger piles

These piles are very useful in soils such as soft alluvium, wet sands and peat soils. They should only be used when a good site investigation is available. The auger on the piling rig has a central core down which a cementitious mortar or fine concrete can be pumped prior to and during removal of the pile spoil on the auger. If, on removal of the auger, it is evident from the soils on the auger tip that the pile has not been formed in suitable bearing strata, then it is necessary to replace the auger and reform a deeper pile.

4.6.3 Design of bored piles

In 1976 Meyerhof determined bearing capacity factors for deep foundations. Similar work was carried out by Berezantsev in 1961 and by Hansen and Vesic, and these factors are listed in Table 4.7.

Ultimate load capacity $Q_u = Q_b + Q_s$

where Q_b = ultimate end bearing component, and Q_s = ultimate skin friction component. Now

$Q_b = q_b A_b = \sigma'_v N_q A_b$

where σ'_v = the effective overburden pressure at the pile toe; N_q = the bearing capacity factor (Table 4.7); A_b = area of pile at base. Now

$Q_s = f_s A_s$

Table 4.7. Bearing capacity factors (after Meyerhof)

ϕ (degrees)	N_c	N_q	N_γ
0	5.14	1.0	0.0
5	6.49	1.60	0.1
10	8.34	2.50	0.40
15	10.97	3.90	1.10
20	14.83	6.40	2.90
25	20.71	10.70	6.80
26	22.25	11.80	8.0
28	25.79	14.70	11.20
30	30.13	18.40	15.70
32	35.47	23.20	22.0
34	42.14	29.40	31.10
36	50.55	37.70	44.40
38	61.31	48.90	64.0
40	75.25	64.10	93.60
45	133.73	134.70	262.30
50	266.50	318.50	871.70

where f_s = average value of skin friction developed over the embedded length of the pile shaft; A_s = surface area of the embedded pile length of the pile shaft. The average value of f_s is given by

$f_s = K_s \sigma'_v \tan \delta$

where K_s = the coefficient of lateral earth pressure; δ = angle of friction between the pile shaft and the surrounding soils. Values for K_s and δ are listed in Table 4.8 (derived by Broms in 1966).

Table 4.8. Typical values for δ and K_s (Broms, 1966)

Pile material	δ	K_s Relative density of soil	
		Loose	Dense
Steel	20°	0.50	1.0
Concrete	0.75 ϕ'	1.0	2.0
Timber	0.67 ϕ	1.50	4.0

ϕ' = Angle of shearing resistance in respect of effective stress values

The values of f_s are limited for pile lengths between 10 and 20 times the pile diameter or pile width. For practical usage f_s maximum is taken as 100 kN/m². Meyerhof determined that q_b is approximately equal to 14 $N D/B$ where N = SPT blow count; B = pile diameter or pile width; D = embedded length of pile in the end bearing strata; f_s is approximately equal to 0.67 $\bar{N}$ kN/m², $\bar{N}$ = the average uncorrected N value over the shaft length considered.

When constructing bored piles in sands and gravels the granular strata will be loosened during removal of the core material. In view of this it is prudent to adopt a cautious approach and use values of ϕ and N_q based on loose soil conditions to determine the ultimate end bearing and skin friction values. Adopting this approach will result in an ultimate bearing capacity lower than that achieved by a driven pile in the same strata but the loosening effect of the boring operations on the base and pile shaft in granular soils can be quite significant.

Example 4.5 Bored piles

A dwelling is to be supported on bored piles which are required because of the closeness of an old existing building. The maximum unfactored line load for the three storey dwelling is 60 kN per metre run. The soil conditions below the site consist of approximately 4.0 m of very loose to loose ash fill which is still undergoing consolidation settlement and, below the fill, a medium dense sand with recorded SPT values of 22. The density of the ash fill is 1300 kg/m³ and all the boreholes revealed dry conditions down to 15.0 m depth. The density of the sand is 1850 kg/m³.

With piles at 4.0 m centres, and using continuously designed ring beams, the maximum working load on each pile will equal $4 \times 1.20 \times 60 = 288$ kN. Ultimate skin friction on pile shaft, $f_s = K_s \gamma_d \tan\delta$. Since boring will loosen the sands, use Table 4.8 for K_s and δ. $K_s = 1.0$, $\phi = 33°$ and $\delta = 0.75 \times 33 = 24.75$. Therefore

Unit skin friction at top of sand $= 1.0 \times 1.30 \times 9.80 \times 4 \times \tan 24.75$
$$= 23.49 \text{ kN/m}^2$$

If we assume piles will be approximately 10 m long:

Unit skin friction at pile toe $= 1.0 \times (1.3 \times 4.0 + 1.85 \times 6)$
$$\times 9.8 \times \tan 24.75$$
$$= 73.64 \text{ kN/m}^2$$

Assuming 400 mm diameter concrete piles:

Total skin friction $= \dfrac{(23.49 + 73.64)}{2.0} \times 6 \times \pi \times 0.40 = 366$ kN

Assuming that during the boring the medium dense sands will be loosened and the value of ϕ will be reduced to 32°. Using the Berezantsev chart (Fig. 4.12):

N_q for $D/B = 6.0/0.40 = 15$ is 33

End bearing resistance $= \dfrac{\pi \times 0.40^2}{4}(1.30 \times 4.0 + 1.85 \times 6) \times 9.8 \times 33$
$$= 662.50 \text{ kN}$$

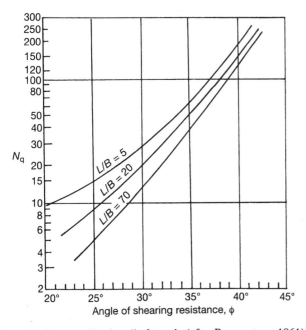

Fig. 4.12 Values of N_q for pile formula (after Berezantsev, 1961).

Therefore

Ultimate resistance $= 662.50 + 366 = 1028.50$ kN

Adopting a combined factor of safety of 3.0, the maximum allowable working load $= 1028.50/3.0 = 342$ kN.

Because the fills are still settling under their own weight, the effect of negative skin friction must be allowed for in the pile design. This value of negative skin friction must be added to the pile working load.

Assume the negative skin friction acts over the top 4.0 m of the pile. The peak value of negative skin friction will not at any time act over the whole length of the pile shaft embedded in the fill and it will therefore be necessary to make a reasoned assessment of the magnitude of the drag-down forces to be used in the design.

Negative skin friction $= \sigma'_v K \tan \phi'_a$.

where σ'_v = effective vertical stress, and $K \tan \phi'_a$ is assumed to be constant for the pile's length.

Unit skin friction at the top of the sand was calculated at 23.49 kN/m². This value can be used as the peak value of the negative skin friction at the base of the fill, as it will be approximately equal. Therefore total negative skin friction on the top 4.0 m of the pile shaft equals

$$\left(\frac{0 + 23.49}{2.0}\right) \times \pi \times 0.40 \times 4.0 = 59.0 \text{ kN}$$

Therefore

Factor of safety on pile $= \dfrac{1028.50}{288 + 59} = 2.96$

This is slightly less than the recommended value of 3.0 but is acceptable because of the low values adopted for the density of the sands.

Using a driven pile

Using a 275 mm × 275 mm precast concrete pile driven through the fills into the medium dense sands ($N = 25$ blows) to a calculated set, a factor of safety of 2.50 can be adopted.

Maximum pile working load $= 288 + 59 = 347$ kN.

Therefore ultimate load capacity requires to be $347 \times 2.50 = 867.50$ kN. For a SPT blow count of 25, $\phi = 36°$. Using the Berezantsev chart (Fig. 4.12), for a 10 m pile length:

$$\frac{L}{B} = \frac{6000}{275} = 21.80$$

Therefore

$$N_q = 60$$

$$Q_s = f_s \times A_s$$

$$f_s = K_s \sigma'_v \tan \delta$$

where $K_s = 2.0$ and $\delta = 0.75 \times 36 = 27$. Therefore

$$Q_s = 2.0 \times \left(\frac{4.0\gamma_1 + 6.0\gamma_2}{2.0}\right) \tan 27 \times 4 \times 0.275$$

$$= 2.0 \times \left(\frac{4.0 \times 13.0 + 6 \times 18.50}{2.0}\right) \times 0.509 \times 1.10 = 91.26 \text{ kN}$$

and

$$Q_b = P_b(N_q - 1)A_b$$

where P_b is the effective overburden pressure at the base of the pile, and A_b is the area of the pile base $= 0.275 \times 0.275 = 0.075$ m². Therefore from the Berezantsev chart

91

$\dfrac{L}{B} = \dfrac{6000}{275} = 21.80$ Therefore $N_q = 60$

$Q_b = 163(60-1) \times 0.075 = 727\,\text{kN}$

$Q_u = 727 + 91.26 = 818\,\text{kN}$

Though this is slightly less than 867.50 kN, the pile will be driven to a calculated set and will most likely penetrate the sands for less than the 6.0 m available.

4.6.4   Set calculations

Using the modified Hiley formula for precast piles,
R_u = ultimate load = $347 \times 2.50 = 867.50$ kN;
E = Transfer energy at pile top = 0.70×10^4,
c = temporary compression of pile and ground per blow, say 12 mm;
s = set blow count.
Therefore

$$s = \dfrac{E}{R_U} - \dfrac{c}{2} = \dfrac{7000}{867.50} - \dfrac{12}{2}$$

$$= 8.06 - 6 = 2 \text{ mm/blow}$$

Therefore adopt a set of 20 mm for 10 blows or less. In practice, f_s is taken as 100 kN/m² maximum.

For driven piles in granular soils there are approximate formulae derived by Meyerhof in 1976 to calculate pile capacities.

$$q_b \approx \dfrac{40ND}{B} \quad \text{but} < 400N \;\left(\text{kN/m}^2\right)$$

where N = SPT blow count, D = embedded length of pile in bearing strata, and B = diameter or width of the pile.

Consider example 4.5:

$Q_b = q_b \times A_b = 400 \times 25 \times 0.275^2 = 756$ kN

and $f_s = N$ kN/m² = the average uncorrected N value over the embedded length of the pile in the bearing stratum. Therefore

$Q_s = 25\,A_s = 25 \times 6 \times 0.275 \times 4 = 165$ kN.

For a bored pile in granular soils

$$q_b = \dfrac{14ND}{B} \text{ kN/m}^2$$

and $f_s = 0.67\overline{N}$ kN/m²

Example 4.6   Working load of precast concrete piles

A 3.0 m thick layer of loose sands and gravels overlie a thicker deposit of dense sands and gravels. SPT tests in the dense sands produced values of 35 from the base of the loose sands to a depth of 12.0 m below ground level at 1.0 m intervals. Using a 275 mm × 275 mm precast concrete pile and adopting a factor of safety of 2.50 determine the maximum allowable working load for the pile.

Ultimate bearing capacity = $Q_u = Q_b + Q_s$

For Q_b, ignore the loose sands and use $q_b = 40\,ND/B$ or $400\,N$ kN/m². Therefore

$$q_b = 40 \times 35 \times \dfrac{D}{0.275} = 400 \times 35 \quad \text{max}$$

$$D = \dfrac{400 \times 35 \times 0.275}{40 \times 35} = 2.75 \text{ m penetration into the dense sands}$$

Therefore

$Q_b = 400 \times 35 \times 0.275^2 = 1058$ kN.

Q_s in loose sands is discounted. Q_s in dense sands = $f_s \times A_s$, where $f_s = N = 35$ kN/m². Therefore

$Q_s = 35 \times 2.75 \times 4 \times 0.275 = 105$ kN

$Q_u = 105 + 1058 = 1163$ kN

Allowable working load $= \dfrac{1163}{2.50} = 465$ kN

In granular strata the end bearing component is much greater than the skin friction component on the sides of the pile. To mobilize this skin friction a significant movement has to occur at the pile toe. In dense granular strata this movement is very small and because of this the factor of safety for a driven pile can be 1.50 for skin friction and 3.0 for end bearing. Applying these factors to Example 4.6 the allowable working load would be

$$\dfrac{105}{1.50} + \dfrac{1058}{3.0} = 70 + 352 = 422 \text{ kN}$$

4.6.5   Dynamic pile formula

The ultimate static resistance of a driven pile can be predicted from the dynamics of the driving operation itself. The kinetic energy imparted by the piling hammer is equated to the work done by the pile in penetrating into the ground. Therefore

Net kinetic energy = Work done during penetration of pile

For a hammer of weight W tonnes falling a drop height of h and causing a penetration or set of s mm, the pile resistance load R_s can be obtained from the formula

$R_s = Wh$ – energy losses

The energy losses are due to the pile and pile cap compression, hammer rebound, and frictional losses in the equipment.

Driving a pile into sands and gravel strata will increase the relative density of the sands and gravels and this has a significant effect on the predictions of load-carrying capacity.

For concrete piles the modified Hiley formula is often applied but the Dutch formula is also often used. The Hiley formula should only be applied to piles which obtain their support in sands and gravels, stiff-to-hard clays or rock. It is not applicable to frictional piles which obtain their support in soft clays by adhesion along their length.

Specialist piling contractors who rely on piling to make their living generally put their trust in the simple drop hammer. These hammers are often considered to be crude and old-fashioned but they are very reliable and just as effective as sophisticated hammers at less cost. Some piling firms have developed their own equipment and use purpose-built hammers not readily available on the open market.

The Hiley formula can also be adopted for steel bearing piles and a factor of safety of 2.0 can be used, where

$$\text{Working pile load} = \frac{R_s}{\text{Factor of safety}}$$

For practical purposes the ultimate load on a pile can be defined as that load which causes a settlement of one tenth of the pile diameter or pile width (Terzaghi and Peck, 1968).

The accuracy of a given dynamic formula can be improved by recalibrating it for a given site against the test load data obtained from static load tests. The formula can then be more confidently used as a guide for selecting final penetrations of those piles which are neither near any tested piles or near ay borehole locations.

Dynamic formulae can be grossly inaccurate: using the Hiley formula, the actual ultimate load obtained by test loading may be between 0.70 and 3.0 times the figure obtained by applying the formula. Pile testing should always be carried out to verify the dynamic formula.

4.6.6 Re-drive tests

These should be carried out on one or more piles at a reasonable frequency rate with a minimum time interval of 12 h. Only if the re-drive final set (mm/blow) is equal to or less than the set of the initial drive can the dynamic formula be adopted. If the re-drive set is greater than the set obtained on the initial drive the formula does not apply, and it will be essential to re-drive the piles until a tighter set is achieved, or to carry out a static load test.

4.6.7 Base-driven steel tube piles

$$R_u = \frac{290 W(1.0 + h)}{s + 12.70}$$

where R_u = ultimate driving resistance in tonnes; W = weight of internal drop hammer in tonnes; h = actual drop of hammer at final set in metres; s = final set (mm/blow). This formula is applicable for:

● drops between 1.20 m and 2.0 m;
● sets less than 5 mm/blow, i.e. 5 blows to 25 mm.

4.6.8 Top-driven steel piles

$$s = \frac{W^2 H}{R(W + p)} \text{ metres penetration per blow}$$

where W = weight of hammer in newtons;
p = weight of pile (unit weight × length) in newtons;
H = effective hammer drop in metres;
R = penetration resistance or ultimate load capacity in newtons;
s = set (penetration per hammer blow) in mm.
Allowing for 30% loss of efficiency:

$$\frac{W^2 H}{W + p} = R \times s$$

where WH = kinetic energy. Therefore

$$\text{set } s = \frac{W^2 H}{R(W + p)}$$

Example 4.7   Steel piles

Working load on pile = 295 kN with pile length = 9.0 m. Factor of safety = 2.50. Therefore

$R = 2.50 \times 295 = 737.50$ kN

Mass of hammer, $m = 3.06$ kN;
weight of hammer, $w = 30$ kN;
weight of pile, $p = 0.648 L$ kN where L = pile length;
effective hammer drop, $H = 0.35$ m.

Velocity of hammer at impact = $2 g H = 2 \times 9.81 \times 0.35 = 2.62$ m/s

$$\text{Kinetic energy} = \frac{mv^2}{2} = \frac{3.06 \times 10^3 \times 2.62^2}{2} = 10\,500.0 \text{ Nm}$$

Reduce this value by 30% for losses = $0.70 \times 10\,500.0 = 7351$ Nm. Therefore

$$s = \frac{W(WH)}{R(W + p)} = \frac{30 \times 10^3 (7351)}{10^3 \times 37.50(30 + 0.648 \times 9)} = 8.34 \text{ mm/blow}$$

Therefore, use three blows of 350 mm for a set of 25 mm.

Example 4.8   Driving precast concrete piles

A 275 mm × 275 mm precast concrete pile is to be used to carry safe working loads of 350–500 kN. The pile is reinforced with eight 12 mm high tensile bars in pairs bundled in each corner. Concrete strength is 50 N/mm². $F_y = 590$ N/mm²; $F_{cu} = 50$; $A'_s = 452$; $A_s = 452$.
Ultimate axial compression load:

$N = 0.40 F_{cu} A_c + 0.75 A_{sc} F_y$
$= 0.40 \times 50 \times 275 \times 275 + 0.75 \times 2 \times 452 \times 590$
$= (1\,512\,500 + 400\,020) \times 10^{-3} = 1912$ kN

For working load of 350 kN, using modified Hiley formula

$$R_u = \frac{E}{s + c/2} \text{ with factor of safety} = 2.25:$$

$R_u = 350 \times 2.25 = 787.50$ kN

Transfer energy at pile top, $E = 0.85 \times 10^4$ kN m; temporary compression of pile and ground, $c = 10$ mm. Set per blow in mm, s, is given by:

$$s = \frac{8500}{787.5} - \frac{10}{2} = 5.80 \text{ mm/blow}$$

Ten blows of hammer give a 58 mm set; therefore use a 4.0 t banut hammer with a 400 mm drop. For working load of 350–500 kN:

$$s = \frac{8500}{500 \times 2.25} - \frac{10}{2} = 2.60 \text{ mm/blow}$$

Therefore ten blows of hammer give a 26 mm set; therefore use a 4.0 t banut hammer with a 400 mm drop, see Table 4.9.

Table 4.9. Hammer transfer energy $\times 10^4$ (Rig: hydraulic Banut type)

Hammer weight (tonnes)	Transfer energy (tonne metres)				
	Hammer drop (mm)				
	300	400	500	600	700
1.50	0.25	0.35	0.45	–	–
3.00	0.55	0.70	0.90	–	
4.00		0.85	1.10	–	
5.0		1.05	1.40		

DUTCH FORMULA

This formula provides an alternative method of determining a pile set using a dynamic formula.

$$s = \frac{W^2 KH}{R_u(W + p)}$$

where s = set in mm/blow;
W = weight of hammer = 35 kN;
K = hammer efficiency = 0.70;
H = hammer drop = 450 mm;
R_u = working load $\times$ 10, 350 $\times$ 10 = 3500 kN;
p = weight of pile = 18 kN.
Therefore

$$s = \frac{35 \times 35 \times 0.70 \times 450}{3500(35 + 18)} = 2.27 \text{ mm/blow}$$

For ten blows this equals 22.70 mm set.

BIBLIOGRAPHY

Berezantsev,V.G. (1961) Load bearing capacity and deformation of piled foundations. *Proc. Fifth International Conference on Soil Mechanics*, Paris, Vol. 2, pp. 11–12.

Broms, B. (1966) Methods of calculating the ultimate bearing capacity of piles: a summary. *Sols (Soils)*, **5**(18/19), 21–31.

BSI (1986) BS 8004: *British Standard code of practice for foundations*, British Standards Institution.

Carter, M. (1983) *Geotechnical Engineering Handbook*, Pentech Press, London.

De Beer, E.E. (1965) Bearing capacity and settlement of shallow foundations on sand. *Proc. Symposium on Bearing Capacity and Settlement of Foundations*, Duke University, pp. 15–33.

Gibbs, H.J. and Holtz, W.G. (1957) Research on determining the density of sands by spoon penetration testing. *Proc. Fourth ICSMFE Conference*, London, Vol. 1, pp. 35–39.

Meyerhof, G.G. (1952) The ultimate bearing capacity of foundations. *Geotechnique*, **2** (4), 301–332.

Parry, R.H.G. (1971) A direct method of estimating settlements in sands from SPT values. Midlands SMFE Society.

Powell, M.J.V. (1979) *House-Builder's Reference Book*, Newnes-Butterworth, London.

Terzaghi, K. and Peck, R.B. (1968) *Soil Mechanics in Engineering Practice*, 2nd edn, John Wiley, New York.

Vesic, A.S. (1966) Tests on instrumented piles. Ogeechee River site. *Journal of Soil Mechanics and Foundation Division, American Society of Civil Engineers*, 96, SM 2.

Chapter 5
Building in mining localities

The following guidance is given for builders and engineers involved with the planning and construction of housing on sites previously undermined by mineral extraction and on sites where future extraction of coal or other minerals will take place after the development has been completed.

There are various techniques for investigating and consolidating old mine workings, securing of old mine shafts, adits etc., and various foundation design options are available to cater for any ground movements likely to arise. Past and current coalmining is the most common cause of subsidence but there are other minerals, such as fireclay, sandstone (Elland flags), chalk, ironstone, salt and gypsum, which can give stability problems below a site. The effects of subsidence from modern longwall extraction methods now in use in the British coalfield can be predicted fairly accurately whereas movements resulting from old shallow mineral workings are not so easily defined and require sound judgements by engineers and geologists experienced in this field based on the available mining and geological data collected.

Building houses on land which is underlain by known shallow coal workings or other mineral workings can result in very expensive development costs. The total costs are difficult to quantify prior to consolidation being carried out because of the lack of information on the volume of material extracted. On some sites it may well be cheaper not to develop certain areas of the site and place public open space over the no-build zones. If shallow workings are discovered in the final phases of a development there will be less properties available to spread the costs.

In known mining areas it is prudent to consult British Coal or other bodies such as the Brine Boards, mineral valuers, and British Geological Survey, before purchasing any land for development. In some localities, planning authorities may lay down conditions in regard to old or future mineral extraction. This has become more frequent since the closure of many mines has allowed mines to flood which results in a diaphragm effect in pushing mine gases such as methane and carbon dioxide to the ground surface. This alone could render a site undevelopable.

Coal, lead, tin, ironstone, fireclay, sandstone, gypsum, salt, chalk, sand, anhydrite and other minerals have been extracted by various methods over the years, but many of the industries related to the minerals have gone into decline. At present the minerals most frequently extracted are coal, gypsum, anhydrite and salt.

Gypsum and anhydrite mines are extensively worked in Cumbria, Yorkshire, Nottinghamshire and Sussex, but the nature of the workings results in very large pillars being left to provide support for the overlying strata. The seams extracted are very thick; often up to 10 m is removed from 30 m thick beds of material.

Chalk was mined in a similar fashion using pillar-and-stall methods in areas of Kent, Bury St Edmonds, Suffolk and Norwich. In some areas the overlying rock strata may have collapsed into swallow holes in the chalk and this material is usually a mass of loose voided material. In areas of swallow-hole activity piling taken below the chalk base is generally required, but if the holes are large and widespread then the site may not be viable owing to the effects of renewed and often unpredictable subsidence on the site infrastructure.

Salt mining can be carried out by mining or by brine pumping and the design of foundations in such areas require special considerations.

5.1 COAL MINING, PAST AND PRESENT

Originally used as a means of obtaining fireclay or ironstone, **bell pits** (Fig. 5.1) were in use from the thirteenth century up to the early 1800s. They were usually found in areas where the thickness of drift (superficial deposits) was relatively thin and the seams were shallow and fairly level. The depth of these pits rarely exceeded 12 m. They consisted of a vertical shaft taken down to the various minerals or coal seam; the shaft was belled out at the bottom to maximize the coal recovery. Quite often, bell pits sunk for coal extraction would encounter bands of ironstone and this was often removed for use. Sometimes shafts were excavated in pairs and interconnected at the bottom. To facilitate ventilation fire

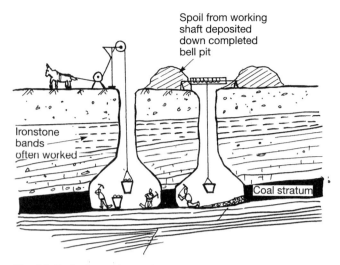

Fig. 5.1 Bell pit workings (after Gregory, 1981).

braziers were suspended over one shaft to create air circulation. When such shafts are being investigated the signs of ashy circles are usually a good indicator of bell pit activity.

Spoil from one shaft was disposed of down the adjacent worked-out shaft in an uncontrolled fashion, resulting in localized soft spots around an area. In certain localities bell pits taken down to a coal seam were worked away from the base of the pit in the form of radiating horizontal adits. In these situations, only the shafts were filled; the horizontal galleries were left open. Such a site was encountered in the Sheffield area and resulted in large volumes of grout consolidation having to be used on a closely spaced grid pattern.

Where bell-pit workings are suspected the most suitable method for locating them is by blading the surface with a wide-face shovel excavator and examining the exposed natural ground for the bell pit outlines. In most cases it will be necessary to remove the layer of spoil which will have been spread around from the workings. In situations where ironstone has also been removed there are generally very thick surface deposits of ironstone and coal measure spoils which make identification of shafts very difficult.

Where bell pits are located, dwellings should be relocated where possible; if the rockhead is reasonably shallow the bell pit can be capped off. Any evidence of horizontal galleries would require the area of the dwellings to be test drilled and possibly grouted. If a bell pit is encountered under a house position and relocation is not possible the fill in the shaft can be tested using SPT equipment to check its state of compaction relative to the natural strata. If it is found to be suitable the most suitable foundation is a stiff edge beam raft or deep heavily reinforced ring beams designed to span 6–8 m across the shaft with a suitable margin at each side.

5.2 COAL SHAFTS

If coal shafts are shown to be present on a site, based on records held by British Coal, then the following procedures must be followed.

The approximate position of the suspected shaft should be surveyed, based on the recorded Ordnance grid reference, and the ground around the peg should be bladed using a wide-face front shovel excavator. In clays or rock strata the backfilled shafts are usually well defined by the different colour of the shaft-fill materials.

If thick fill deposits are found to exist over the suspect area then it will be necessary to remove this fill down to the natural stratum. This is generally only economic to depths of 2–3 m. Any fill in excess of these depths may require a more sophisticated mine shaft detection system to be adopted. This may be done using a closely spaced borehole pattern over a pegged-out grid or by using proton magnetometer or similar resistivity techniques. These are not always successful, as the coal measure fills around shafts can contain minerals which affect the search instrumentation. If a recorded shaft cannot be located then it may be necessary to zone off an area of land on which nothing should be built.

Once a shaft has been located its grid co-ordinates should be determined; many shafts' positions have been lost on re-excavation because surface fills can obliterate the shaft outline. If the shaft is filled it must be capped off at rockhead level or on to strata which would be acceptable to British Coal surveyors. The general requirements are that the shaft cap be twice the shaft diameter and be designed to support the depth of fill over and any surcharge loads from large vehicles.

Table 5.1. Mine shaft cap designs for various shaft sizes

Shaft dia. (m)	Cap size (m)	Cap depth (mm)	Reinforcement details			
			Mild steel		High-yield	
			Bar diameter (mm)	Distance between centres (mm)	Bar diameter (mm)	Distance between centres (mm)
1.0	2 × 2	450	20	250	16	250
1.50	3 × 3	450	20	200	16	200
2.0	4 × 4	500	20	175	20	250
3.0	6 × 6	500	25	200	20	175
4.0	8 × 8	600	25	150	20	150
5.0	10 × 10	600	32	175	25	150
6.0	12 × 12	750	40	200	32	200

All reinforcement to be in top and bottom and spanning in both directions

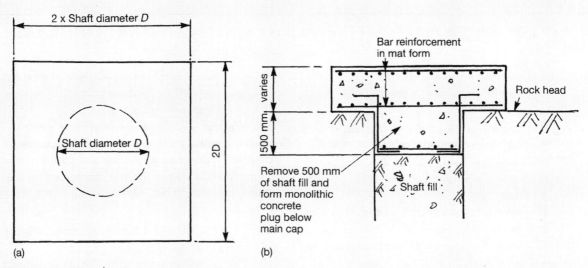

Fig. 5.2 Typical mine shaft cap: (a) plan; (b) typical section.

Table 5.1 gives a range of cap designs for varying shaft diameters and fill depths. Figure 5.2 shows a typical cap arrangement.

When shafts are encountered very close to dwellings (Fig. 5.3) it is important to ensure that the house foundations are deepened so that a 45° line of repose passes below the bottom of the cap excavation. If this results in the house

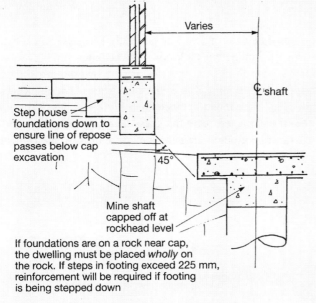

Fig. 5.3 Deepen foundations if close to mine shaft cap.

foundation being on rock then the dwelling must be placed *wholly* on the rock.

Where rockhead is in excess of 3–4 m then the shaft filling will need to be investigated and a borehole outside the shaft will be required to determine rockhead depth. The shaft should be drilled to its base and through any platforms often placed at intervening seam levels.

Many shafts were filled by depositing large objects down the shaft, such as mine bogies, pit props, and conveyor belts, to form a makeshift plug. Sometimes a timbered or steel platform was constructed at a level just below the surface or at a

higher seam horizon to support the shaft filling. When test drilling such shafts it is important to carry out full safety precautions such as the use of steel bridging beams supported well beyond the shaft, the use of full safety harness for drilling operatives, and regular checks for mine gases (especially methane and carbon dioxide). Such shafts can be made safe by drilling and pressure grouting the shaft infill and capping off at ground level.

Where a shaft is found not to have been filled below a temporary platform level, the usual method for stabilizing the shaft is to drill and fill with pea gravel from the shaft base or use a mixture of PFA and cement grout at the base of the shaft workings and top up with pea gravel. Once the shaft has been stabilized, houses should be kept at least an 8 m radius away from the centre of the shaft. Though it is technically possible to build over a deep capped shaft there is still the potential risk of migrating mine gases and there are recorded fatalities from such situations.

5.3 SHALLOW MINEWORKINGS

Old shallow mine workings less than 30 m deep can be a major consideration when considering a site's stability (Fig. 5.4). Where the stratum overlying mine workings is competent rock and is thick, the stability is achieved by arching and bulking of the galleries.

Past mining was generally carried out using pillar-and-stall systems of extraction with extraction rates varying according to the mine geology and roof conditions. Where insufficient rock cover exists, the roof of the mine galleries can collapse and result in crown or plump holes at the surface.

Old mine workings are notoriously ill-documented; in many cases, no plans existed. Workings were often entered from an adit driven into the hillside to intercept the coal. Where the seam was deeper, shafts were sunk and the coal was removed in tunnel fashion with substantial coal pillars left on a crude grid basis. The size and positioning of these

---

GUIDELINES FOR THE DESIGN OF REINFORCED CONCRETE CAPS OVER FILLED DISUSED SHAFTS

In accordance with British Coal criteria.

1. The cap shall be designed to support the following loading criteria:
   (a) a 20 tonne point load placed centrally above the shaft at final ground level; *or*
   (b) a superimposed load of 33 kN/m² over the full cap surface whichever is the more onerous condition; *plus*
   (c) allowance for overburden from cap surface to final ground level; *plus*
   (d) allowance for suction pressures created on collapse of shaft fill; *plus*
   (e) allowance for reverse bending caused by the collapse of the shaft walls; *plus*
   (f) an allowance for any loads which the cap designer considers particular to the intended position or after-use of the shaft site.
2. The cap should be founded at rock level and should be square centred on the shaft with sides equal to twice the diameter of the shaft.
3. The top 500 mm of fill should be removed from the shaft and replaced by concrete poured at the same time as the cap to form a shear key.
4. The minimum cap thickness should be 450 mm.
5. Recommended good practice on the choice of concrete type, mixing, placing and curing of the concrete and fixing of the steel reinforcement and shuttering should be observed, subject to the following.
   (a) Concrete should have a minimum 28 day strength of 30 N/mm².
   (b) The minimum cover to reinforcement should be 75 mm.
   (c) Partial substitution of cement content by PFA or blast furnace slag is recommended.

Where the depth to rockhead is excessive the cap should be founded on solid ground (not made-up ground) or supported on piles driven or bored down to rockhead level.

---

pillars were often dictated by the roof geology and thickness of seam available. Where seat earth or fireclay was encountered above or below a coal seam these materials were also removed, resulting in increased extraction thicknesses. In some localities, where fireclay was the principal mineral being mined, any coal removed was an added bonus.

Where test drilling reveals that old workings have collapsed and virtually closed up, or where there is evidence of back stowing, then grouting up is difficult and it is usual in these situations to provide a stiff edge beam raft foundation designed to span 3 m and cantilever 1.5 m.

Where workings exist with known voids or broken ground then the thickness of competent rock cover over the workings must be determined. The generally accepted rule is that where the competent rock cover exceeds ten times the extracted seam thickness then no major crown holing should result at the surface: this is the **ten-times seam thickness rule** (Fig. 5.5). If the competent rock cover is less than ten times the extracted seam thickness then generally the workings must be grouted up using a mixture of PFA and cement

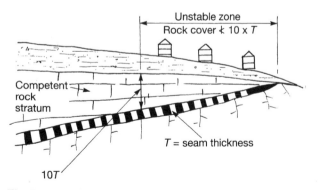

Fig. 5.4 Shallow pillar and stall workings.

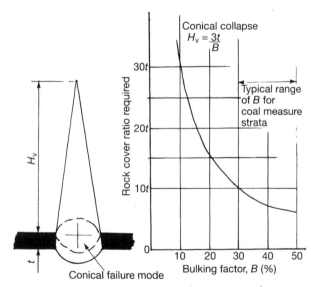

Fig. 5.5 Conical failure mode of shallow coal workings.

grout placed under pressure. Once the workings are stabilized, foundations to houses can be heavily reinforced strip footings or pseudo edge beam rafts. Where grouting is not required it is still prudent to reinforce strip footings, especially in marginal areas.

There are occasions when the ten times seam thickness rule does not apply: for instance where massive bedded sandstones overlie the seams. It can be demonstrated that the bulking factor of such rocks would result in a reduced rock ratio being adequate.

The seam thickness is the theoretical measurement but, in reality, the base of the seam was often covered with a layer of stone debris and dust, and often heaved up because of the pressures on the adjacent pillars. It is reasonable therefore to conclude that the rule is conservative, but its use is reasonable bearing in mind that geological faulting and similar discontinuities may exist.

Maximum height of void migration is given by:

$$H_v = \frac{nt}{B-1}$$

where

n = 1, 2, or 3 for rectangular, prismatic and conical collapse modes;
t = the extracted seam thickness;
B = the bulking factor = volume of collapsed strata/volume of *in-situ* strata.
For highly weathered coal measure shales B is equal to 1.1 to 1.3.
Assuming a conical collapse mode for the roof then

$$Hv = \frac{3t}{(1.3-1.0)} = 10t$$

5.4 DRILLING INVESTIGATIONS

Where shallow workings are suspected, a drilling investigation is required using rotary flush drilling techniques. The pattern of drilling is based on the outcrop position of the seam and is generally carried out on a domino-five pattern. The primary aim is to establish the seam depth and the amount of rock cover. If the first few holes indicate that sufficient rock cover exists, even allowing for the dip of the seam, then it is pointless spending more money on drilling more holes. If the drillings reveal that there is insufficient rock cover then further holes must be drilled to establish whether the seam has been worked.

If 20 holes are drilled on a grid pattern and no broken ground or voids are found then it is reasonable to conclude that the seam has not been worked. If, however, one or two holes reveal voids or broken ground then it must be assumed that the seam has been worked throughout the site.

An important point to remember when carrying out a drilling investigation is to ensure that the ground levels are always recorded at each borehole position. Failure to obtain these levels will make it impossible to produce a cross-section between holes showing the seam horizon.

In some localities there may be more than one shallow seam with one seam outcrop overlapping the other, resulting in two unstable zones which will require grouting up.

5.5 STABILIZING OLD WORKINGS

The most common solution is to fill the workings with a PFA cement down predrilled holes, generally on a 3 m grid basis. The grid should extend at least 2 m beyond the building line.

Where the seam is thick and extensively voided, it will require the installation of a grout curtain around the perimeter of the grid. The grout mix proportions will need to be modified by the addition of sand or fine gravels to produce a grout with low slump characteristics so preventing the grout from flowing away. This method is also used for seams which have a substantial dip. Once the grout curtain is installed, the infill holes can be grouted up with the standard mix.

5.5.1 Collapsed workings

Where old mine workings have partially collapsed, and any voids are less than 0.5–1.0 m, a grout curtain can be formed on the perimeter of the grid using a mix known as a **pancake mix**, as it is placed in a series of grout layers. The constituent parts of the mix are:

Cement	1 part
Sand	2.5 parts
PFA	10.5 parts
Bentonite	0.05 parts
Water	

For perimeter curtain walls in seams with voids exceeding 1.0 m depth, a cement:PFA:sand mix of 1:6:1 should be used with bentonite added to limit the spread of the grout during injection. Placed in a progressive series of grout layers these mixes give a good cohesivity and limit segregation with a compressive strength of about 1 N/mm^2 at 28 days.

Grout injection should always commence at the deepest side of the seam dip with the end returns being grouted up the dip in advance of any infill grouting. The higher section of the curtain wall should be left ungrouted until the central infilling has been completed. In flooded seams this operation would displace the water ahead of the grout and force it out of the last holes.

After grouting the efficiency of the operations should be checked by drilling holes and carrying out grout or water injection. Water injection tests generally prove to be inconclusive as complete filling of all voids is impossible.

5.5.2 Special conditions

Steeply sloping sites

When the dip of the seam is too steep for normal grout cones to form in a perimeter wall, then a barrier can be formed using 'porcupines'. These are spring steel splines which open out in the workings. Drill holes need to be at least 100 mm in diameter. Pea gravel and grout are injected down a secondary line of holes to form the perimeter curtain wall.

Open workings

Where workings are accessible and not flooded, and provided they are safe to enter, they can be surveyed below ground. The workings can be sealed off at predetermined locations using grout-filled sand bags followed by infill grouting through the sandbagged walls.

Multiple seams

Where several coal seams have been worked at shallow depths and the outcrops overlap, the collapse of a lower seam can affect the seams above. The lower seam must therefore be grouted up after the upper seam has been stabilized.

5.6 FOUNDATIONS IN AREAS WITH SHALLOW WORKINGS

In marginal cases where the coal seam may have already collapsed, or where the existing rock cover is supplemented by thick stiff clays or shale deposits, the use of a stiff edge beam raft often provides a suitable foundation. The edge beams are designed to span 3 m and cantilever 1.5 m at the corners. Brick reinforcement in the bed joints at critical levels in the superstructure is also a prudent and cost-effective measure.

This form of construction is not recommended where void migration could occur at the surface resulting in serious subsidence and possible a risk to human safety. Grouting with PFA / cement is the only safe option with reinforced footings following on. Generally the reinforced footings should be 300 mm thick reinforced with a layer of B503 fabric mesh top and bottom.

Where workings are very shallow and are grouted up there is a greater risk of residual subsidence as crown holing may have occurred prior to consolidating the workings. In these situations an edge beam raft is the most suitable foundation. The use of such rafts is recommended where the rock cover is less than five times the seam thickness.

5.7 ACTIVE MINING

When a site is to be developed for housing in areas where British Coal or other mining firms are currently working coal or other minerals, or there are seams available which may be undermined at some future date, the following procedures should be carried out.

- Consult with British Coal or the relevant mining firm and determine what their plans are in respect of mineral extraction below and adjacent to the site in question. Where British Coal are the owners of the land being considered they generally impose conditions to the sale of the land. One of these conditions is that the developer must submit foundation proposals to British Coal for their comment prior to commencing work on site.
- Seek advice and guidance from geological plans, British Geological Survey, mineral valuers, National House Building Council engineers and local authorities.

5.8 FUTURE MINING

British Coal hold vast reserves of deep mined coal and coal suitable for extraction by opencast methods. A lot of the deep-mined coal held in reserve could never be mined economically. British Coal generally work on a five-year plan of forward projection and if there are coal seams below a particular site the British Coal mining report will normally take either one of two forms.

- If a particular area is within their forward plan, the mining report will indicate which seams they intend to work, and the approximate commencement date, and will generally recommend that developers consult British Coal regarding precautions that may be required to mitigate subsidence effects to houses and the services.
- If the site is not in the five-year plan, the report will normally point out that there are seams of coal in reserve which may be mined in the future but there are no plans to extract coal at present. This does not preclude the possibility of the site being included in a future plan.

The background to the question of liability for subsidence damage is vague. British Coal have declared that, in general, they will never comment on a developer's proposals, because this may result in a liability to pay compensation for extra over normal development costs. They appear to prefer to make no comment and pay out statutory compensation for any damage they cause to those properties affected. Presumably this is a commercial decision which would indicate that it is cheaper to repair some houses on a site following extraction than to pay compensation to developers for increased foundation costs for all the houses.

It has been recognized for a long time by engineers, surveyors and insurance companies that this policy is not always in the best interests of the property owner. Responsible developers in active mining localities generally take precautions against subsidence damage and such costs obviously are paid for by the eventual property owner. Only on large prestigious buildings or sensitive buildings such as nuclear power stations or dams would British Coal consider making comments as it is most likely that the cost of subsidence damage would far exceed the cost of taking adequate precautions.

When there are no definite proposals to remove coal from below a site then in strict legal terms a developer is not obliged to take any precautions. Failure to take token measures may result in adverse sales, because property buyers in mining areas are aware of the potential damage. The provision of a strip foundation with some light mesh is a prudent and cost-effective measure which will also increase the foundation strength on marginal or varying ground conditions.

Where there are definite proposals to mine under a site and British Coal have issued or intend to issue a 'withdrawal of support' notice, the developer should seek engineering advice and provide suitable foundations. It is considered by most engineers to be good practice not to build while coal is being removed from or adjacent to a site. There are sound valid reasons for this policy; partly constructed buildings are less able to tolerate the tension and compressive forces induced at ground level by subsidence. A building shell without floors or roof is more likely to suffer damage from mining strains than one which is structurally complete. In addition, any damage caused may not be so evident: walls out of plumb, for example, or shear failure on the bed joints in walls. British Coal policy is that buildings should be erected during a 'construction window'; failure to follow this advice may well result in counter-claims by British Coal surveyors when assessing claims for damage at some later date.

5.9 MITIGATING THE EFFECTS OF MINING SUBSIDENCE

It is usual to take precautions against the effects of mining subsidence when coal extraction is likely to occur under or close to a site. The British Coal subsidence engineer should always be contacted as he will be in possession of all the relevant information, such as widths of panel extraction and direction of workings.

When the coalfield is currently being exploited, certain conditions should be ascertained prior to development. The critical conditions for low-rise housing are as follows.

1. What are the working depths of the seams being extracted? These should not be within 225 m of the surface.
2. What is the total subsidence likely to be at one pass of the working face? This is generally about 80% of the seam thickness extracted and should not exceed 825 mm.
3. What is the rate of advance of the coal face? This should be about 1.50 m per day.
4. What is the angle of incidence (Fig. 5.6)? This should be a minimum of 20°.

Once this information is available it is possible to estimate the likely ground strains using the *Subsidence Engineers Handbook* (British Coal, 1975). The layout of the proposed development can be adjusted to take account of site conditions such as geological faults and special foundations.

5.9.1 Longwall mining (advancing system)

Where subsidence strains are to be catered for it is common policy to design foundations that ride out the subsidence wave and also to introduce some flexibility into the foundations and superstructure. The usual foundation for low-rise housing is a flat-bottomed raft placed on a 1200 gauge polythene dpm laid over a 150 mm thick bed of compacted coarse sand. This reduces the friction between the raft base and the ground below.

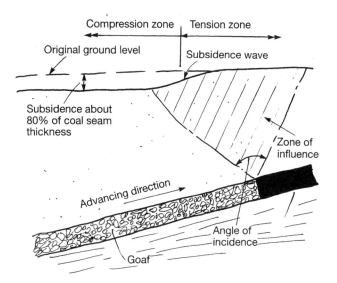

Fig. 5.6 Longwall mining.

When the subsidence wave occurs the resultant ground tensions and compressions are transmitted to the underside of the foundation by friction. Where the natural ground is a good bearing stratum the flat raft on a polythene and sand cushion is effective. The coefficient of friction between the sub-base and the underside of the raft is usually taken as 0.66 in the tension calculations. However, where poor ground conditions require downstand edge beam rafts to be used, then the tension forces in the ground will be locked in, producing a greater tension in the raft slab. The coefficient of friction should therefore be increased to 1.0 to take account of this. Figure 5.7 shows a typical mining raft using a precast or timber ground floor construction.

5.9.2 Designing buildings for future mining subsidence

The most suitable method of controlling and reducing damage to houses arising from active mining subsidence is to introduce **flexibility** into the structural shell and foundations. This can be achieved by adopting some or all of the following recommendations.

1. Use a flat reinforced flexible raft foundation laid on a slip membrane and sand cushion, and do not build over known geological faults.
2. Ensure that any holes for service entry/exit pipes are suitably sized to give adequate clearances.
3. Design housing layouts so that any long blocks are positioned with their width in the direction of the subsidence wave. If this is not possible, ensure that the block length is suitably divided by movement joints at least 50 mm wide. Spacing of these joints can be every second unit or 12 m centres whichever is the less (Fig. 5.8).
4. Where possible, avoid steps in or between dwellings. If they cannot be avoided then provide a movement joint at that position.
5. Provide bands of brick reinforcement in brickwork bed joints at selected positions in the superstructure.
6. Only build foundations and structural shells during a 'construction window'.

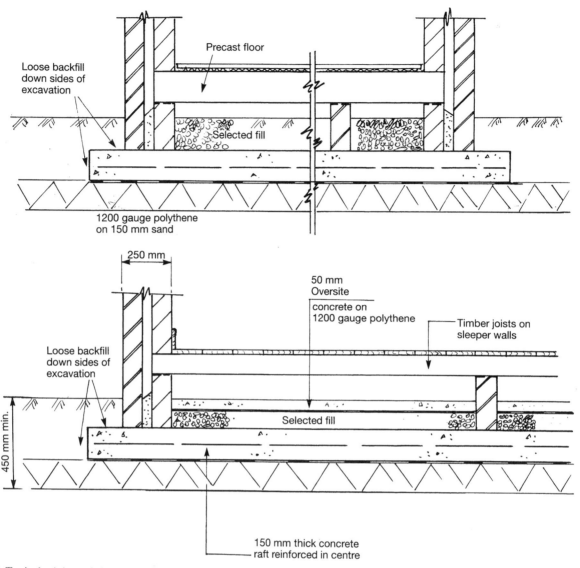

Fig. 5.7 Typical mining raft for small dwellings.

7. Avoid basement constructions. If this is not possible, provide a flat-bottomed raft slab over the whole basement area and ensure that a compressible material such as low-density polystyrene (Claymaster, for example) is placed down the sides of the external walls. This material should be at least 75 mm thick and the backfilled working space should consist of loose stone or sand backfill. Where an external damp-proofing membrane is used the polystyrene will form a protective backing.

8. Avoid piled foundations if longwall mining is likely to take place. If the ground is so poor that piles are the only technical solution, then the development will have to take place after the mining subsidence wave has passed under the site. Failure to follow this advice could result in piles shearing and tie beam junctions failing in tension. If the project cannot be delayed then the piling system will need to be designed to have the piles' end bearing established through the site overburden on to a suitable stratum using a prebored oversized hole which can be filled with compressible materials such as foam, polyurethane, plastic or viscous bituminous mixtures

after a steel tube, concrete shell or concrete precast pile has been formed.

9. Take one third of the imposed loads in the foundation design and make no allowance for wind or snow loads.

10. Take the coefficient of friction under the raft slab as 0.66.

11. Allowable reinforcement stress to be 310 N/mm^2 and the permissible compressive stress in concrete to be 14 N/mm^2.

12. Design the tension in the raft steel based on half of the building's mass.

5.9.3 CLASP system of construction

In general, structures built in active mining areas should be either completely flexible or completely rigid. The completely rigid building will have a foundation strong enough to cater for the horizontal strains and curvature effects and will be designed to allow tilting to take place without damage to the superstructure. Some allowance is usually made for jacking points to enable the building to be jacked back to level. This is a very expensive approach to adopt.

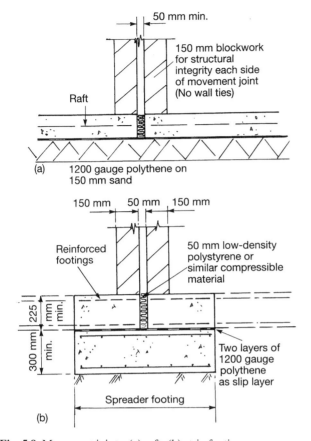

Fig. 5.8 Movement joints: (a) raft; (b) strip footings.

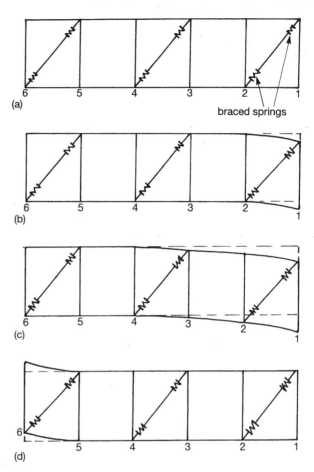

Fig. 5.9 CLASP system of braced springs.
(a) Building on level ground prior to extraction of coal.
(b) Following coal extraction the ground below column 1 subsides. The load on column 1 overcomes the resistance of the wind brace springs allowing panel 1–2 to take a lozenge shape. All the stanchions remain vertical because the majority of the spring-braced units are still on supported ground.
(c) Bracing in panel 3–4 tilts as a result of springs in 5–6 balancing springs in 1–2. Hence 5–6 is lozenged and lozenging of 1–2 is relatively reduced.
(d) The structure has almost completely subsided and is settling back level at a lower horizontal position. Panel 5–6 is still affected by the subsidence wave and is lozenged. Panels 1–2 and 3–4 are overcoming the spring resistance.

A flexible structure, designed to follow the ground movements and with a foundation which will accommodate the surface strains, is a more practical and economic solution. One example of a lightweight flexible structural system which was able to tolerate a considerable magnitude of mining subsidence was the CLASP system (Consortium of Local Authorities Special Programme). It was used extensively in active mining areas for schools, hospitals and similar low-rise buildings. Usually the buildings were two or three storeys.

The system consisted of a superstructure of lightweight steel frame with pinned joints at all connections to allow freedom of movement. Resistance to wind forces was provided by the use of diagonal lattice bracings with coiled spring inserts to allow the frame to take up a lozenge shape as the subsidence wave passed below the structure. The coiled spring bracing provided for the frame to return to vertical once the mining wave had passed. Internal and external cladding consisted of timber or precast panels fastened at the top corners only and free to slide against the adjacent component parts. Special window frames were used to allow for movement of the glazing without cracking. Services into the buildings had flexible joints, designed for abnormal movements.

Foundations consisted of a simple flat reinforced concrete raft laid on a polythene dpm over a 100 mm layer of sand. The raft was split into small panels using movement joints.

Irregular-shaped buildings were designed so that additional reinforcement was provided in the rafts at the changes in plan.

Though the design and construction of this system was economic it went into decline as a result of cutbacks in the mining industry, problems with carbonation of the precast panels, and also problems encountered when these types of buildings experienced a fire. Figure 5.9 indicates how the braced spring system worked. These principles are still applicable for structures such as light factory frames.

5.9.4 Mining rafts

Raft foundations need to be reinforced to cater for tensile stresses. The raft foundation should be kept as high as possible; if levels permit, a more suitable solution is to place the raft above the existing ground level but ensure that any clays have adequate frost cover (Fig. 5.7).

Where possible, the dwellings should be of rectangular plan form and projections are best avoided. If projections are unavoidable the raft will have to be designed for the condition in which the extension 'flops' down ahead of the main building. This design method is covered in more detail in Example 5.2.

When movement joints are provided in a building, a double gable wall arrangement should be adopted at party walls to ensure that each dwelling forms its own box. This is illustrated in Fig. 5.8.

Example 5.1   Calculating the reinforcement in a mining raft

Calculate the subsidence load at a particular cross-section of a raft slab being considered for a two storey building 20 m by 12 m wide (Fig. 5.10).

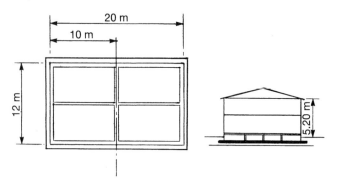

Fig. 5.10 Example 5.1: plan and section of two-storey dwelling.

It is usual to consider sections at one half and one quarter points only in each direction. The subsidence load is taken as the weight of the building plus one third of the imposed loadings on floors and roof. The frictional coefficient of the sand below the raft is taken as 0.67.

SUBSIDENCE LOAD

Roof

	(kN/m²)
Imposed load = 0.33 × 0.75 kN/m²	0.25
0.33 × 0.25	0.08
Dead load =	1.25
Total	1.58

First floor

Imposed load = 0.33 × 1.50	0.50
Dead load =	0.50
Total	1.0

Ground floor

Imposed load = 0.33 × 1.50	0.50
Dead load = 0.15 × 24	3.60
Total	4.10

External walls

3.75 kN/m² × 5.20 = 19.5 kN/m

Party walls

3.0 kN/m² × 6.50 = 19.5 kN/m

Spine wall

1.50 kN/m² × 2.50 = 3.75 kN/m

Total wall loading

	(kN)
20 × 2 × 19.50 =	780
12 × 2 × 19.50 =	468
12 × 1 × 19.50 =	234
20 × 1 × 3.75 =	75
Total =	1557

Unit load / m² = 1557/20 × 12 =	6.48 kN/m²
Plus roof	1.58 kN
Plus first floor	1.0
Plus ground floor	4.10
Total =	13.16

Total subsidence load = 13.16 × 20 × 12 = 3158 kN

Permissible steel stress taken as 310 N/mm²

$$\text{Tension in slab at centre} = \frac{0.50 \times 3158 \times 0.67}{12} = 88 \text{ kN/m}$$

$$\text{Tension steel required} = \frac{88 \times 10^3}{310} = 283 \text{ mm}^2/\text{m}$$

Use fabric reinforcement A393 in centre of raft or a layer of A193 top and bottom of raft. Concrete grade 25N with a minimum cement content of 290 kg/m³ using 20 mm aggregate (Fig. 5.11).

5.9.5   Irregular-shaped units

When irregular plan shapes occur on housing sites, additional reinforcement must be provided to cater for the eccentric forces set up. These forces result from the extended sections trying to 'flop' down ahead of the main body of the raft as the subsidence wave passes below the site.

When the subsidence wave occurs it first induces tension in the ground. This can occur under the building from any direction. When projecting areas are built on, the tension wave attempts to detach the projecting area from the main dwelling section, and additional reinforcement is required in the longitudinal and transverse directions.

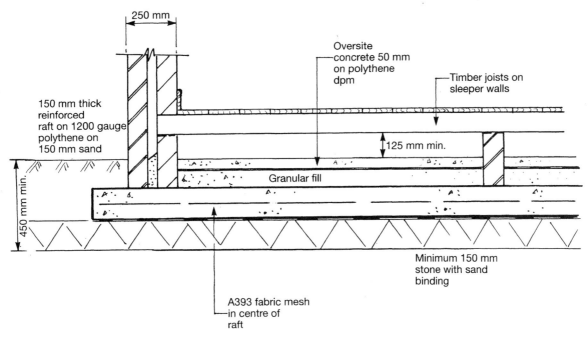

Fig. 5.11 Example 5.1: mining raft details.

Subsidence in longitudinal direction

In this situation (Fig. 5.12) the subsidence wave is approaching from right to left, and acts first on the small projecting area shown cross-hatched. This in effect causes a pull from left to right which must be resisted by additional reinforcement at points A and B.

The area of steel required to resist the eccentric moment is given by

$$A_{el} = \frac{\mu W_s \bar{x}}{310(b-0.9)} \times 10^3 \ \text{mm}^2$$

where μ = coefficient of friction on ground = 0.67;
W_s = subsidence load for shaded area (kN);
$\bar{x} = b/2 - c/2$ (m);
b = width (m).

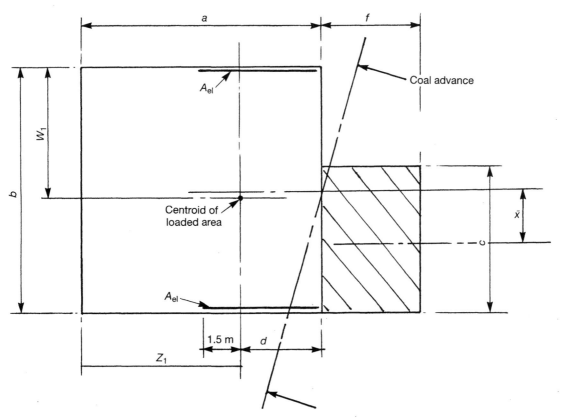

Fig. 5.12 Example 5.2: irregular-shaped units, subsidence in longitudinal direction. $x = b/2 - c/2$; d = distance over which moment acts.

Building in mining localities

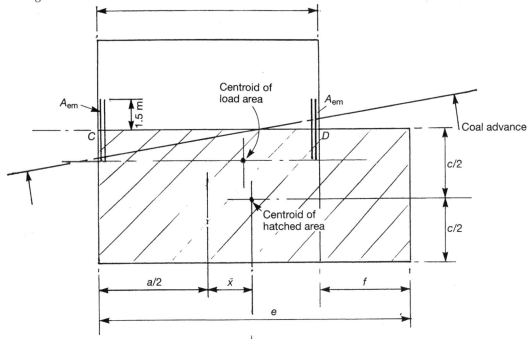

Fig. 5.13 Example 5.2: irregular shaped units, subsidence in transverse direction. $x = e/2 - a/2$.

Subsidence in transverse direction

In Fig. 5.13 the subsidence wave creates a downward pull on the hatched area which must be resisted by additional tension steel at points C and D. These forces can be caused from either direction.

Area of additional steel is given by

$$A_{em} = \frac{\mu W_s \bar{x} . 10^3}{310(a - 0.9)} \times 10^3 \; mm^2$$

where $\bar{x} = e/2 - a/2$; a = length (m).

Effect at the projection

The subsidence wave attempts to detach the projection in a downward pull which requires additional reinforcement at points E and F (Fig. 5.14).

The area of additional steel is given by

$$A_{ep} = \frac{\mu W_s \times 0.5 f \times 10^3}{310(c - 0.9)}$$

The areas of additional steel, A_{el}, A_{em} and A_{ep}, should be made up using 12 mm high tensile bars or 20 mm high tensile bars:

$$\text{Number of bars} = \frac{A_{el}}{113} \; or \; \frac{A_{el}}{314}$$

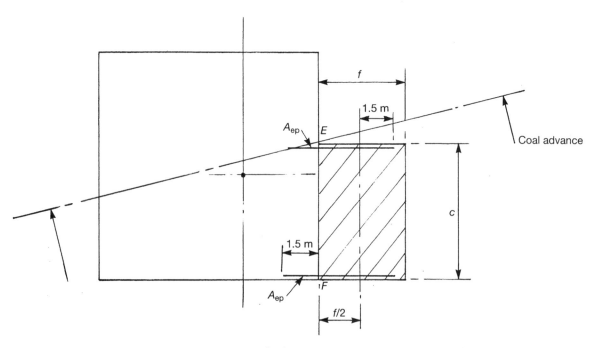

Fig. 5.14 Example 5.2: irregular-shaped units, effect at the projection.

106

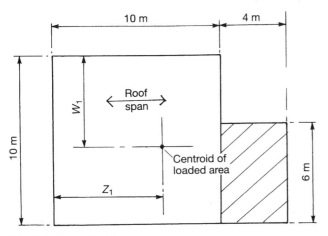

Fig. 5.15 Example 5.2.

Example 5.2   Designing the reinforcement in an irregular-shaped dwelling

A bungalow is being built in an active mining area and it has a projection of 4 m by 6 m at one corner. Calculate the additional reinforcement required to supplement the normal reinforcement in the main body of the raft (Fig. 5.15).

Calculate Z_1 for dead load plus one third imposed load.

Roof $1.25 + \dfrac{1}{3} \times 1.0 = 1.58$ kN/m^2

Raft $3.60 + \dfrac{1}{3} \times 1.50 = 4.10$ kN/m^2

Walls $= 4.5$ kN/m^2

Element		Area		Unit load		Load area		L_a		Moment
10×10	=	100	×	5.68	=	568	×	5	=	2840
4×6	=	24	×	5.68	=	136.32	×	12	=	1636
10×2.5	=	25	×	4.50	=	112.50	×	0	=	—
10×2.5	=	25	×	4.50	=	112.50	×	10	=	1125
6×2.5	=	15	×	4.50	=	67.50	×	14	=	945
$10 \times 2.5 \times 2$	=	50	×	4.50	=	225.00	×	5	=	1125
$4 \times 2.5 \times 2$	=	20	×	4.50	=	90	×	12	=	1080
				Total	=	1311.82				8751

Therefore

$$Z_1 = \frac{8751}{1311.82} = 6.67 \text{ m}$$

Calculate W_1 (taking moments about top wall).

Element		Area		Unit load		Load area		L_a		Moment
10×10	=	100	×	5.68	=	568	×	5	=	2840
4×6	=	24	×	5.68	=	136.32	×	7	=	954.24
10×2.5	=	25	×	4.50	=	112.50	×	0	=	0
14×2.5	=	35	×	4.50	=	157.50	×	10	=	1575
4×2.5	=	10	×	4.50	=	45	×	4	=	180
$10 \times 2 \times 2.5$	=	50	×	4.50	=	225	×	5	=	1125
$6 \times 1 \times 2.5$	=	15	×	4.50	=	67.50	×	7	=	472.50
				Total	=	1311.82			=	7146.74

Therefore

$$W_1 = \frac{7146.74}{1311.82} = 5.44 \text{ m}$$

Calculate A_{el}.

			(kN)
Raft 6×4	= 24×5.68	=	136.32
Walls $4 + 4 + 6$	= $14 \times 2.5 \times 4.50$	=	157.50
		W_s =	293.82

Therefore

$$A_{el} = \frac{0.66 \times 293.82 \times 10^3 \times (5-3)}{310 \times (10-0.90)} = 137 \text{ mm}^2$$

Therefore using T12 bars 137/113 = Two T12.

Calculate A_{em}.

$$\bar{x} = \frac{e}{2} - \frac{a}{2}$$

Therefore

$$A_{em} = \frac{0.66 W_s \bar{x}}{310(a-0.90)} \times 10^3$$

					(kN)
W_s raft =	6×14	=	84×5.68	=	477.12
Walls	=	$3 \times 6 \times 2.5$	= 45×4.50	=	202.50
Walls	=	$2 \times 4 \times 2.5$	= 20×4.50	=	90.00
Walls	=	10×4	= 40×4.50	=	180.00
				W_s =	949.62

Therefore

$$A_{em} = \frac{0.66 \times 949.62 \times (7-5) \times 10^3}{310 \times (10-0.90)} = 445 \text{ mm}^2$$

Using T20 bars 445/314 = Two T20

Calculate A_{ep}.

W_s as for A_{el} = 293.82 kN. Therefore

$$A_{ep} = \frac{0.66 \times 293.82 \times 10^3 \times 0.5 \times 4}{310 \times (6-0.90)} = 245 \text{ mm}^2$$

Use three T12 mm bars as shown in Fig. 5.16.

5.9.6   Designing strip footings in active mining areas

Where a good bearing stratum exists on a site it is sometimes possible to design strip foundations to mitigate the effects of mining subsidence on small buildings. Various assumptions require to be made relative to the ground conditions, as follows.

- Normal single- or two-storey dwellings of traditional brick and block construction are to be built.
- The removal of support in total or in part from below the strip foundations must be assumed to equate to about half the span of the house. The strip foundations are to be

107

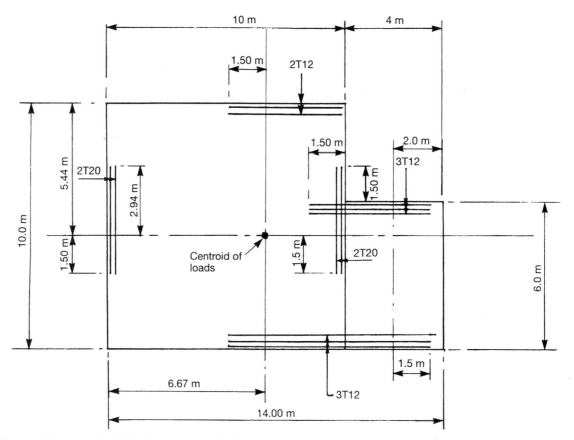

Fig. 5.16 Extra reinforcement details for irregular-shaped units.

considered as a series of interlocked fixed beams using a bending moment coefficient of $WL^2/12$.

In order to reduce the amount of ground strain transmitted into the foundations it is prudent to cast the foundations level all round on a 75 mm thick compacted layer of sand or gravel blinded off with fine stone dust or, alternatively, a 75 mm thick layer of blinding concrete with the top surface trowelled off smooth to receive a 1200 gauge polythene slip membrane. The strip foundations are cast on top of the polythene and they should contain sufficient reinforcement to cater for the tensions in the foundation.

If the site is sloping then the top of the mass concrete make-up should be kept level, with the slip membrane and footing being added on top as indicated in Fig. 5.17.

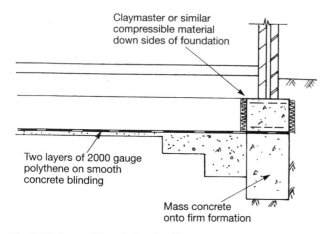

Fig. 5.17 Stepped foundation details.

Calculations for reinforced strip footings:

Working stress in reinforcement in tension = 310 N/mm^2

Working stress in concrete in compression = 4.2 N/mm^2

Elastic modulus ratio = $E_s/E_c = 15$

Depth of beam to neutral axis = h

Depth of beam to centre of reinforcement = d_1

Example 5.3 Reinforcement mesh

Loadings on external walls = 35 kN/metre run

Loadings on internal walls = 18 kN/metre run

Assume fully fixed situation

Maximum bending moment $= \dfrac{35 \times 3.40^2}{12} = 33.71$ kNm

using a 600 mm wide beam with top and bottom reinforcement mesh. Therefore

$$\frac{h}{d-h} = \frac{f_c E_s}{f_s E_c}$$

$$\frac{h}{d-h} = \frac{4.20 \times 15}{310} = 0.203$$

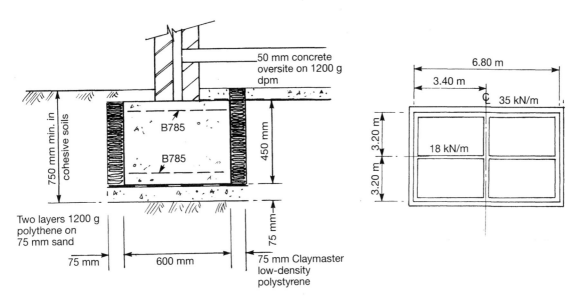

Fig. 5.18 Example 5.3: thick reinforced strip footings.

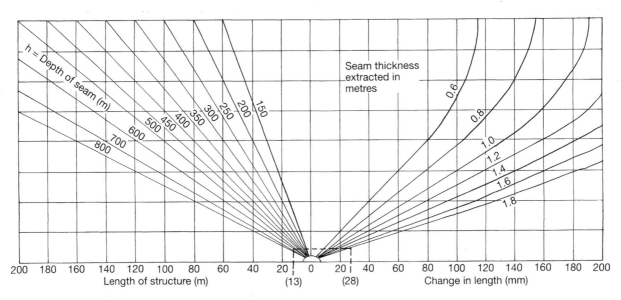

Fig. 5.19 Nomogram for the determination of change in length of superstructure for various seam depths and thicknesses of coal extraction.

Therefore

$h = 0.168\,d$

With $d = 400$ mm and beam depth = 450 mm:

$h = 0.168 \times 400 = 67$ mm

Moment $M = f_s A_s \left(d - \dfrac{h}{3} \right)$

$A_s = \dfrac{33.71 \times 10^6}{(400 - 67/3)310} = 288$ mm^2

Using B785 mesh then 500 mm wide sheet provides $0.50 \times 785 = 393$ mm^2. Therefore provide B785 top and bottom in all foundations.

5.9.7 Movement joints

When long terrace blocks have to be sectioned into shorter block lengths it is important that the size of movement joint

is adequate to allow for the change in length of the structure and the effect of ground curvature at the surface. These joint widths can be obtained by using the nomograms in Figs 5.19 and 5.20.

Example 5.4 Calculating movement joint sizes

A terrace block on a housing development is to be split up with movement joints at 13 m centres. The height of the housing is 6.0 m average. Mining is proposed in the future of a 1.80 m thick coal seam at a depth below ground level of 600 m. Determine the minimum joint thickness to accommodate the mining strains.

From Fig. 5.19:

Length of structure = 13 m

109

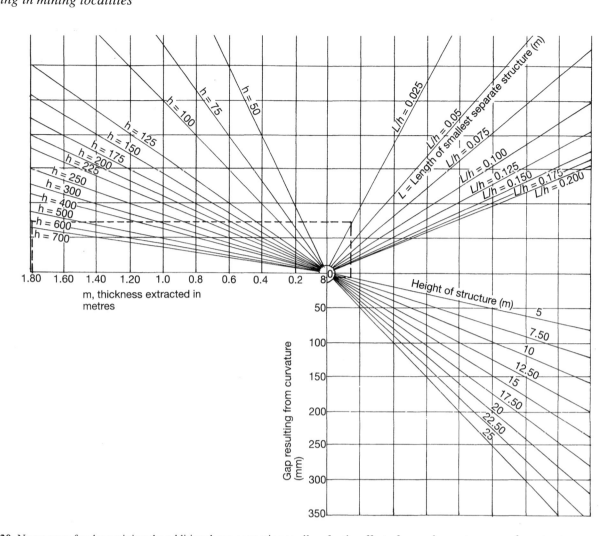

Fig. 5.20 Nomogram for determining the additional gap correction to allow for the effect of ground curvature on surface structures.

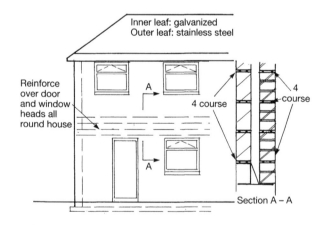

Fig. 5.21 Location of brick reinforcement.

This length is selected on the baseline and is projected up to intercept the appropriate depth of seam line. At this intersection

transpose a line horizontally across to intersect the appropriate seam thickness line. Transposing a vertical line down on to the base line will give the change of length of the structure.

From Fig. 5.20:

$$\frac{\text{Length of building}}{\text{Depth of seam}} = \frac{13}{600} = 0.022$$

Select the seam thickness of 1.80 m on the baseline and transpose vertically to intercept the depth of seam line and transpose from this intersection a line horizontally to intersect the appropriate L/h line.

At this intersection transpose down vertically to intersect the appropriate height of structure line. This intersection is then taken horizontally to the curvature base line.

The two gaps obtained from the nomograms are added together and rounded up to the nearest 5 mm. In this example the change in length comes to 28 mm and the change in length due to surface curvature is 8 mm, giving a total of 36 mm. In this case a 40 mm or 50 mm joint would be adequate.

SPECIFICATION FOR BRICK REINFORCEMENT

1. All brick reinforcement shall be of proprietary manufacture and shall comply with all the relevant British Standards.
2. All reinforcement shall be high-yield steel with a minimum yield stress of 460 N/mm².
3. Reinforcement shall be fabricated of two bars of 3.5 mm minimum diameter spaced a minimum distance of 65 mm apart linked together by steel laterals at 450 mm maximum centres.
4. Reinforcement to be stainless steel when used in the external leaf of a brick cavity wall. For internal walls and the inner leaf of an external cavity wall the reinforcement should be galvanized steel to BS 729.
5. Only one layer to be used in a mortar bed joint, except at laps, where bars shall be placed adjacent.
6. Minimum side and end cover to be as laid down in the manufacturer's requirements, but never less than 15 mm.
7. Minimum lap length to be 300 mm.
8. All reinforcement shall be free of mud, grease, or ice, and should be stored in a proper manner to ensure that the galvanized steels do not get scratched or damaged.

GLOSSARY OF MAIN MINING TERMS

Adit	Access into coal workings, driven at a shallow angle. Often worked into the side of a hillside and sometimes referred to as a **drift mine**.
Bulking	Process of increase in volume of mineral either when mined or of strata which have collapsed into the old workings.
Cap	A thick reinforced concrete slab placed over a shaft at rockhead and designed to support the overburden up to ground level.
Crown hole	1. Surface depression resulting from the collapse of strata into old mine workings.
	2. Surface depression formed as a result of loss of fine soils into rock fissures opened up by deep mining.
Discontinuity	Fracture or break in soil stratification in the form of geological faults or joints.
Dip	Angle at which a coal seam is inclined to the horizontal.
Fault	A fracture zone along which there has been vertical displacement of strata.
Goaf	The caved area of mine workings where the roof has collapsed following coal extraction.
Longwall mining	A modern method of coalmining involving total extraction along a working face generally about 200 m long.
Outcrop	The face of a rock stratum where it emerges at ground level or at the junction of the drift materials and rockhead.
Pillar	A block of mineral left unworked to support the overburden above the coal seam.
Stowing	Packing of waste material into worked-out mine galleries.

111

BIBLIOGRAPHY

Bell, F.G. (1975) *Site Investigations in Areas of Mining Subsidence*, Newnes-Butterworth, London.

British Coal (1975) *Subsidence Engineer's Handbook*.

British Coal (1982) *The Treatment of Disused Mine Shafts*.

CLASP (1957) Consortium of Local Authorities Special Programme.

Gregory, O. (1981) Defining the problems of disused coal mine shafts. *Proc. Conference on Disused Mine Shafts*, Royal Institution of Chartered Surveyors, London, Vol. 4.2, pp. 4–15.

Healy, P.R. and Head, J.M. (1984) *Construction over abandoned mine workings*, CIRIA Special Publication No. 32, Construction Industry Research and Information Association, London.

Institution of Civil Engineers (1977) *Ground Subsidence*, Thomas Telford, London.

National Building Studies (1951) *Mining subsidence effects on small houses*. Special Report No. 12, HMSO, London.

Tomlinson, M.J. (1980) *Foundation Design and Construction*, 4th edn, Pitman.

Chapter 6
Sites with trees

6.1 FOUNDATION DESIGN

Since the UK droughts in 1947 and 1976, research work by the Building Research Establishment, National House Building Council and other organizations has resulted in a better understanding of the effects of trees on soils that are classed as shrinkable clays. On housing developments, existing trees and new planting improve the amenity of the urban landscape visually, and provide wind and sound breaks between areas.

However, the sheer size of certain tree species, which are often taller than the adjacent housing, brings them into conflict for the available space. If trees are to be retained close to houses, then adequate provision must be made in the foundation design to enable the trees and housing to co-exist.

In 1980 the British Standard Code of Practice BS 5837 *Trees in Relation to Construction* officially recognized that most buildings are likely to come into close proximity with trees during their useful life. Where the presence of shrinkable clay has been established it is advisable to take precautionary measures in the design of a building's foundation.

The following guidelines and procedures demonstrate the design philosophy required if subsidence damage is to be prevented.

1. First determine the broad nature of the soil strata, e.g. is it cohesive or granular?
2. Determine the species and potential mature heights of the existing trees and their relative positions on a detailed site survey. Tables 6.1(a) and 6.1(b) give a range of tree species with their mature heights and water demand characteristics. Tree heights in parentheses are those used in the NHBC Standards Chapter 4.2 2003 and stated as the 'average mature heights to which healthy trees of the species may be expected to grow in favourable ground and environmental conditions'.

   Bear in mind that the NHBC guidance, while it is the best available, is based on insured risk and it is not uncommon to find trees that are taller than the NHBC Chapter 4.2 heights and which are likely to be very healthy specimen trees. Obviously in these situations, it would be prudent to give due respect to such existing specimen trees. The author has encountered many occasions where foundations have had to be placed deeper than the NHBC tables, based on the *actual* heights of the mature trees. Figures 6.1, 6.2 and 6.3 illustrate some common trees, which have low, moderate and high water demand characteristics.
3. Classify the soil. If clay soils are found, determine by laboratory testing the Atterberg limits, i.e. the liquid limit, plastic limit, plasticity index and the natural moisture content.

   NHBC Chapter 4.2 was modified in April 2003 to take account of those clay soils that contain medium and fine sand particles smaller than 425 microns. For these clays the Modified Plasticity Index can be used.
   Modified Plasticity Index

   $$I'_p = I_p \times \frac{\% \text{ of sample less than 425 microns}}{100\%}$$

   Generally economies in the design of foundations will probably only occur in glacial tills and mixed soils. Where the plasticity index is not known, then the high category of 40% and greater must be used. These soil categories are listed in Table 6.4
4. Examine the soil strata *in-situ* to determine whether the soils are desiccated. This is easily recognized: the clay cannot be moulded by hand and comes out of the excavation bucket in small cubical pieces. In these situations precautions against ground heave will most probably need to be taken.
5. Establish the depths of the clay stratum and record the presence and depth of any granular strata or rock strata in the trial pits.
6. Establish the depth of any standing water table.
7. Where there is a group of trees, one or more trees in the group may predominate. The tree that has the greatest

Table 6.1(a).   Tree species – Broad leafed
Mature heights are specimen heights. Figures in brackets are heights from N H B C Chapter 4.2 Appendix B

Botanical name	Common name	Water demand	Mature height (m)
Crateagus monogyma	Common hawthorn	High	14 (10)
Crateagus oxyacantha	Hawthorn		14 (10)
Eucalyptus	Gum		– (18)
Populus alba	White poplar		26 (15)
Populus canescens	Grey poplar		35 (25)
Populus nigra	Black poplar		30 (28)
Populus nigra Italica	Lombardy poplar		30 (25)
Populus tremula	Aspen poplar		20 (12)
Populus alba	White poplar		26 (22)
Quercus borealis	Red oak		35 (24)
Quercus cerris	Turkey oak		39 (24)
Quercus ilex	Holm oak		28 (16)
Quercus robur	Common oak		28 (20)
Salix alba	White willow		(24)
Salix caprea	Goat or pussy willow		(5)
Salix fragilis	Crack willow		(24)
Salix vitellina	Weeping willow		22 (16)
Ulmus glabra	Wych elm		29 (18)
Ulmus procera	English elm		36 (24)
Ulmus procera wheatley	Wheatley elm		30 (22)
Acer campestre	Field maple	Moderate	24 (13)
Acer negundo	Box elder maple		15 (12)
Acer platanoides	Norway maple		27 (18)
Acer pseudoplatonus	Sycamore		34 (22)
Acer saccharinum	Silver maple		31 (22)
Aesculus carnea	Horse chestnut		38 (20)
Ailanthus altissima	Tree of Heaven		30 (20)
Alnus cordata	Italian alder		27 (18)
Alnus glutinosa	Common alder		26 (18)
Alnus incana	Grey alder		25 (18)
Castanea sativa	Sweet chestnut		36 (20)
Fagus spp.	Beech		(20)
Fagus sylvatica	Common beech		41 (20)
Fraxinus excelsior	Common ash		45 (23)
Fraxinus ornus	Manna ash		24 (18)
Juglans regia	Common walnut		23 (18)
Laurus nobilis	Bay laurel		15 (10)
Malus aldenhamensis	Flowering crab		9 (7)
Malus tschonskii	Pillar apple		15 (9)
Platanus acerifolia	London plane		45 (26)
Prunus spp.	Plum		(10)
Prunus avium	Wild cherry or Gean		18 (17)
Prunus dulcis	Almond		(8)
Prunus laurocerasus	Cherry laurel		(8)
Prunus padus	Bird cherry		15 (10)
Prunus serrulata	Japanese cherry		12 (9)
Prunus spinosa	Blackthorn or sloe		(8)
Pyrus communis	Common pear		15 (12)
Robinia pseudoacacia	False acacia		(18)
Sorbus aria	Whitebeam		(12)
Sorbus aria — Lutescens	Whitebeam		20 (13)
Sorbus aucuparia	Rowan		20 (9)
Sorbus commixta	Japanese rowan		16 (10)
Sorbus intermedia	Swedish whitebeam		15 (9)
Tilia cordata	Small-leafed lime		30 (22)
Tilia euchlora	Crimean lime		18 (15)
Tilia europaea	Common lime		45 (22)
Tilia petiolaris	Pendant silver lime		27 (20)
Tilia platyphyllos	Large-leaved lime		31 (20)
Tilia tomentosa	Silver lime		27 (20)

Table 6.1(a). *Continued*

Botanical name	Common name	Water demand	Mature height (m)
Betula pendula	Silver birch	Low	– (14)
Betula verrucosa	Birch		– (14)
Carpinus betulus	Common hornbeam		30 (17)
Corylus avellana	Hazel		– (8)
Ficus carica	Fig		9 (8)
Gleditsia triacanthos	Honey locust		42 (14)
Ilex aquifolium	Common holly		22 (12)
Laburnum anagyroides	Laburnum		(12)
Liriodendron	Tulip tree		32 (20)
Magnolia spp.	Magnolia		(9)
Morus spp.	Mulberry		(9)
Sambucas nigra	Elder		10 (10)

Table 6.1(b). *Coniferous species*

Botanical name	Common name	Water demand	Mature height (m)
Cupressocyparis leylandii	Leyland cypress	High	30 (20)
Cupressus macrocarpa	Monterey cypress		25 (20)
Chamaecyparis lawsoniana	Lawsons cypress		30 (18)
Araucaria araucana	Monkey puzzle or Chile pine	Moderate	30 (18)
Cedrus spp.	Cedar		(20)
Larix deciduas	European larch		45 (20)
Larix kaempferi	Japanese larch		37 (20)
Picea sitchensis	Sitka spruce		40 (18)
Pinus sylvestris	Scots pine		35 (20)
Sequiadendron giganteum	Wellingtonia		50 (30)
Taxus baccata	English Yew		25 (12)
Pseudotsuga menziesii	Douglas fir		25 (20)

potential for damage should be selected when carrying out an assessment, i.e. the potentially largest or the one with the highest water demand.

8. Make allowance for the geographical location of the site (Fig. 6.5).
9. If necessary seek the advice of a chartered structural or civil engineer or qualified arboriculturist for any recommendations regarding the precautions necessary for the building foundations.

For sites where foundation depths are likely to exceed 2.50 metres, or where the site slopes are greater than 1 in 7, it is an NHBC requirement that a chartered structural or civil engineer must assess the foundations design.

With the above information it should be possible to obtain from NHBC Chapter 4.2 2003 sufficient information to decide on the foundation types required and the required foundation depths. For simple situations involving one or more trees, NHBC Chapter 4.2 Appendix D enables one to arrive at foundation depths at the point on the building closest to the trees and at various distances away from the trees.

There are other special factors that often need to be considered in terms of an overall foundation design for the following situations:

- Sites that are extensively wooded, which will have a substantial number of trees removed and on which some large species will be retained
- Buildings that are to be constructed over or within the zone of influence of existing trees, hedges and shrubs
- Sites where the strength of the clay soils reduces significantly with depth and where deeper foundations would be undesirable. This occurs in areas like Humberside where the over-consolidated clay crust lies over soft to very soft glacial deposits and peat lenses
- Sites that have rock at very shallow depths. While the building foundations should not pose a problem, it is important to ensure that the external drains and services are adequately protected from tree roots running on top of the rock.
- Sites where future tree removal will be required, or later phases of a development may require precautions to be adopted in the earlier phases
- Sites on clay fills with special foundations.

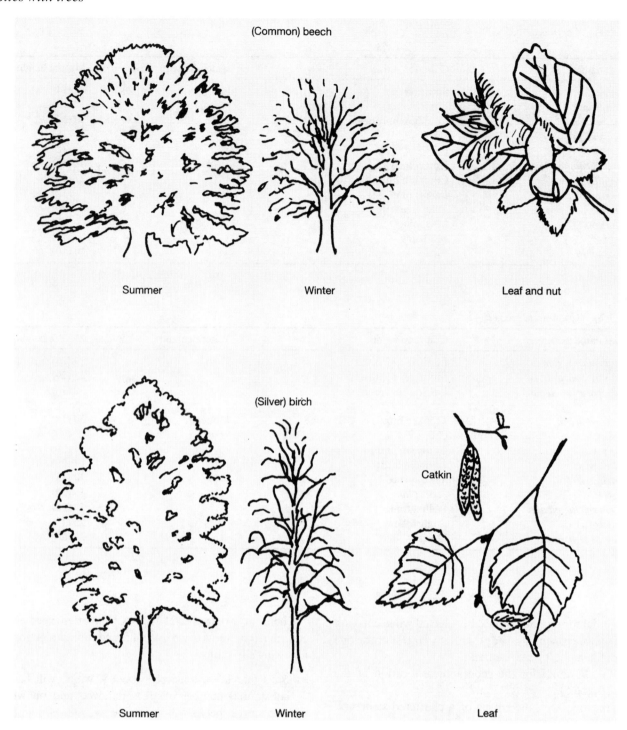

Fig. 6.1 Tree identification: low water demand trees.

Table 6.2. Zone of tree influence (NHBC Chapter 4.2 2003)

Water demand category	Zone of influence
High	1.25 × mature height
Moderate	0.75 × mature height
Low	0.5 × mature height

Example 6.1   Foundations on clay soils with young poplar trees

A dwelling is to be constructed on a site in Leeds, which has a row of small black poplar trees 6 metres high and 12 metres from the gable wall. Soil tests show that the clays have the following characteristics;

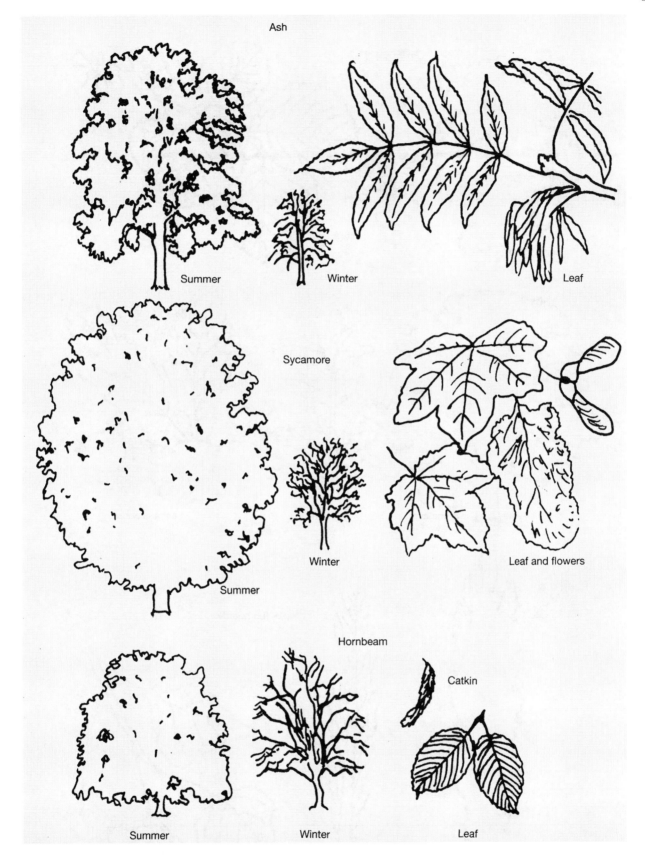

Fig. 6.2 Tree identification: moderate water demand trees.

- Liquid limit   53%
- Plastic limit   26%
- Plasticity index = Liquid limit − plastic limit
  = 53 − 26 = 27%
- Natural moisture content = 26%.

The clay soils are therefore in the medium range in terms of volumetric change potential, between 20% and 40% as indicated in Table 6.3. The black poplars are a high water demand tree and have an average mature height of 28 metres if applying the NHBC criteria. Therefore Table 17 in Chapter 4.2 Appendix D

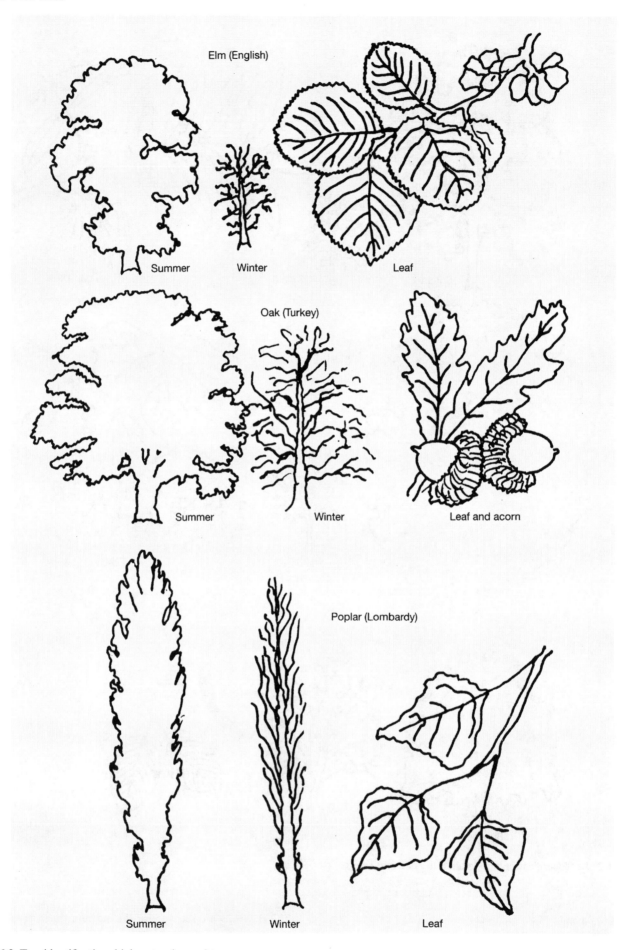

Fig. 6.3 Tree identification: high water demand trees.

Table 6.3. Example 6.1 Required foundation depths taken from NHBC Chapter 4.2 Appendix D Table 17

Distance from trees (m)	Table 17 Depths	Climatic variation from Fig. 6.5	Final depth (m)
12	2.30	From Fig. 6.5	2.15
13	2.25	this shows a	2.10
14	2.20	permitted	2.05
15	2.10	reduction of	1.95
16	2.05	150 mm to the	1.90
17	2.00	table depths	1.85
18	1.95	for the	1.80
19	1.90	Leeds area	1.75
20	1.80		1.65

applies for medium plasticity soils and high water demand species.

From the table a foundation depth of 2.30 metres is applicable with a reduction of 150 mm for climatic variation, giving a final foundation depth of 2.150 metres. By using Table 17 it is possible to determine the foundation depths for the remaining foundations, as shown in Table 6.3

The foundations should be stepped up in accordance with recommended good practice using 300 mm maximum steps with a minimum overlap of 600 mm. These depths are indicated in Fig. 6.4, which shows the site location and position of the trees relative to the dwelling.

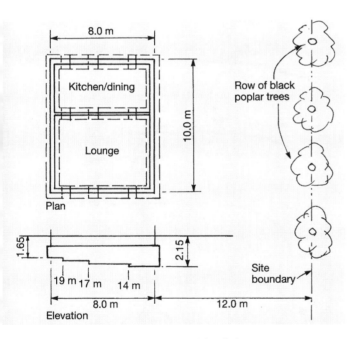

Fig. 6.4 Example 6.1: site location with existing trees.

As the black poplar trees are young trees they have not yet fully extended their root system and the clay soils under and close to the dwelling are unlikely to be desiccated. In this situation the use of concrete trench fill is acceptable. It would be prudent to provide a 1000 gauge polyethylene sheet membrane down the external face of the trench fill to act as a slip membrane in case a dry spell of weather causes the external clays to dry out and move down. This will help to mitigate any eccentric draw-down forces on the foundation sides.

Note

When using deep trench fill foundations, the strata below the base of the foundation should be checked for at least a further 1.0 metre to ensure that the bearing strata remains competent. In this example trial pits or hand auger borings taken to a depth of 3.25 metres should be adequate. Any deep trial pits should be located at least 3.0 metres from the walls of the building.

6.1.1 Climatic variation

High rainfall reduces the moisture deficits caused by trees, hedges and shrubs, with cool damp weather reducing the amount of water lost by the trees in transpiration. This combination therefore reduces the amount of clay shrinkage. As the driest and hottest weather conditions in the UK usually prevail in the south-east of England, the greatest risk to foundations occurs in that area. The south-east of England also contains the highly shrinkable clays.

To cater for the climatic change from the south-east to the north and west the NHBC introduced 50-mile zoning contours from London and these are shown in Fig. 6.5. For every 50 miles from London a reduction of 0.05 metres (50 mm) may be applied to foundation depths derived from the Appendix D tables.

Table 6.4. Volume change potential (NHBC Chapter 4.2 2003)

Modified plasticity index	Volume change potential
40% and greater	High
20% to less than 40%	Medium
10% to less than 20%	Low

The above table can be used without modifying the clay plasticity index. However, sufficient samples should be taken across the site to provide confirmation that the test results are representative of the soil volume change potential for the site. Never average out the results, as on large sites there can exist pockets of differing strata.

6.1.2 Distances between trees and foundations

Direct damage can result to a building if it is too close to a tree. The future physical growth of the trunk and roots can create large lateral and vertical forces, which will affect the walls and foundations.

BS 5837 and NHBC Chapter 4.2 Appendix F provide guidance on the safe distances buildings can be located near trees. Where these safe distances cannot be achieved, precautions to allow for future growth must be taken by:

- Reinforcing foundations adequately to resist lateral forces
- Reinforcing walls and foundations slabs to allow for bridging over the tree roots, allowing sufficient clearance for future growth

Sites with trees

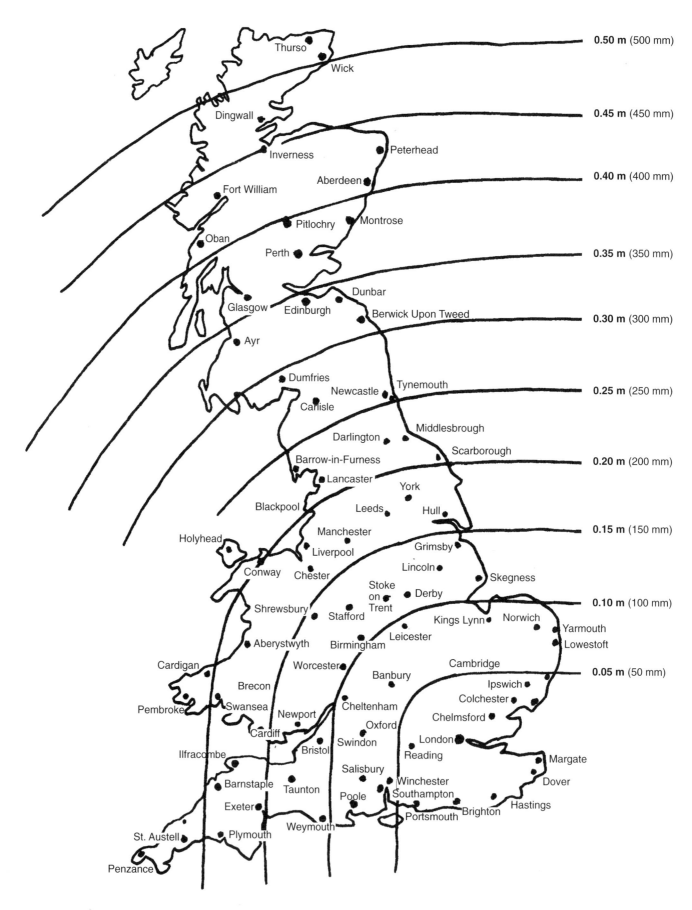

Fig. 6.5 Reductions in foundation depths due to climatic variations (NHBC Chapter 4.2 Appendix 4.2 E).

- Lay paving on a flexible base to allow for movement
- Free-standing boundary walls should have footings that span over roots and the brickwork should be reinforced with careful use of movement joints
- Drains and services must be adequately designed and protected and allow for ground movement. Any service pipes passing through walls or foundations should have sufficient clearances to take account of the potential ground movement.

6.1.3 Foundation depths related to proposed tree and shrub planting

Developers often construct foundations prior to getting agreement with the local planning authority on proposed landscaping. This puts buildings at risk, if trees or shrubs are planted that:

- Are too close to the buildings, or
- The species selected are inappropriate for the site in terms of future water demand, mature height and distance from the buildings.

Table 6.5. Minimum foundation depths allowing for restricted new planting (NHBC Chapter 4.2 2003)

Volume change potential	Minimum depth (m)
High	1.50
Medium	1.25
Low	1.0

Foundation depths in Table 6.5 above are based on agreed planting schedules to exclude trees within the distances from foundations shown in Table 6.6.

Foundation depths shown in Table 6.7 are based on zone limits agreed in the new planting schedules to exclude trees within the zone of influence shown in Table 6.2.

New landscaping and planting schedules should, where possible, always be agreed with the local planning authority before the site is purchased. There may be

Table 6.6. No tree-planting zone for minimum depth foundations (NHBC Chapter 4.2 2003)

Water demand	No tree planting zone
High	1.0 × mature height
Moderate	0.5 × mature height
Low	0.2 × mature height

Table 6.7. Minimum foundation depths outside zone of influence (NHBC Chapter 4.2 2003)

Volume change potential	Minimum depth (m)
High	1.0
Medium	0.9
Low	0.75

existing trees on the site that the local planning authority wish to retain and which could have a large influence on the development costs.

Any proposed landscaping should not compromise the foundation designs on the development.

Foundation depths related to new shrub planting

Some shrubs have considerable potential to cause damage to building foundations. Such species as *Cotoneaster*, *Pyracantha* and climbers such as ivy, virginia creeper and wisteria can be particularly damaging. Foundation depths relating to new landscaping should be based on the following criteria:

- Foundation depths shown in Table 6.5, in which case there are no restrictions where the shrubs can be planted
- Foundation depths shown in Table 6.7, with limits agreed in the planting schedules to exclude shrubs within the distances from foundations shown in Table 6.8.

Table 6.8. No shrub planting zone for minimum depth foundations (NHBC Chapter 4.2 2003)

Volume change potential	No shrub zone (m)
High	3.0
Medium	2.50
Low	2.0

6.1.4 Measurement of foundation depths

Once the foundation depths have been determined, the next step is to examine the existing ground levels on the site and the proposed ground levels in conjunction with any new planting proposals.

Where ground levels are to remain unaltered the foundation depths should be measured from ground level at the excavation. Where ground levels are reduced or increased, the foundation depths should be measured as shown in the following figures.

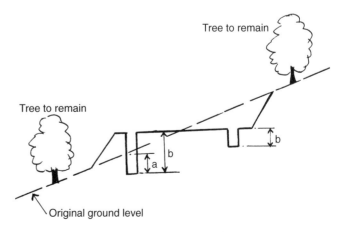

Fig. 6.6 Where trees or hedges are to remain.

Sites with trees

Use what is the lower of:

(a) Foundation depth based on the appropriate tree height (see Fig. 6.9)
(b) Foundation depth based on the mature height of the tree.

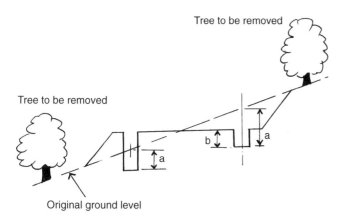

Fig. 6.7 Where trees or hedges are removed.

Use what is the lower of:

(a) Foundation depth based on the appropriate tree height (see Fig. 6.9)
(b) Minimum foundation depth (see Table 6.7).

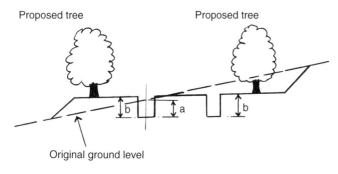

Fig. 6.8 Where new trees or hedges are proposed.

Use the lower of:

(a) Minimum foundation depth (see Table 6.7)
(b) Foundation depth based on the mature height of the tree.

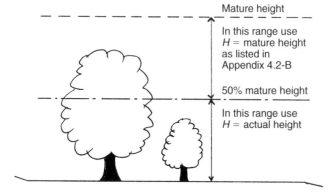

Fig. 6.9 Tree height H to be used for particular design cases.

This guidance should be used when:

- Deriving foundation depths where trees have been removed (use tree height at time of removal)
- Checking the appropriate level from which depths should be measured when trees remain and ground levels are increased (use tree height at time of construction relative to original ground level)
- Determine whether heave precautions should be provided (use tree height at time of construction).

6.2 BUILDING ON WOODED SITES

The design of foundations for buildings that are to be built over areas where previous mature trees or hedges existed is not an exact science. Rates of hedge and tree growth vary considerably; they are dependent on such factors as exposure and variations in soil conditions. Because of this it is wise to adopt recommendations based on common sense and well-judged criteria.

The most useful published guidance is contained in the NHBC Chapter 4.2 April 2003 *Building Near Trees*.

If the NHBC guidance is followed when evaluating a wooded site for development, the critical parameters and data required to make sensible and commonsense judgements can be determined.

There are five generic solutions for sites where clay soils have been desiccated by previous significant tree growth:

1. The use of piles and concrete ring beam foundations with precautions for heave
2. Deep trench-fill concrete footings with voided suspended ground floors and compressible sheet materials down the sides of the footings
3. Deep strip footings with loose stone backfill and a voided suspended ground floor
4. Stiff raft foundations placed on a thick cushion of well-compacted Type 1 granular fill to a DoT specification
5. The use of deep-pad and stem foundations with precautions for heave.

6.2.1 Piled foundations

This solution is generally used in areas where clays have become heavily desiccated to depths in excess of 3.0 metres and it is required that foundations are taken down to a moisture-stable level. Where large trees are within 4.0 metres of a proposed building and the trees and clays are such that foundation depths are impracticable, it is advisable to adopt a piled ring beam solution.

Where bored piles are used with a proprietary sleeve the piles should contain sufficient reinforcement to a depth of at least 3.0 metres or to the depth of the zone of influence, whichever is the greater.

This reinforcement must be linked to the ground beams and the pile should extend a minimum distance of 750 mm below the zone of influence or to a minimum depth of

6.0 metres or to suitable load-bearing strata, whichever is the deeper.

Heave precautions for pile and ground beam foundations

If piling is the only practical solution, the piles and ground beams should be designed to withstand the effects of clay heave. This is usually catered for by providing special sleeved piles and using low-density expanded polystyrene sheets such as Claymaster around and below the ground beams, with a fully voided suspended ground floor construction (Fig. 6.10). Table 6.9 indicates the void dimensions required down foundation sides and below ground beams and floor slabs.

Table 6.9. Minimum void dimensions for foundations, ground beams and suspended ground floors

Volume change potential	Against side of foundation and ground beam Void dimension (mm)	Under ground beam and *in-situ* suspended concrete floor Void dimension (mm)
High	35	150
Medium	25	100
Low	0	50

Note

When using compressible materials to achieve the required void space, the materials should be able to compress to accommodate the heave. The actual thickness of compressible material should be established from the manufacturer's recommendations and is generally in the order of twice the void dimension shown.

Compressible materials should never be used below thin in-situ concrete floor slabs because there is insufficient dead weight to resist the upward heave forces.

For void formers the void dimension is the remaining void after collapse. The actual thickness of void former required should be determined from the manufacturer's recommendations.

6.2.2 Deep trench-fill concrete foundations

Of all the solutions this is the one most preferred by builders. Unfortunately it is also the one most likely to give rise to structural damage later due to inadequacies in the design and construction process. The following problems can result:

- The width of the excavated trench can be uneven, resulting in inverted ledges that, when subjected to the vertical heave pressures, can lift upper levels of mass concrete from the lower levels (Fig. 6.11)
- Compressible sheet materials such as low-density polystyrene (Claymaster) must be used on the inside face of the foundation when the foundation depths exceed 1.50 metres based on the appropriate tree heights or where the clays are known to be desiccated

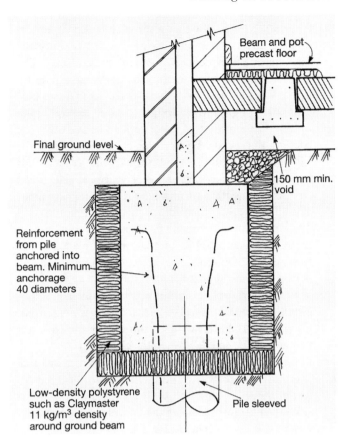

Fig. 6.10 Ground beam on piles.

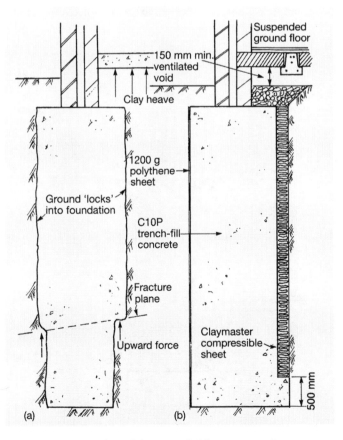

Fig. 6.11 Installation of deep trench-fill concrete and Claymaster: (a) incorrect method; (b) correct method.

- It is the author's view that a slip membrane should be used on the external face of the trench to reduce the 'locking-in' effect between the concrete and clay. The NHBC Chapter 4.2 does not stipulate this as a requirement, despite past evidence of recorded insurance claims. It is logical when designing bored concrete piles to carry an amount of the pile load using the 'skin friction' that occurs between the concrete pile and the surrounding ground. Why should the reverse not happen when clay heave takes place on the deep concrete foundations, which have a larger surface area? The effect is more likely following construction after a dry summer and before the building over is completed. It is also exacerbated if foundations are placed in the summer and building over is not completed in the period between September and March

- While compressible materials are not required on deep internal foundations as the lateral forces are balanced, provision should be made for slip membranes to cater for the locking-in effect when the clays heave

- If trench-fill concrete is adopted only on two or three sides of a dwelling, a dam effect can be created, resulting in a rapid groundwater build-up and a subsequent re-hydration of the internal clay dumpling, which will result in clay heave

- The deep trench-fill can act as a barrier to tree roots so preventing the tree from reaching its water source. Dumpling heave is then potentially likely

- NHBC recommends a maximum depth of 2.50 metres for deep trench-fill concrete. Any foundation deeper than this must be designed by a qualified chartered structural or civil engineer.

- It is important that the ground below the base of the trench-fill is shown to be competent for at least another 1.0 metre.

6.2.3 Deep strip footings with loose stone backfill

This solution (Fig. 6.12) is more suited to areas where the clay soils are in the low shrinkage category, i.e. plasticity index less than 20%. These clay conditions are generally found north of the River Humber but there are exceptions where in certain localities, high-plasticity clays can be encountered. Such clays can be found in Yorkshire (Kirbymoorside), Wakefield, Cleveland (South Bank and Great Ayton) and even as far north as Jarrow.

Such foundations are usually no deeper than 1.50 metres, resulting in 1.20 metre deep trenches, which are safe enough to lay bricks in without the need for shoring. By using trench-block masonry with loose stone backfill down the sides of the footing scarcement, the effects of horizontal and vertical swelling can be significantly reduced. A fully suspended voided ground floor is recommended if there is any likelihood of trees affecting the clays below the floor area.

6.2.4 Stiff raft foundations on a thick cushion of granular fill

For sites likely to be affected by dumpling heave, or sites where the soil strengths decrease with depth this can be an effective foundation solution, for the following reasons:

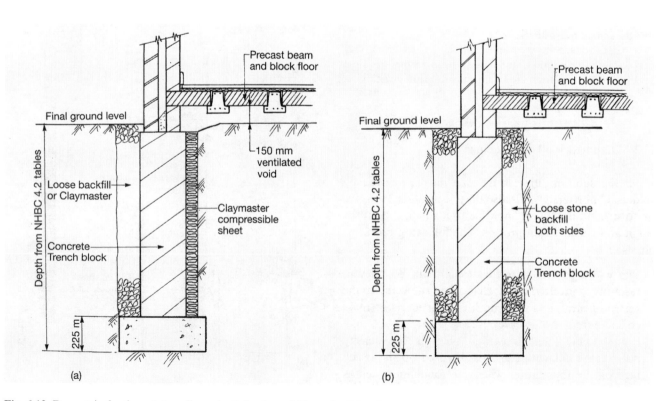

Fig. 6.12 Deep strip footings: (a) medium plasticity clays; (b) low plasticity clays.

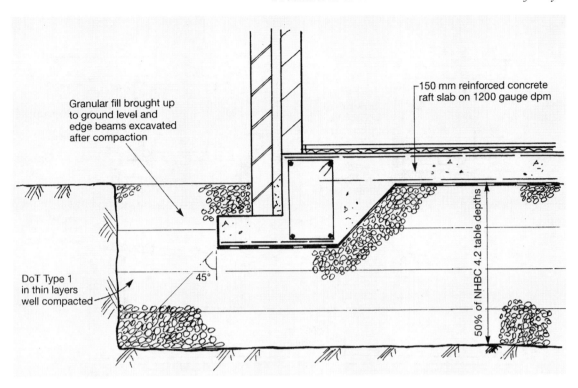

Granular fill brought up to ground level and edge beams excavated after compaction

150 mm reinforced concrete raft slab on 1200 gauge dpm

DoT Type 1 in thin layers well compacted

45°

50% of NHBC 4.2 table depths

Fig. 6.13 Raft foundations on a thick stone cushion. NHBC recommends maximum depth of stone fill to be 1.25 metres but this may be increased if filling is supervised and laid to a correct engineer's specification.

- It usually works out cheaper than deep trench-fill foundations and suspended ground floors
- It is simpler to construct and does not require any shoring
- It kills two birds with one stone. By removing the desiccated inner clay dumpling and replacing it with D of T Type 1 sub-base the stiff foundation is better able to accommodate seasonal variations as the remaining trees grow (Fig. 6.13)
- The raft can be made stiff enough to cater for any differential movements likely to occur in the underlying strata
- This method is most suitable in those locations where the clay soils become weaker with depth. Such areas as Hull, Grimsby and Grangemouth have firm to stiff over-consolidated clay crusts about 1.50 to 2.0 metres thick underlain by soft alluviums and peat beds
- Where used as an alternative to deep trench-fill, NHBC require the depth of stone fill to be 50% of the trench-fill depths from the Chapter 4.2 tables, with a maximum permitted depth of 1.25 metres of stone up-fill. Deeper depths of stone would be acceptable if laid and compacted to an engineer's specification and adequately supervised and tested on site
- This method is also suitable on sites where clay fills are present and there are existing trees that require the foundations to be designed to comply with NHBC Chapter 4.2. Provided the clay fills are suitable to support a raft, the excavation of the clay fills

below the raft and replacement with a thick cushion of compacted granular fill can only improve the ground conditions and help to spread the loads more evenly.

6.2.5 Deep pad and stem foundations

This solution is very similar to the ring beam and piling solution but is often adopted when only a few dwellings are involved to avoid the high costs in piling. The pads can be excavated without shoring and the stems can be formed using precast concrete manhole rings as formers. Heave precautions will be required if the soils are desiccated and these are indicated in Fig. 6.14.

6.3 PRECAUTIONS TO TAKE WHEN THERE IS EVIDENCE OF CLAY DESICCATION

1. Provide low-density polystyrene such as Claymaster on the inside face of trench-fill concrete foundations. Provide a polyethylene slip membrane on the external face of the concrete trench fill to break the bond between the clay and the concrete. Where the existing trees are within a distance of twice the foundation depth, the use of Claymaster is recommended on both sides of the trench-fill concrete.
2. Provide a fully suspended voided ground floor construction, i.e. no sleeper walls, off over-site concrete.

125

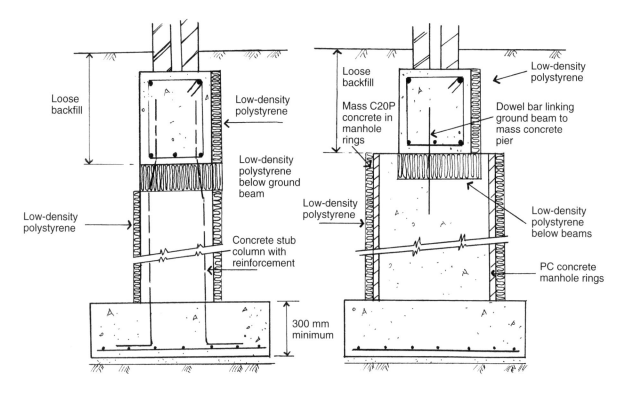

Fig. 6.14 Pad and stem foundations showing heave precautions.

3. Check to ensure that the foundation stratum below the foundation base level remains competent for a distance of at least twice the foundation width.
4. Trench-fill concrete foundations can be adopted up to a depth of 2.50 metres without having to be engineer-designed.
5. All foundation excavations should be concreted straight after excavation and the foundation and super-structure should be constructed as soon as possible to counteract the effects of ground heave.
6. All lighter buildings such as porches and garages should be founded at the same depth as the main dwelling foundations.

6.3.1 Suspended floors

These can be either timber joists or precast pot and beam systems with a ventilated void below, in accordance with Table 6.10.

6.3.2 Drainage and services

If the construction of trench-fill concrete foundations has interrupted the natural run-off of surface and ground-water, then a dam effect could be created. To prevent this it may be necessary to provide French drains to intercept any groundwater and to transfer it into the existing storm drainage system.

6.3.3 Protection to drainage

Where new drainage is likely to be affected by tree root action, the following precautions should be taken:

- The drains should be provided with a concrete surround. On flexible drainage this will necessitate providing Flexcell board between the joints at the spigot and socket connections
- Provide mass concrete barrier walls. This is often required on sites where rock is very close to the surface.

Table 6.10. Minimum void dimensions under precast concrete and timber ground floors

Soil heave potential	Precast concrete floors	Suspended timber
	Void dimension (mm)	Void dimension (mm)
High	225	300
Medium	175	250
Low	125	200

Notes
(i) The measurement is from the underside of the precast beams to the ground floor and includes the 75 mm ventilation allowance.
(ii) The measurement from the underside of the timber joists to ground level includes 150 mm ventilation allowance.

Example 6.2 Strip foundations on clay soils with mature and semi-mature trees

A proposed housing development in Leeds, consisting of a large two-storey masonry construction, is to be built in the grounds of a former hospital complex. The new building covers a large plan area as shown in Fig. 6.15 and is surrounded on all sides by mature and semi-mature trees (see the tree schedule in Table 6.11). In addition, several existing trees will have to be removed from within the building footprint. The soils have been tested following a trial pit investigation and found to be in the medium plasticity category.

The clay soils have a high bearing capacity and are suitable for strip foundations. Using NHBC Chapter 4.2 determine the most suitable foundations and the required foundation depths. Also provide details of the most suitable ground-floor construction.

The tree heights in brackets in Table 6.11 are the average mature heights referred to in the NHBC Chapter 4.2 2003 Appendix B. When considering the trees that are to be lopped, use the height of the tree at the time of removal.

Table 6.11. Example 6.2: Tree schedule

Tree No.	Tree species	Height (m)	Comments
1	Sycamore	20 (22)	Prune
2	Sycamore	18 (22)	Retain
3	Sycamore	18 (22)	Retain
4	Sycamore	20 (22)	Retain
5	Sycamore	18 (22)	Prune
6	Birch (silver)	10 (14)	Prune
7	Birch (silver)	10	Remove
8	Sycamore	16	Remove
9	Beech	20 (22)	Retain
10	Sycamore	22 (22)	Retain
11	Ash (Manna)	14 (18)	Retain
12	Ash (Manna)	14 (18)	Retain
13	Sycamore	16 (22)	Retain

DETERMINATION OF FOUNDATION DEPTHS

Southern wall

The nearest sycamore tree to the south-east corner is 10.50 metres away. Using Chapter 4.2 Table 18:

Depth required = 1.32 metres by interpolation

$$\text{Less} \quad \frac{0.15}{1.17} \text{ m for climatic variation}$$

Therefore make foundation 1.20 metres deep.

Check south-west corner

Using Table 19 as birch are low water demand species:

Birch 14 metres high: depth required for 5 metres distance = 1.10 − 0.15 = 0.95 metres
Sycamore 22 metres high: depth required for 14 metres distance = 1.10 − 0.15 = 0.95 metres
Birch tree to be removed is within the footprint and is 10 metres high = 1.35 − 0.15 = 1.20 metres.

With minimum foundation depth of 900 mm the zone of influence of the sycamores will be 14 metres.

The zone of influence of the lopped birch will be 5.0 metres. Precautions will need to be taken for clay heave within a 5.0 metre radius of Tree No. 7. This will be best catered for by using a 50 mm sheet of Claymaster, a low density polystyrene on the inside face of the foundations and by providing a voided suspended ground floor over the area shown shaded in Fig. 6.15.

Check north-east corner

Predominant tree is the ash No. 12.

From Table 18 for a 9 metre distance the depth required = 1.30 − 0.15 = 1.15 metres.

The sycamore No. 8 that is to be lopped at 16 metres requires a depth of 1.90 − 0.15 = 1.75 metres.

The zone of influence of Tree No. 8 will be 10 metres. Precautions will need to be taken to cater for clay heave within a 10 metre radius of Tree No. 8 as previously described.

Eastern wall

Tree No.	Distance (m)	Depth required (m)
No. 10	6.0	1.60 − 0.15 = 1.45
	7.0	1.55 − 0.15 = 1.40
	9.0	1.40 − 0.15 = 1.25
	10.0	1.35 − 0.15 = 1.20
	13.0	1.15 − 0.15 = 1.00

North-west corner

Sycamore tree No. 13 is at a distance of 10 metres. Depth required = 1.35 − 0.15 = 1.20 metres.

As this corner has to be 1.75 metres deep to cater for Tree No. 8 the distance from Tree No. 13 before the depths can be the 900 mm minimum will be 14 metres.

6.3.4 Precautions against clay heave

See Fig. 6.16. As there is more than 50% of the floor area that could be affected by clay heave, it would be simplest to specify a voided floor throughout the flats. If a concrete beam and block system is used, no over-site concrete will be required but a 175 mm ventilated void will have to be provided.

Floor finishes can be 18 mm chipboard on 25 mm polystyrene with a 500 gauge polyethylene vapour barrier.

If a timber floor is preferred then a 50 mm concrete over-site on a 1200 gauge polyethylene dpm will be required with a 250 mm minimum ventilated void space.

Chapter 4.2 Table 9 requires a 25 mm void to the inside face of the foundations. A 50 mm thick sheet of Claymaster will give 25 mm allowing for 50% compression. This should be placed down the inside face of the under-building along the walls within the extent of the shaded

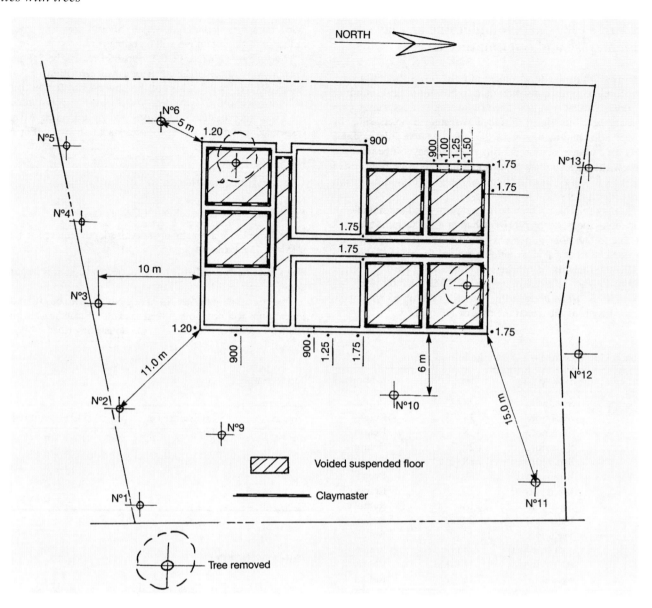

Fig. 6.15 Example 6.2: site layout and tree schedule.

area. On the external face of the walls, loose stone backfill should be placed in the foundation scarcement.

Example 6.3 Foundations on a site with mature trees on stiff clay soils overlying soft alluvium

A site in Hull is to be developed for low-rise housing. The site investigation has revealed firm to stiff clays down to a depth of 2.50 metres underlain by a thin band of peat 150 mm thick, which in turn is underlain by a thick layer of soft silty grey clay extending to a depth of 6.0 metres below ground level. The site survey shows that on the northern and southern boundaries, there are many mature black poplars on the adjoining land.

The laboratory tests show that the upper clays have a high bearing capacity of 110–150 kN/m². The soft silty clays are shown to have an allowable bearing capacity of 40 kN/m² with a maximum settlement of the order of 45–50 mm.

The dwellings are to be two-storey traditional brick and block construction with maximum line loads on the front and rear walls of 35kN/metre run. Determine a suitable foundation type for the houses and garages on the site. The plasticity index of the upper clays is 33% and the distance from the poplar trees to the houses was measured at 10 metres.

Using Chapter 4.2 Appendix D Table 17:

Depth of foundation = 2.40 m − 0.15 = 2.25 metres

If trench-fill foundations are used to a depth of 2.25 metres then the soft clays will be over-stressed and settlements of a high magnitude will result. Because of the prevailing ground conditions, the most suitable foundation solution will be a stiff raft placed on a thick cushion of well-compacted granular fill. The thickness of the stone cushion will have to be 50% of the trench-fill depths derived from Table 17, i.e. 1.20 metres. The sub-base will need to be laid in 200 mm layers and given six passes of a Bomag 65 vibratory roller (Fig. 6.17). The raft must be designed to span 3.0 metres and cantilever 1.50 metres at the corners.

128

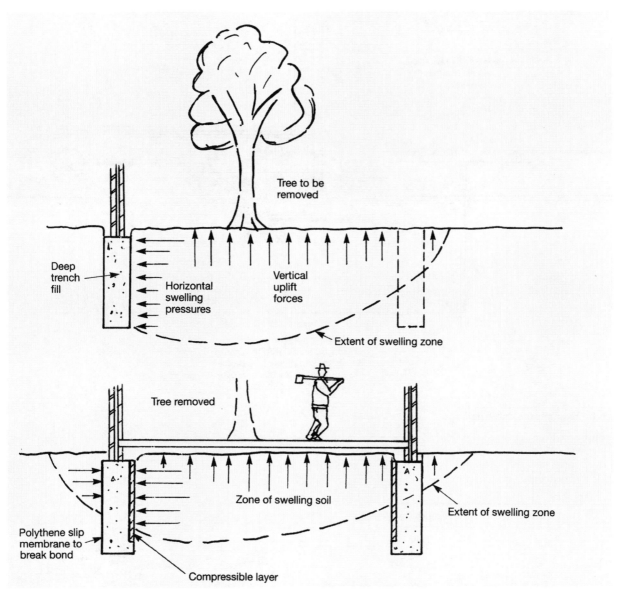

Fig. 6.16 Example 6.2: dumpling heave.

Example 6.4   Site investigation and foundation design on a heavily wooded site

A builder is to construct a large bungalow on a site that is heavily wooded. The layout of the dwelling and driveway will result in several large trees being removed, in addition to several large shrubs.

A tree survey has been carried out by a local arborist for planning purposes and this is scheduled in Table 6.12. The layout of the site is as shown in Fig. 6.18. The engineer is required to have a site investigation carried out and is to make recommendations for foundations to the bungalow.

SITE INVESTIGATION

Four boreholes were drilled at locations around the proposed bungalow to depths of 4.0 metres. Samples of clay were taken at various depths in each borehole and submitted to a soils laboratory for testing. The laboratory test results are shown in Table 6.13.

Ground conditions

The borehole details shown in Fig. 6.19 indicate that very similar ground conditions exist over the site. Below the topsoil, firm sandy clay was evident except in borehole No. 4 where soft clays were encountered from 0.80 metres to 1.20 metres down. At varying depths between 0.75 metres and 1.80 metres the sandy clays changed to uniform deposits of a stiff boulder clay.

Tree roots

Active tree roots and fine fibrous rooting were found within the upper clays at the following depths:

129

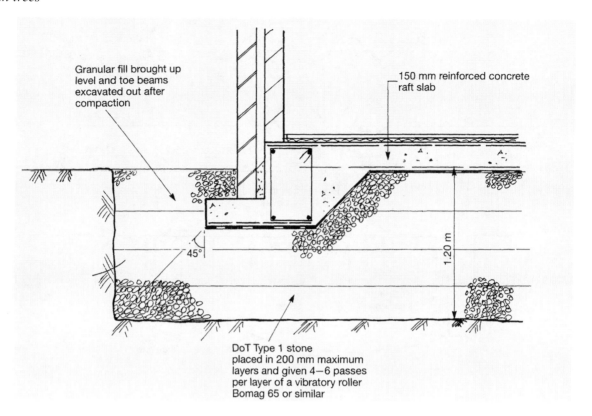

Granular fill brought up
level and toe beams
excavated out after
compaction

150 mm reinforced concrete
raft slab

45°

1.20 m

DoT Type 1 stone
placed in 200 mm maximum
layers and given 4—6 passes
per layer of a vibratory roller
Bomag 65 or similar

Fig. 6.17 Example 6.3: raft details.

Borehole No.	Depth (m)
1	1.30
2	0.90
3	1.20
4	1.50

LABORATORY TESTING

Samples of the clay were tested for natural moisture content, liquid limit, plastic limit, plasticity index and soluble sulphates. The natural moisture contents were shown to range from 14% to 24% and the results indicated that there was a reduction in moisture content with depth.

The plasticity index of the upper clays was in excess of 20% and this puts these clays into the Medium plasticity category as defined in the NHBC Chapter 4.2. Tests on the samples of stiff clays revealed plasticity index values of 12%, which puts them into the Low plasticity category

FOUNDATION DESIGN

The boreholes indicate slight variations in the upper sandy clays, which are generally of a firm consistency except for borehole No. 4 where the soils were very weak down to depths of 1.20 m. All the boreholes, however, revealed stiffer clays at depths of 0.75 m and 1.80 m with the clays showing a lower natural moisture content. Active tree roots from the existing trees were evident at depths of 1.40 m.

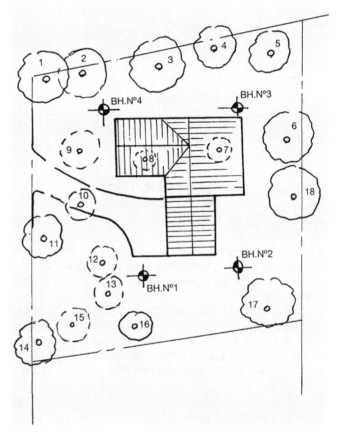

Fig. 6.18 Example 6.4: site plan. See Table 6.4 for the tree schedule.

Table 6.12. Example 6.4: Tree schedule

Tree No.	Tree species	Height (m)	Water demand	Comments	Foundation depths (m)
1	English oak	15 (24)	High	Retain	1.50
2	Sycamore	15 (22)	Moderate	Retain	
3	Horse chestnut	18 (20)	Moderate	Retain	
4	Horse chestnut	18 (20)	Moderate	Retain	
5	Horse chestnut	16 (20)	Moderate	Retain	
6	Ash	16 (23)	Moderate	Retain	1.08
7	Rhododendron bush			Removed	
8	Rhododendron bush			Removed	
9	Ash	10	Moderate	Removed	
10	Ash	12	Moderate	Removed	
11	Ash	14 (23)	Moderate	Prune	
14	English oak	15 (20)	High	Retain	
16	Holm oak	16 (16)	High	Retain	
17	English oak	15 (20)	High	Retain	1.50
18	Ash	12 (23)	Moderate	Retain	1.08

Note: The figures in brackets are the NHBC Chapter 4.2 tree heights.

Table 6.13. Example 6.4: laboratory test results

Borehole No.	Sample No.	Depth	Moisture content (%)	Liquid limit	Plastic limit	Plasticity index	SO₃ (g/l) 2:1
1	2	1.0	24	46	21	25	
4	5	1.30	22	46	24	22	
1	4	1.50	16	31	16	15	
3	5	1.30	18	30	16	14	0.23
2	4	1.35	15	32	21	11	0.16

It is recommended that standard strip footings be used on the site. To ensure that the existing trees are catered for, these footings will need to be deepened to various depths into the stiffer boulder clay stratum.

It was evident that root penetrations were confined within the upper clays at a maximum depth of 1.40 metres. The foundation depths can be determined by reference to Tables 20 and 21 in NHBC Chapter 4.2 Appendix D.

Foundation depths (see Fig. 6.20)

From the tree survey and site survey the required foundation depths can be found using the Low shrinkage category Tables 20 and 21. The foundation depths are indicated in Table 6.12 for the particular tree species. These depths should be checked on site during excavations, as it is important to ensure that:

1. The same stratum is evident at formation level; and
2. No root growth is present at the formation level of the foundation. If roots are evident the foundation should be deepened at least 300 mm below the lowest roots.

Removal of existing trees

The removal of existing mature trees and large bushes will result in a potential for ground heave of the upper clays. To avoid vertical and lateral pressures on the foundation sides it is recommended that a 50 mm thick sheet of Claymaster, a low-density polystyrene, is placed on the inside face of the foundation. In addition to uplift forces, the external clays can subside because of future moisture abstraction. When this occurs, trench-fill concrete can be subjected to an eccentric force as a result of the 'locking-in' adhesion between the clay and the concrete. To minimize this effect a 1200 gauge polyethylene sheet should be placed on the external face of the trench-fill concrete excavation prior to concreting.

Should existing trees be closer than twice the foundation depth then it would be prudent to use Claymaster on both sides of the trench-fill concrete.

Ground floor construction

Because of the risk of dumpling heave it is a requirement of NHBC Chapter 4.2–S4 that ground floors be of the voided fully suspended type. A precast concrete beam and block floor with a 175 mm ventilated void or full-span timber joists with a 250 mm ventilated void above the concrete over-site level.

Concrete mix for foundations

The samples tested revealed low soluble sulphate levels. It is recommended that C20P Ordinary Portland cement concrete be used in the trench-fill concrete foundations.

Boring Method					Sheet: 1 of 1	Job. Nº: 1641
Hand Auger Bores Nºs 1 – 4						
				Ground Level: 1060 AUD	Site: Leeds St Huddersfield	

Borehole Nº				Description of Strata	Depth (m)	Legend
	BH	Nº2		Topsoil with roots	0.30	
				Clay (Firm) – Light brown and grey mottled with silty partings with roots and fibrous rooting evident.	0.90	
				Clay (stiff) – Dark brown silty clay with stone inclusions	1.50	
				Clay (very stiff) – Dark brown with weathered shale inclusions	1.90	
	BH	Nº4		Topsoil	0.30	
				Clay (soft) – Light brown mottled possibly fill material	0.80	
				Clay (soft) – Grey, silty with an abundance of tree roots	1.20	
				Clay (firm) – Mottled yellow grey with some roots present	1.50	
				Clay (stiff to very stiff) – Dark brown with weathered shale inclusions. Very stiff to hard at base of hole	2.30	
	BH	Nº1		Topsoil	0.30	
				Clay (firm) – Light brown and grey mottled and silty partings with roots and fibrous rooting evident	0.90	
				Clay (stiff) – Dark brown, silty with stone inclusions	1.50	
				Clay (very stiff) – Dark brown with weathered shale inclusions	1.90	
	BH	Nº3		Topsoil	0.30	
				Clay (firm) – Brown with inclusions of small sandstone pieces and weathered shale. Roots evident	0.70	
				Clay (firm) – Dark grey mottled with shale inclusions. Many root fibres visible	1.20	
				Clay (firm) – Mottled sandy clay with inclusions of weathered shale. Root at top of band	1.65	
				Clay (stiff) – Dark brown with inclusions of weathered shales. Stiff becoming hard at base of hole	2.30	

Fig. 6.19 Example 6.4: borehole logs 1–4.

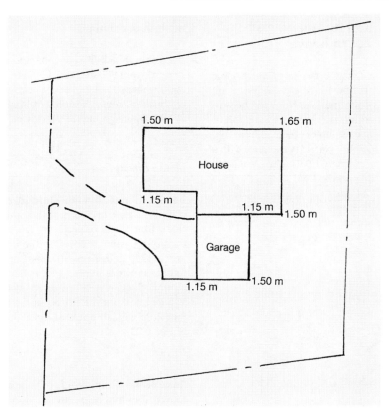

Fig. 6.20 Example 6.4: foundation depths below finished ground level (metres).

CHECKLIST – To be used in determining foundation solutions to cater for trees

1. Establish the shrinkage classification of the bearing stratum and its *in-situ* condition.
2. Determine the actual height of the trees on site and their potential mature height.
3. Take account of any proposed tree and shrub planting scheme.
4. Examine the site topography. Any major slopes or site re-grading could have a significant influence on the foundation depths. Remember that where trees are lower than the dwelling, the depths required are measured from the ground level in accordance with NHBC Chapter 4.2 Figures 5–7.
5. Determine which trees are to remain and the position and height of trees to be removed.
6. Determine the geographical reduction due to climatic variation north of London.
7. Determine from NHBC Chapter 4.2 Appendix D the required depths of foundations less the geographical reduction.
8. Where deep foundations step up away from the trees, give careful consideration to the depth and positioning of steps.
9. Check on the potential for clay heave. Take precautions in accordance with NHBC Chapter 4.2–D8.
10. Examine the position of existing trees relative to the proposed buildings and ensure that the trees are not damaged by construction activities. Follow the recommendations given in BS 5837 (1991).
11. Check that any drainage is suitably protected from the ingress of tree roots. Where possible provide a concrete cut-off wall or surround the drain in concrete.
12. Consider what other foundation solution could be adopted. Deep trench-fill concrete solutions are not suitable on sites where the soil conditions become weaker with depth.
13. Where trees are to be retained and some are to be removed, it may be more appropriate to adopt piled foundations or a raft foundation on a thick cushion of stone.

Example 6.5 Foundation design on a site with mature trees, subject to heave

Development is proposed on a fairly level site of three large detached houses in Wakefield, as shown in Fig. 6.21. There are two mature holm oak trees on the site and Tree No. 1 will need to be removed as it suffers from a serious fungal disease.

Plasticity index tests carried out on samples obtained from trial pits show the clays to be in the Medium plasticity category with a plasticity index of 31%. Bearing capacity of the clays are good and 600 mm wide foundations are proposed.

The two oak trees will affect Plot No. 4 and foundation depths will require to be in accordance with NHBC Chapter 4.2 Appendix D. In addition precautions will need to be taken to cater for potential clay heave following the tree removal.

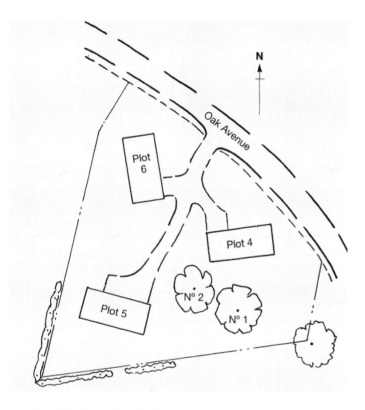

Fig. 6.21 Example 6.5: site layout.

FOUNDATION DEPTHS

Holm oaks have a mature height of 16 metres and are a high water demand tree. For medium plasticity soils refer to NHBC Chapter 4.2 Table 17. The distance from Plot No. 4 is 8 metres minimum for Tree No. 2 and the Table 17 depth required is therefore 2.20 metres. The allowable geographical reduction is 0.150 and this makes for a foundation depth of 2.20 − 0.150 = 2.05 metres.

At 16 metres distance the foundation depth will need to be 1.35 − 0.15 = 1.20 metres.
At 19 metres distance the foundation depth will need to be 1.05 − 0.15 = 0.90 metres.

Therefore provide a standard-width foundation at a minimum depth of 900 mm at distances in excess of 19 metres from Tree No. 2, as indicated in Fig. 6.22.

Steps in foundations

By using trench-fill concrete the foundation bottom can be stepped from 900 mm to 2.05 metres using 250 mm steps. Avoid making steps at or close to junctions or wall intersections. Steps in trench-fill concrete should overlap not less than twice the depth of the step height or 1.0 metre, whichever is the greater.

Heave precautions

Examination of the foundation excavation has confirmed that the clays are desiccated for a distance of 15 metres radius from the tree centre. It will be necessary therefore to provide a 50 mm low-density polystyrene sheet on the inside face of the foundations and a 1200 gauge polyethylene slip membrane on the external face, as shown in Table 6.14. In addition a fully suspended voided ground floor will be required with a ventilated air space of 175 mm under a precast beam floor or 250 mm below a timber joist floor with air bricks at 2.0 metre centres maximum, as shown in Fig. 6.23.

Table 6.14. Example 6.5: minimum void space to sides of foundations and below ground floor beams

Soil shrinkage category	Void dimension (mm)	
	Below ground beams	Against sides of foundation
High plasticity index > 40%	150	35
Medium plasticity index 20–40%	100	25
Low plasticity index < 20%	50	0

6.4 FOUNDATIONS IN GRANULAR STRATA OVERLYING SHRINKABLE CLAYS

Quite often, sites have thick surface deposits of sands and gravels and these will not be adversely affected by tree growth. However, occasionally the sands and gravels may be underlain by clay deposits at shallow depth. In such cases the foundation depths can be reduced to 600 mm subject to adequate bearing capacity and provided the depth of the sands or gravel stratum is greater than 0.75 times the depth of foundation required in a clay soil from the tables in NHBC Chapter 4.2 Appendix D.

This condition is valid only when the thickness of the granular stratum below the foundation is greater than the foundation width, as indicated in Fig. 6.24.

When applying this rule the following points should be confirmed:

1. The soil conditions across the whole building need to be confirmed by trial pit investigation.
2. No tree roots larger than 5 mm diameter should be present in the foundation bottom.

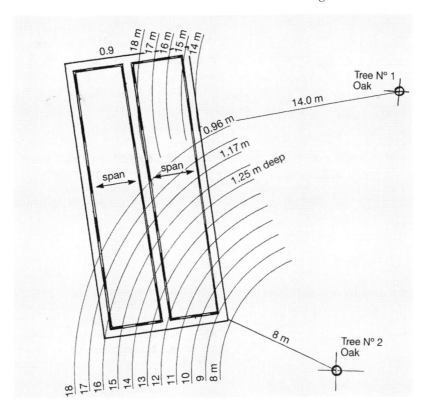

Fig. 6.22 Example 6.5: foundation depths to Plot No. 4.

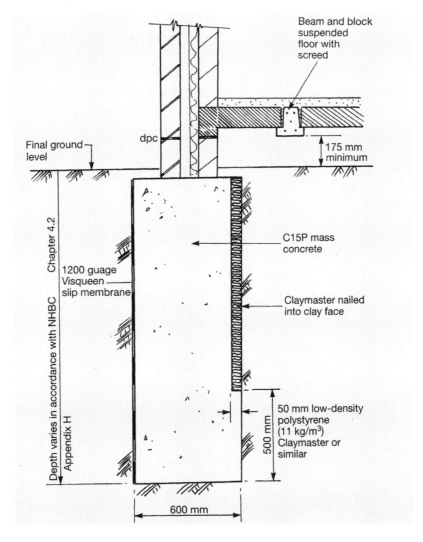

Fig. 6.23 Example 6.5: trench-fill foundation with heave precautions.

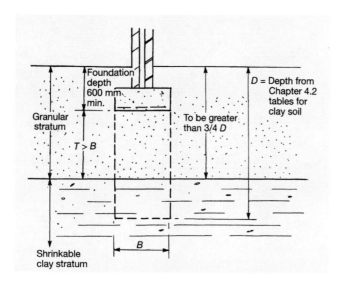

Fig. 6.24 Foundation bearing on a granular stratum overlying shrinkable clays.

3. No root balls or masses of fibrous rooting should be left in place beneath a footing or ground floor slab.

If roots are evident at the base of the excavation then the foundations should be deepened to a minimum of 300 mm below the lowest 5 mm roots and the thickness of the granular material below the foundation should be re-checked to ensure that it exceeds the foundation width.

BIBLIOGRAPHY

Arboriculture Advisory and Information Service. Forest Research Station, Farnham, Surrey.

BRE (1980) Building Research Digests 240–242: *Low-rise Buildings on Shrinkable Clay Soils*, Parts 1–3, Building Research Establishment, Watford.

BRE (1985) Building Research Digest 298: *The Influence of Trees on House Foundations in Clay Soils*, Building Research Establishment, Watford.

BSI (1991) BS 5837 *Trees in Relation to Construction*, British Standards Institution.

BSI (1981) BS 5930: *Code of Practice for Site Investigations*, British Standards Institution.

BSI (1990) BS 1377: *Methods of Testing for Civil Engineering Purposes*, British Standards Institution.

NHBC (2003) *NHBC Standards*, Chapter 4.2 Building Near Trees.

Phillips, R. (1978) *Trees in Britain, Europe and North America*.

Chapter 7
Developing on sloping sites

7.1 STABILITY OF SLOPES

Generally, slope failures in mixed soils, i.e. with cohesive and frictional strengths, occur in the form of a **rotational slip**. The failure usually exhibits heaving of the ground mass at the toe of the slope with back-scar cracking at the crest (Fig. 7.1).

Checking the stability of existing slopes is now generally carried out using **effective stress analysis** to determine the factors of safety. If water pressures are evident the factor of

safety can be determined for the drained or undrained condition of the slope. If water levels exist within the slope then the porewater pressures must be determined by field measurements using piezometers or, in highly permeable soils, standpipes.

Effective stress analysis was developed by Bishop (1955) for checking existing slopes. For new embankments and dam constructions Bishop developed a **rigorous method**, as the scale of such projects required a more exacting appraisal for

---

SYMBOLS

The following symbols have been used in this chapter:

a	Shear span (mm)
A	Horizontal cross-sectional area of section (mm^2)
A_s	Cross-sectional area of primary reinforcing steel (mm^2)
β	Angle of slope of retained material (degrees)
b	Width of section (mm)
c	Lever arm factor
d	Effective depth (mm)
ϕ	Angle of shearing resistance of the retained material (degrees)
μ	Angle of internal friction between base concrete and formation (degrees)
f_k	Characteristic compressive strength of masonry (N/mm^2)
f_{kx}	Characteristic flexural strength (tension) of masonry (N/mm^2)
f_s	Stress in the reinforcement (N/mm^2)
f_v	Characteristic shear strength of masonry (N/mm^2)
f_y	Characteristic tensile sgrength of reinforcing steel (N/mm^2)
γ_m	The moist bulk density of the backfill material

γ_s	The bulk density of the natural soils
g_a	Design vertical load per unit area (N/mm^2)
M_d	Design moment of resistance (N/mm or kNm)
N	Design axial load (N or kN)
Q	Moment resistance factor (N/mm^2)
Q_k	Characteristic imposed load (N or kN)
s_v	Spacing of shear links (mm)
t	Overall thickness of wall or column (mm)
t_{ef}	Effective thickness of a wall or column (mm)
t_f	Thickness of flange in a pocket-type wall (mm)
τ_f	Partial safety factor for load
τ_m	Partial safety factor for the material
τ_{mv}	Partial safety factor for shear strength of masonry
V	Shear force due to design loads (N)
v	Shear stress due to design loads (N/mm^2)
Z	Section modulus of section considered (mm^4)
z	Lever arm (mm)
τ_{mm}	Partial safety factor for comprehensive strength of masonry
τ_{ms}	Partial safety factor for strength of steel reinforcement
V_h	Design shear stress.

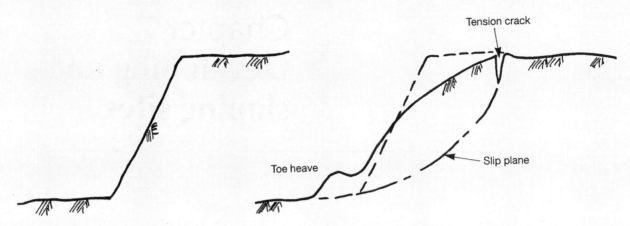

Fig. 7.1 Rotational slip: cohesive strata.

economy. On sites where fill has been deposited over an existing slope the rigorous method of analysis is recommended if the failure is likely to be rotational.

The normally accepted factor of safety for mixed strata is about 1.25 or greater. In view of the low stress parameters used in the analysis it is economic if the factor of safety does not exceed 1.50.

Before the recent developments of computer programs for carrying out slope stability checks, calculations were carried out manually and were very laborious. With the computer software now available it is possible to examine many cross-sections of a slope and many potential slip circles relatively quickly, for various groundwater levels.

When commencing a slope stability check the site investigation should be examined to try and establish the type of likely failure mode (Fig. 7.2). If the indications are that any slip would be rotational, the first object is to try and locate the centre of the most critical circle. This can only be achieved by trial and error with the various sections being

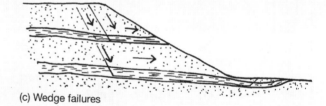

(a) Surface slide

(b) Shallow translational slide

(c) Wedge failures

(d) Rotational slip surface failure

Fig. 7.2 Different failure modes of slopes.
(a) *Surface slide*. Generally occurs in granular strata. Mode of failure occurs when groundwater levels rise within the slope.
(b) *Shallow translational slide*. A subsurface slide can result at the base of a thin layer of weak stratum, especially if excavations are carried out along the base of the slope.
(c) *Wedge failures*. These usually occur where the slope is made up of varying strata bands. Some slopes may contain many weak bands.
(d) *Rotational slip surface failure*. This is the usual type of failure in cohesive strata, and is usually deep-seated. The base of the critical slip circle is usually controlled by any underlying hard stratum.
(e) *Complex slope failure*. Slopes with complex soil strata do not give rise to rotational slip circle failures. The usual method of analysis is by the Janbu method which involves taking strip segments along the slope, similar to that described in Bishop's rigorous method.

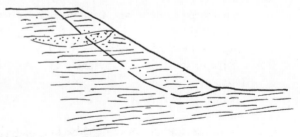

(e) Complex slope failure

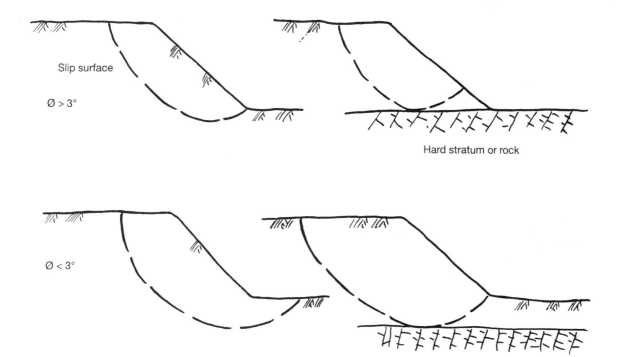

Fig. 7.3 Types of rotational slip.

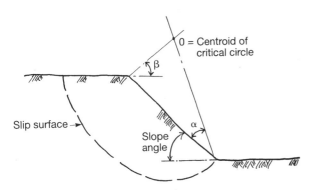

Fig. 7.4 Fellenius construction for centre of critical circle.

Table 7.1. Fellenius critical circle method

Slope	Angle of slope (degrees)	Angle α (degrees)	Angle β (degrees)
1:0.58	60	29	40
1:1	45	28	37
1:1.50	33.79	26	35
1:2	26.57	25	35
1:3	18.43	25	35
1:4	11.32	25	37

analysed and factors of safety obtained. Once sufficient circles have been tried it is possible to draw the contours of the factor of safety values which generally form an elliptical pattern, the centre of which should eventually coincide with the centre of the critical circle.

Obviously it is essential to determine the centre of the first trial circle as accurately as possible. For soils where ϕ is less than 3° the critical slip circle usually passes through the toe of the slope if the slope angle exceeds 53°, unless there is a layer of rock or stiffer strata at the base of the slope. For cohesive deposits with small or zero values of ϕ, the critical circle is generally deeper and extends in front of the toe of the slope. This again is affected should rock or stiffer strata be present at the base of the slope (Fig. 7.3).

For purely cohesive deposits Fellenius (1936) developed a method for determining the centre of the critical circle as shown in Fig. 7.4 and Table 7.1. The table lists Fellenius's values for α and β for various slope angles.

This method is not suitable for mixed soils and it was therefore adapted by Jumikis (1962) for use with mixed soils. To use the Jumikis construction for $c-\phi$ soils it is still necessary to obtain the centre of the Fellenius circle O_1 and then to construct the line A–O_1 as shown in Fig. 7.5.

These methods are only to be used to obtain a series of trial slip circles that are sensibly located. They become less reliable when the soil conditions cease to be homogeneous or when the slopes are not even, but they can give an indication of the likely factors of safety.

Effective stress Analysis

These methods of analysis are now in general use, and the formula

$$\text{Factor of safety } F' = \frac{1}{W \sin \alpha}\left[c'L + W(\cos \alpha - r_u \sec \alpha)\tan \phi\right]$$

where L is the length of the clip circle, should enable existing slopes to be analysed quickly.

Rigorous method. In this method

$$F = \frac{1}{W \sin \alpha}\left[c'b + W(1 - r_u)\tan \phi\right] \frac{\sec \alpha}{1 + \tan \phi'.\tan \alpha / F}$$

139

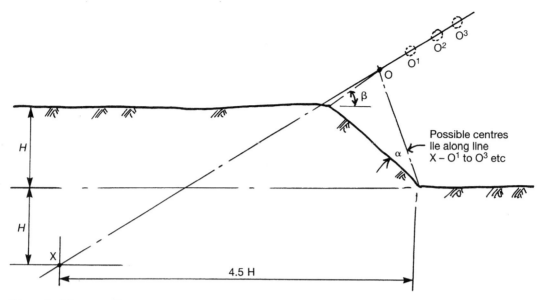

Fig. 7.5 Jumikis method for c–$\emptyset$ soils.

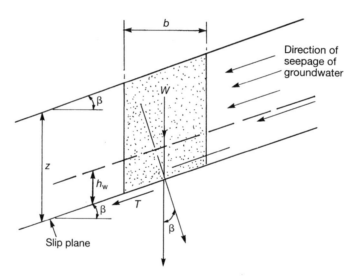

Fig. 7.6 Determination of r_u. $r_u = h_w\gamma_w/\gamma z$; $W = \gamma z\, b \cos \beta$; $T = W \sin \beta$.

where

α = angle between the vertical and the line drawn from the circle centre through the centre of the segment on the slip circle;

Z = height of segment (m);

c' = effective stress (cohesion);

b = width of segment (m);

W = weight of segment (kN);

r_u = pore pressure ratio;

ϕ' = effective stress (angle of shearing resistance);

F = factor of safety.

The slope to be checked is first divided into a suitable number of segments, and the pore pressure ratio r_u is determined at the mid-point of the segment base.

DETERMINATION OF R_U

See Fig. 7.6. As $r_u = u/\gamma Z$, where $u = \gamma_w\, h_w$, then $r_u = h_w\, \gamma_w/\gamma_z$. The various values for r_u can be determined for each segment.

With the rigorous method a first approximation is tried using an assumed value for the factor of safety in the expression

$$\frac{\sec \alpha}{1 + \tan \phi' \cdot \tan \alpha / F}$$

For greater accuracy the final columns should be recalculated using the F value obtained from the first approximation.

Case Study 7.1   Sloping site with clay fills over boulder clay

A site investigation has been carried out on a section of a housing development in Yorkshire. The investigation has revealed that clay fills have been deposited over an existing slope which consists of boulder clays. The slope has a watercourse at its base. A slope stability analysis is required using Bishop's rigorous method and the soil parameters to be used are shown in Table 7.2. These parameters for the filled ground were determined from the *in-situ* SPT tests (according to Peck, Hanson and Thornburn, 1953).

The values for the sandy silty clays have been considered to be typical values for the boulder clays of the glacial deposits in that area of Yorkshire.

Profiles drawn through Plots 6, 10, 20 and 27 with assumed groundwater levels at various depths have been used for checking the various slip circles (Figs 7.7–7.10).

The following data need to be determined and tabulated for the sections being considered:

Table 7.2. Case Study 7.1: soil parameters

Stratum	Bulk density (kg/m^3)	Effective stress parameters	
		c'	ϕ'
Clay fill	2000	0	30
Sandy silty clay	2000	0	30

Height of slice considered, Z
Width of slice, b
Weight of slice, W
Groundwater height above slip plane, h_w

Assume $c' = 0$ and $\tan \phi' = 0.57$. The results are listed in Tables 7.3–7.8.

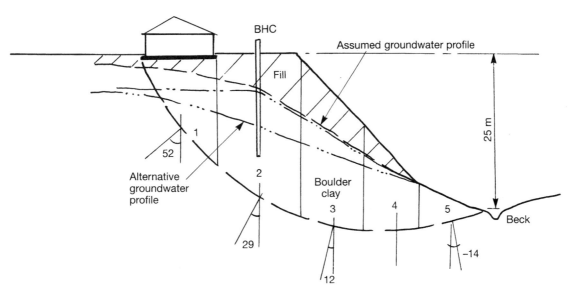

Fig. 7.7 Case study 7.1: plot 6, section 1. Scale 1:50.

Table 7.3. Lower water level

Slice no.	z (m)	b (m)	W (kN)	h_w (m)	α (deg)	r_u	$\sin \alpha$	$W \sin \alpha$	Eqn (1)	$\sec \alpha$	$\tan \alpha$	Eqn (2)	(1) × (2)
												$F = 1.20$	$F = 1.20$
1	11	12.0	1920	2.0	52	0.09	0.78	1513	1008	1.62	1.28	1.00	1008
2	18	14.0	5040	10.0	29	0.28	0.48	2444	2068	1.14	0.55	0.87	1799
3	16	10.0	3200	10.0	12	0.31	0.20	666	1274	1.02	0.21	0.91	1159
4	11	9.0	1980	7.0	0	0.31	0	0	788	1.0	0	1.00	788
5	5	10.0	1000	3.0	−14	0.30	−0.24	−242	404	1.03	−0.24	1.20	485
							Total	4381				Total	5239

Equation (1): $W(1 - r_u) \tan \phi'$. Equation (2) : $\dfrac{\sec \alpha}{\dfrac{1 + \tan \phi' \tan \alpha}{F}}$

$F = 5239/4381 = 1.19$

Case Study 7.1 (*contd.*)

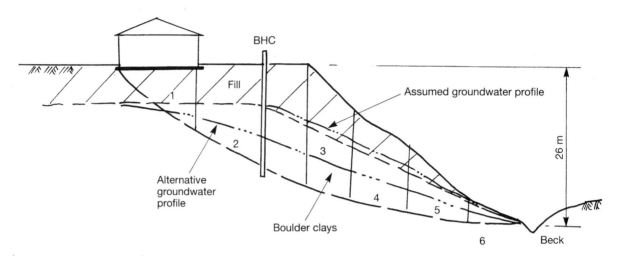

Fig. 7.8 Case study 7.1: plot 10, section 2.

Table 7.4. Higher water level

Slice no.	z (m)	b (m)	W (kN)	h_w (m)	α (deg)	r_u	$\sin \alpha$	$W \sin \alpha$	Eqn (1)	$\sec \alpha$	$\tan \alpha$	Eqn (2)	$(1) \times (2)$
												$F=1.20$	$F=1.20$
1	6.0	13.0	1560	2.5	36	0.21	0.60	940	714	1.25	0.75	0.92	656
2	15.0	18.0	5400	9.5	27	0.32	0.45	2451	2127	1.12	0.51	0.90	1914
3	16.0	7.0	2240	10.5	19	0.32	0.33	729	870	1.07	0.34	0.91	792
4	12.5	9.0	2250	9.0	13	0.36	0.23	505	832	1.03	0.23	0.93	772
5	7.5	9.5	1428	5.0	7	0.33	0.12	174	548	1.02	0.13	0.95	521
6	1.5	8.0	230	1.0	3	0.33	0.05	12	92	1.0	0.05	0.97	89
							Total	4811				Total	4744

$F = 4744/4811 = 0.986 < 1.0$

Table 7.5. Lower water level

Slice no.	z (m)	b (m)	W (kN)	h_w (m)	α (deg)	r_u	$\sin \alpha$	$W \sin \alpha$	Eqn (1)	$\sec \alpha$	$\tan \alpha$	Eqn (2)	$(1) \times (2)$
												$F=1.20$	$F=1.20$
1	6.0	13.0	1560	0.0	36	0.00	0.60	940	900	1.25	0.75	0.92	828
2	15.0	18.0	5400	4.0	27	0.13	0.45	2451	2712	1.12	0.51	0.90	2441
3	16.0	7.0	2240	5.0	19	0.16	0.33	729	1085	1.07	0.34	0.91	987
4	12.5	9.0	2250	9.0	13	0.16	0.23	505	1090	1.03	0.23	0.93	1015
5	7.5	9.5	1428	5.0	7	0.17	0.12	174	683	1.02	0.13	0.95	648
6	1.5	8.0	230	1.0	3	0.00	0.05	12	138	1.00	0.05	0.97	134
							Total	4811				Total	6053

$F = 6053/4811 = 1.26$

Case Study 7.1 (*contd.*)

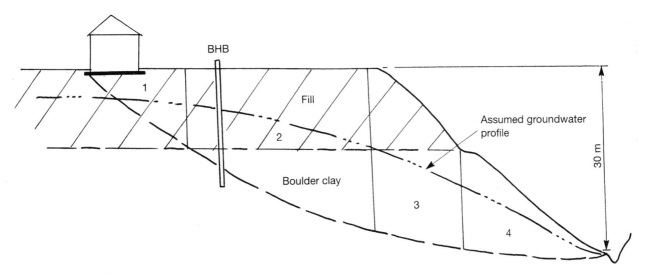

Fig. 7.9 Case study 7.1: plot 20, section 3.

Table 7.6.

Slice no.	z (m)	b (m)	W (kN)	h_w (m)	α (deg)	r_u	$\sin \alpha$	$W \sin \alpha$	Eqn (1)	$\sec \alpha$	$\tan \alpha$	Eqn (2)	$(1) \times (2)$
												$F=1.20$	$F=1.20$
1	7.0	16.0	2240	1.5	37	0.11	0.61	1344	1150	1.25	0.752	0.92	1056
2	20.0	30.0	12000	11.0	25	0.29	0.42	5076	4985	1.10	0.463	0.90	4498
3	22.0	14.0	6160	11.5	14	0.27	0.24	1491	2630	1.04	0.251	0.918	2416
4	8.0	24.0	3840	4.0	4	0.25	0.07	269	1662	1.00	0.071	0.97	1616
							Total	8180				Total	9585

$F = 9585/8180 = 1.171$

Case Study 7.1 (*contd.*)

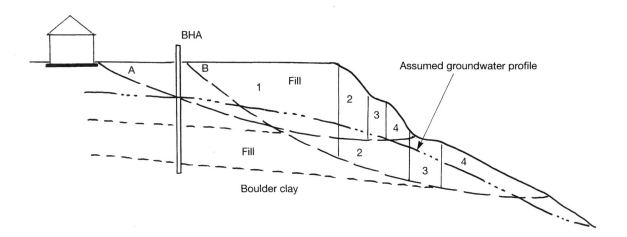

Fig. 7.10 Case study 7.1: plot 27, section 4, circle A.

Table 7.7. Circle A

Slice no.	z (m)	b (m)	W (kN)	h_w (m)	α (deg)	r_u	$\sin \alpha$	$W \sin \alpha$	Eqn (1)	$\sec \alpha$	$\tan \alpha$	Eqn (2)	$(1) \times (2)$
												$F=1.20$	$F=1.20$
1	8.5	44.0	7482	1.5	18	0.091	0.31	2310	3930	1.05	0.33	0.93	3666
2	10.0	4.5	898	1.0	5	0.052	0.09	79	490	1.0	0.08	0.968	477
3	7.0	3.0	419	0	3	0	0.05	22	240	1.0	0.05	0.98	237
4	3.0	5.0	301	0	2	0	0.03	12.0	175	1.0	0.03	0.99	170
							Total	2421				Total	4550

$F = 4550/2421 = 1.87$

Table 7.8. Circle B

Slice no.	z (m)	b (m)	W (kN)	h_w (m)	α (deg)	r_u	$\sin \alpha$	$W \sin \alpha$	Eqn (1)	$\sec \alpha$	$\tan \alpha$	Eqn (2)	$(1) \times (2)$
												$F=1.20$	$F=1.20$
1	10.0	28.0	5600	2.0	30	0.11	0.52	2880	2907	1.17	0.60	0.91	2640
2	12.5	12.0	3000	6.0	18	0.23	0.31	931	1317	1.10	0.33	0.91	1198
3	7.5	5.5	820	5.0	14	0.33	0.23	183	320	1.03	0.23	0.93	293
4	4.5	20.0	1805	1.0	6	0.11	0.11	188.0	925	1.01	0.11	0.96	890
							Total	4182				Total	5021

$F = 5021/4182 = 1.20$

Case Study 7.1 (*contd.*)

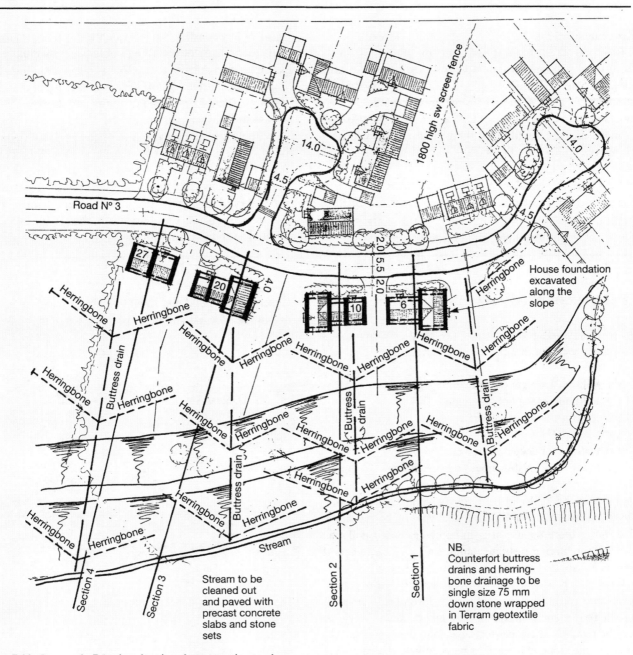

Fig. 7.11 Case study 7.1: plan showing slope-retention works.

Table 7.9. Case study 7.1: typical factors of safety for the varying sections

Section	Relevant borehole	Slip circle	Groundwater profile	Factor of safety
Plot 6 Section 1	C	As shown	High	< 1.0
Plot 10			Low	1.26
Plot 10	C	As shown	High	0.986 < 1.0
Plot 20	B	As shown	As shown	1.17
Plot 27	A	A	As shown	1.87
		B	As shown	1.20

Conclusions

Typical factors of safety for the varying sections are indicated in Table 7.9. From these results it can be seen that, generally, the overall stability of the site is reasonable for long-term drained conditions. However, at the sections taken for plots 6 and 10 the slope stability depends significantly on the position of the groundwater in the slope.

In borehole C a standing groundwater level was recorded at a depth of 6.0 m below ground level and this may be the typical case. If piezometer readings confirm these levels then instability exists in the slope and remedial measures need to be taken.

Case Study 7.1 (*contd.*)

Recommendations

Additional piezometers should be installed on the sections for plots 6 and 10 to establish and monitor the level of the groundwater more accurately. Once more reliable data are available, a more exacting analysis can be carried out. If unstable conditions still exist then the dwellings will have to be repositioned or the slopes stabilized by cutting them back to a safer angle.

In addition, the slope should be drained using deep counterfort buttress drains placed down the slope. Shallow herringbone drains should be used to collect the surface runoff. These drains should be installed several months ahead of any construction work on the dwellings. During the construction of the dwellings polythene sheeting can be pegged out on the top of the slope as a temporary measure to improve the surface runoff. The watercourse at the base of the slope should be cleared of debris and the base and sides should be lined with concrete slabs.

When constructing the dwellings the foundations will need to be taken through the upper fills into the firm natural clays below. The side wall foundations should be positioned across the slope contours and the front and rear walls should be placed on ground beams spanning from side to side. No deep excavations should be carried out parallel to the slope contours.

7.2 DEVELOPING ON SLOPING SITES

On steeply sloping sites there may be a risk of ground movement resulting from:

- the additional weight of the dwellings;
- the additional weight of any regrade fill;
- changes in the groundwater level or alterations to the surface runoff;
- excavations for deep drainage etc.;
- removal of any existing trees and vegetation.

On such sites a full and comprehensive site investigation should be carried out to determine the types of ground strata. This may require the use of deep boreholes and installation of water-monitoring instrumentation such as piezometers if there is a suspected stability problem.

Where such investigations reveal potential slip planes, great care must be exercised when designing a housing layout. To prevent a landslip developing, retaining structures may be required at critical sections along the site slopes. In addition, the slope may need to be positively drained using herringbone and deep counterfort drains to maintain any standing water at a safe depth and to prevent the build up of porewater pressures in the clay strata.

7.2.1 Additional weight of dwellings

See Fig. 7.12. Some slopes may exist for years in a state of incipient failure, i.e. on the point of slipping. Any man-made interference, such as adding extra weight on the crest, may precipitate movement. If houses are built too close to the top of a slope the additional weight could precipitate a slip circle failure if the slope stability is critical. There must be no risk of a slip, and an adequate factor of safety must be achieved. This may be achieved by simply positioning the houses a sufficient and safe distance from the top of the slope.

Generally, no buildings should be closer than a 1 in 3 line drawn from the base of the slope to the top of the slope.

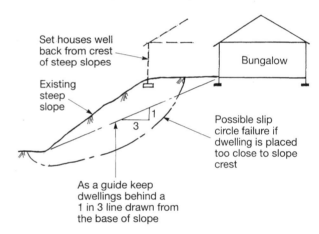

Fig. 7.12 Siting dwellings on a slope.

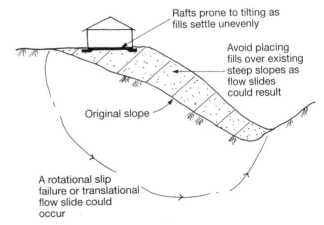

Fig. 7.13 Avoid placing fill on slopes.

7.2.2 Additional weight of regrade fill

See Fig. 7.13. Care must be exercised when placing fill materials on an existing slope as the natural drainage paths can close up and an increase in porewater pressure could precipitate a surface flow. In addition, the extra weight could trigger off a rotational slip. Before carrying out such works the slope stability should be examined and an adequate factor of safety should exist.

7.2.3 Changes in the groundwater level or surface runoff

See Fig. 7.14. Excavation of deep drainage at the base of a slope could result in a draw-down of the water table. In cohesionless strata this could cause settlement of buildings on or above the slope. Should any existing drainage become blocked a rise in the groundwater will occur and there could be a build up of porewater pressures in any clay soils, resulting in a slip (Fig. 7.15).

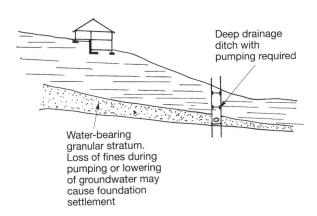

Fig. 7.14 Effects of lowering groundwater levels.

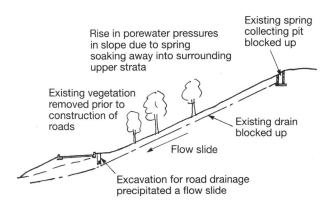

Fig. 7.15 Failure of slope at Chapel Hill, Pocklington, Yorkshire.

7.2.4 Excavations for deep drainage

Excavations for services at the toe of a slope or across a slope could precipitate a slope failure. There may be a weak soil layer along which upper strata could slide.

Removal of weight from the toe of the slope could give rise to a slip circle failure mode or a translational failure mode if the soils are weak.

7.2.5 Removal of trees and vegetation

See Fig. 7.16. When existing trees and shrubs are removed from a slope there will be a rise in the groundwater regime. In cohesive strata this could result in a strength reduction with a reduced factor of safety. Drainage of the slope is therefore a prudent measure.

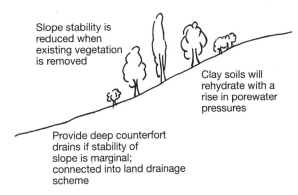

Fig. 7.16 Removing vegetation from existing slopes.

Excavations for foundations on a sloping site may result in parts of the dwelling being on different strata if foundations are level. In these circumstances it is usual to take the foundations on the top side of the slope down through any soft strata to found in similar strata (Fig. 7.17).

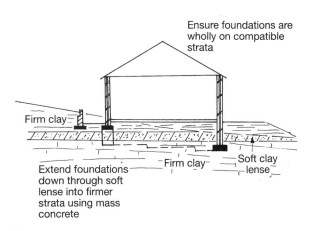

Fig. 7.17 Avoid changes in bearing strata.

Where raft foundations are being proposed on a sloping site because of weak bearing capacity, the formation for the raft should be benched into the slope in order that the stone thickness under the raft remains fairly uniform (Fig. 7.18).

147

Developing on sloping sites

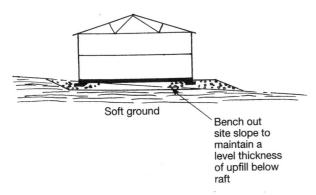

Fig. 7.18 Bench slopes to avoid differential settlements under rafts.

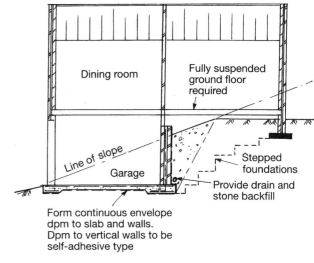

Fig. 7.19 Split-level development on steep slopes.

7.2.6 Split-level housing

See Fig. 7.19. Where a site is steeply sloping it is often advantageous to design split-level units on the site. This can be achieved by utilizing a retaining wall internally or by providing an external retaining wall separate from the dwelling. When considering this type of design the workmanship and construction detailing must be of a high standard to prevent water ingress.

IMPORTANT POINTS TO NOTE ON SPLIT-LEVEL HOUSING

1. Ensure that the vertical and horizontal damp-proof membranes form a complete envelope with adequate laps at joints. These membranes must not pass through the retaining wall; otherwise a slip plane will be created which will weaken the wall's resistance to sliding, and overall stability.
2. Provide adequate cavity trays at the top of any retaining walls if cavity wall construction is being used above. These should terminate at least 150 mm above the finished ground level.
3. Ensure that there is adequate subfloor ventilation to voided areas and adequate provision of drainage sumps to prevent a build up of water.
4. Most important: provide a land drain at the base of a retaining wall connected to the site drainage system.
5. Do not place soakaways on the high side of a split-level dwelling. All such drainage should be piped down to soakaways positioned at the lowest point of the site.
6. If using drained cavity construction ensure that the cavities are well ventilated.
7. Where possible, try to keep the natural slope of the land below ground-floor level, as this cuts down on expensive retaining walls. This may require the ground floors to be fully suspended.

7.3 RETAINING SYSTEMS

Retaining systems are generally required when the slope to be supported is at a steeper angle than its natural angle of repose. There are many types of retaining systems suitable for housing sites.

7.3.1 Gravity type retaining systems

This usually consists of mass brickwork, mass concrete, or a combination of both (see Fig. 7.20).

7.3.2 Cantilever walls: reinforced concrete or brickwork

This is usually the most economic retaining system. The masonry designs are usually concrete-filled cavity walls with the brick or block walling acting as a shutter. Bar or mesh reinforcement is cast in the base and the cavity is filled in heights not exceeding 900 mm. Generally, the two skins of brickwork are tied together using the twist-type fishtail tie. See Fig. 7.21. Alternatively, a reinforced concrete wall can be constructed using formwork.

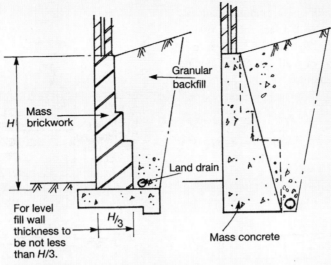

Fig. 7.20 Mass brick and concrete retaining walls.

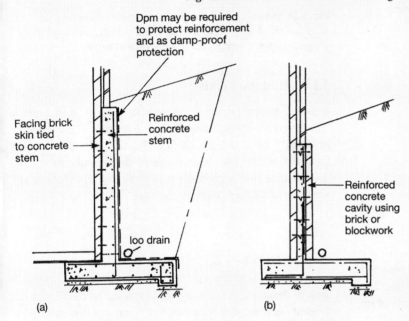

Fig. 7.21 Reinforced retaining walls: (a) reinforced concrete; (b) reinforced brickwork.

7.3.3 Gabions, crib walling and reinforced earth

(a) Gabion walls

These are usually constructed of wire-mesh baskets filled with fairly large stones just laid on a prepared formation. They are generally 2.0 m long by 1 m square in section and the retained heights are achieved by building up the units as shown in Fig. 7.22 to produce a front-face batter.

For increased durability the wire baskets can be plastic-coated or galvanized, and it is now possible to obtain baskets made out of high-density polyethylene polymers. The baskets have a central diaphragm in the length to enhance the stiffness and are wired together using 2.5 mm diameter steel wire. The base of a gabion wall is usually one half of the retained height.

(b) Crib walling

These are constructed using precast concrete segments or treated timber segments. They generally consist of a series of stretcher segments interlocked with transverse headers. The

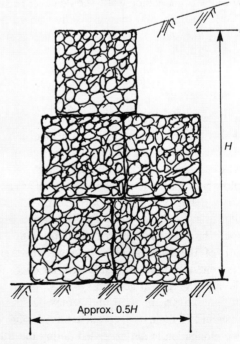

Fig. 7.22 Gabion walls.

149

space between is filled with free-draining granular material. The walls are usually constructed with a face incline of 1 in 6 and can be erected very quickly. See Fig. 7.23. Because of its flexible construction this type of wall can accommodate relatively large differential settlements.

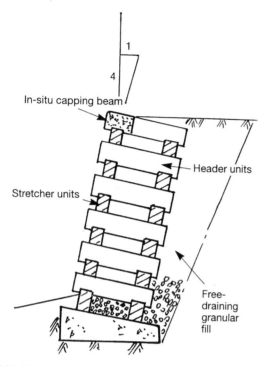

Fig. 7.23 Crib walls.

The concrete base to these walls is usually flat-bottomed with the top of the base the same angle as the wall face. The base width can vary from 0.50 *H* to 1.0 *H* and heights of up to 6.0 m can be built.

Important: Crib or gabion walls should never be subjected to surcharge loading, and masonry skins should never be used to face the units. Any bulging of the gabions will damage the masonry.

(c)   Geotextile walls

See Fig. 7.24. These are designed on the principle that a mass of soil can be given tensile resistance in a specific direction. The idea of using materials to strengthen soils and bricks was in use as far back as biblical times.

The pioneer of the modern technique known as **reinforced earth** was Vidal in 1966. The reinforced earth-retaining wall is designed as a gravity structure. The fill material behind the wall should be granular and free-draining, with not more than 10% of the material passing a 63 μm sieve. The polymers generally used are manufactured by Netlon Ltd under the name of Tensar geogrid systems. In selecting the grade of geotextile to be used, the geotextile strain of the material should be limited to a predetermined value in order that creep elongation is not exceeded during the life of the wall.

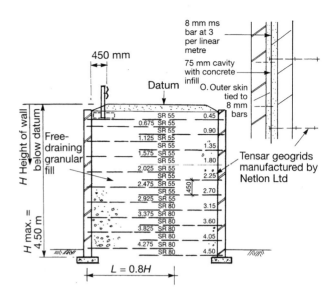

Fig. 7.24 Retaining wall using geotextile grids. General arrangement: double-sided wall up to 4.5 m high, showing Tensar geogrid reinforcement layout (type, level and length).

7.3.4   Steel sheet piling

This is generally used for retaining river banks using cantilevered sections or an anchored section. The anchorage is usually obtained by casting the tie rods into a mass concrete anchor block. Because of the amount of ground disturbed behind such walls it is generally wise to pile any structures which are close to the wall.

7.4   DESIGNING RETAINING WALLS

Before designing a retaining wall system, the stability of the ground mass must first be checked to ensure that there is no mechanism which could result in a global failure: i.e. a lateral slide due to a rise in groundwater levels.

Causes of failure in retaining walls:

- Failure due to an inadequate bearing capacity below the wall base.
- Failure by overturning. Though a wall may be bearing on rock, its overall stability must be checked out to ensure that there is an adequate factor of safety against overturning.
- Failure by sliding. Often a problem on clay soils; a downstand toe may be required to develop some passive resistance.
- Failure due to a complete rotational landslip.
- Failure resulting from inadequate design, bad workmanship or poor-quality materials.

7.4.1   Active pressure on walls

For cohesionless soils, applying Rankine's theory with retained surface level (Fig. 7.25):

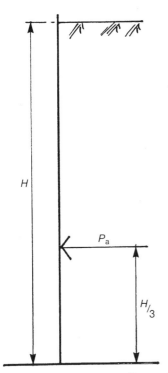

Fig. 7.25 Example 7.1: horizontal pressure diagram (level fill).

$$P_a = K_a \gamma \frac{H^2}{2}$$

$$K_a = \frac{1 - \sin \phi}{1 + \sin \phi}$$

where γ = soiled density; H = retained height of fill; ϕ = angle of shearing resistance of the soils.

Applying Rankine's theory with retained surface inclined β degrees (Fig. 7.26):

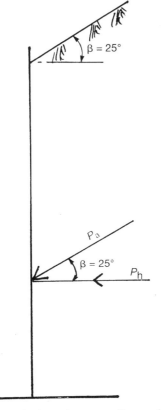

Fig. 7.26 Example 7.1: horizontal pressure diagram (sloping fill).

$$P_a = K_a \gamma \frac{H^2}{2}$$

$$\text{where } K_a = \cos \beta \left[\frac{\cos \beta - \sqrt{(\cos^2 \beta - \cos^2 \phi)}}{\cos \beta + \sqrt{(\cos^2 \beta - \cos^2 \phi)}} \right]$$

This pressure will act on the wall parallel to the surface of the soil: i.e. at an angle of β degrees to the horizontal. The horizontal component will therefore equal $P_a \cos \beta$.

Example 7.1   Thrust on a retaining wall

A retaining wall 3.0 m high retains sloping ground at an angle of 25°. The fill behind the wall is granular with an angle of shearing resistance $\phi = 35°$. Determine the horizontal thrust if (a) the fill behind the wall is level and (b) the soil is sloping at 25°.
Soil surface horizontal:

$$K_a = \frac{1 - \sin 35}{1 + \sin 35} = \frac{1 - 0.57}{1 + 0.57} = 0.273$$

Maximum active pressure:

$$P_a = \gamma H K_a = 19 \times 3 \times 0.273 = 15.56 \text{ kN/m}^2$$

$$\text{Total active pressure} = \frac{15.56 \times 3}{2} = 23.34 \text{ kN/m}$$

Soil surface sloping at 25°

$$
\begin{aligned}
K_a &= \cos 25 \left[\frac{\cos 25 - \sqrt{(\cos^2 25 - \cos^2 35)}}{\cos 25 + \sqrt{(\cos^2 25 - \cos^2 35)}} \right] \\
&= 0.906 \left[\frac{0.906 - \sqrt{(0.906^2 - 0.819^2)}}{0.906 + \sqrt{(0.906^2 - 0.819^2)}} \right] \\
&= 0.363
\end{aligned}
$$

Therefore

$$\text{Total active pressure} = \frac{0.363 \times 19 \times 3^2}{2} = 31.03 \text{ kN/m}$$

Horizontal thrust $= 31.03 \cos 25 = 31.03 \times 0.906 = 28.11$ kN/m

Figure 7.26 shows the line of thrust on the wall due to a 25° slope.

7.4.2   Surcharge loading

This is additional load imposed on the ground surface being retained by the retaining wall. These loads can be due to construction plant, vehicles, or stacked materials.

For a uniform load the designer may consider surcharge loads as equivalent to an extra height of fill to be retained. Therefore the equivalent height h_e is given by:

$$h_e = \frac{W_s}{\gamma}$$

when retained fills are level, where γ = unit weight of retained soil; W_s = intensity of surcharge loading/unit area. Total pressure due to surcharge loading, $P_s = K_a \gamma h_e$.

Example 7.2 Pressures and bending moments on a retaining wall

A vertical retaining wall is 3.0 m high and retains the ground below the car park area of a split-level bungalow. The density of the granular fill can be taken as 19 kN/m³. The angle of shearing resistance ø = 30°. The surcharge loading is to be taken as 5 kN/m². Determine the resultant pressures on the wall and the bending moment in the wall stem.

Applying Rankine's theory:

$$K_a = \frac{1 - \sin 30}{1 + \sin 30} = \frac{0.50}{1.50} = 0.33$$

Active pressure at base of wall:

$$P_a = 0.33 \times 19 \times 3.00 = 19.00 \text{ kN/m}^2$$

Surcharge pressure:

$$P_s = 0.33 \times 5.0 = 1.65 \text{ kN/m}^2$$

The pressure diagram is shown in Fig.7.27.

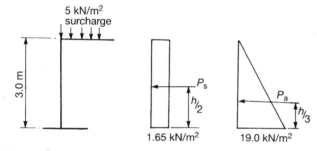

Fig. 7.27 Example 7.2: pressure diagram for surcharge conditions.

Total pressure $= p_s + p_a$

$p_s = 1.65 \times 3.0 = 4.95$ kN/m

$p_a = 19.00 \times \dfrac{3.0}{2} = 28.50$ kN/m

Total pressure $= 33.45$ kN/m

Bending moment on stem $= 4.95 \times \dfrac{3}{2} + 28.50 \times \dfrac{3}{3} = 35.92$ kN m

7.4.3 Passive resistance (granular soils)

The passive resistance P_p is given by

$$P_p = K_p \gamma d$$

where

$$K_p = \frac{1 + \sin \phi}{1 - \sin \phi} = \tan^2\left(45° + \phi/2\right)$$

and d is the depth of soil in front of the wall which can develop passive resistance.

On clay soils where ø = 0:

$$p_p = \frac{\gamma H}{K_2} + \frac{2c}{K_2} = \gamma H + 2c$$

where K_2 is a soil pressure coefficient dependent on ø, but c is best limited to 50 kN/m².

For a mixed soil with values of ø the Rankine formulae were developed by Bell and for horizontal surfaces in front of a wall

$$P_p = \gamma H \tan^2\left(45 + \text{ø}/2\right) + 2c \tan\left(45 + \text{ø}/2\right)$$

The main forces acting on a retaining wall are the horizontal pressures imposed by the retained materials and surcharge loads. Most designers assume a granular fill behind a wall with consideration given to any hydrostatic pressures if present. In firm clays this is a conservative approach as such clays can in theory stand vertical without support.

When designing retaining walls it is worth noting that the overturning moments are proportional to the fill height cubed. Double the fill height and the overturning moments increase eightfold.

The bending stress at any point on wall $= \pm \dfrac{M}{Z}$ where $Z = \dfrac{bd^2}{6}$

(The + and − signs indicate tension and compression.) Therefore

$$f_b = \frac{\dfrac{K_a \gamma H^3}{6}}{\dfrac{bd^2}{6}} = +\frac{K_a \gamma H^3}{bd^2}$$

This bending stress will produce tensile stresses on the earth face and compressive stresses on the internal face of the wall. There will also be a uniform compressive stress vertically due to the self-weight of the wall and other axial loads. These stresses combine to give

$$f_b = \frac{W}{A} \pm \frac{M}{Z} = \frac{W}{A} + \frac{M}{Z} \text{ or } \frac{W}{A} - \frac{M}{Z}$$

where W = total axial load above section being considered; A = cross-sectional area of section being considered; M = bending moment. It can be seen that if there is a high preload stress from axial loads above, which are equal to or larger than the bending stress, then no tension will develop in the wall.

7.5 CANTILEVERED RETAINING WALLS

These can be designed using various construction methods:

- mass brickwork or blockwork;
- reinforced cavity brick or block walls;
- reinforced pocket walls;
- reinforced concrete walls;
- brick walls with geotextile polymer grids.

7.5.1 Mass brick or block walls

Many builders prefer this type of wall as it is a type they are familiar with. They are generally expensive to construct; they can be built using a rule of thumb of the thickness being at least one third of the retained height for a level surface. They can also be designed with an adequate factor of safety to the requirements of BS 5628 Part 1 (1978).

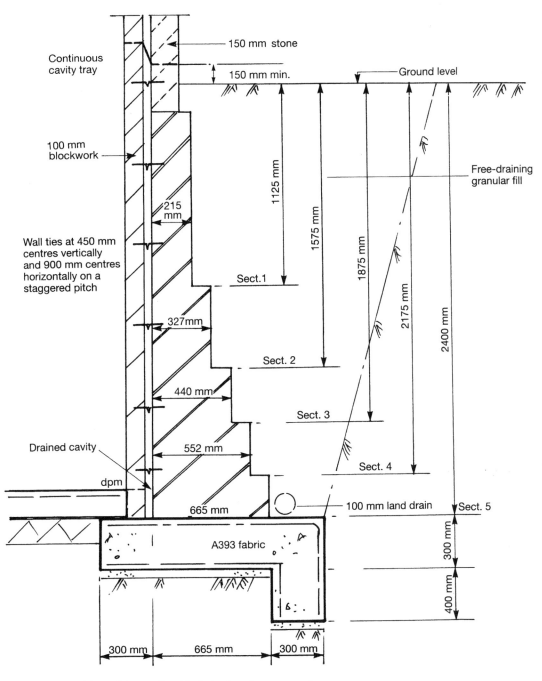

Fig. 7.28 Example 7.3: mass brick retaining walls with drained cavity construction.

Example 7.3 Thickness of retaining wall

A retaining wall (Fig. 7.28) is to be constructed in 25.0 N brickwork and is to retain earth for a height of 2.40 m. A wall above imposes a line load from the roof and first floor. Determine the wall thicknesses for the final condition carrying full axial load. Take density of fill = 19 kN/m^3 and $\varnothing$ = 35°, with density of brickwork taken as 22.0 kN/m^3. Assume a surcharge load of 1.50 kN/m^2. Use bricks with a porosity less than 7% and having a crushing strength greater than 25 N/mm^2 laid in a 1:3 cement:sand mortar. Concrete base to be 35 N/mm^2 mix with a minimum cement content of 300 kg/m^3. Base to be founded on a stratum with a safe bearing capacity of 150 kN/m^2.

INNER LEAF

Roof	= 2.0 × 2.25	=	4.50 (kN/m)
First floor	= 2.50 × 1.50	=	3.75
Self-weight of wall	= 1.80 × 5.50	=	9.90
			18.15

OUTER LEAF

Self-weight of wall = 3.50 × 2.80 = 9.80 kN/m

153

Section 1

$h = 1.125$ m $\qquad t = 215$ mm

$$M_d = \left(1.40 \times 0.27 \times 19 \times \frac{1.125^3}{6}\right) + \left(1.60 \times 0.27 \times 1.50 \times \frac{1.125^2}{2}\right)$$

$$= 2.114 \text{ kN m}$$

Design vertical load per unit area, g_a, is given by:

$$g_a = \frac{9.80 + (0.9 \times 22.0 \times 1.125 \times 0.215)}{215}$$

$$= \frac{9.80 + 4.78}{215} = \frac{14.58}{215} = 0.067 \text{ N/mm}^2$$

Section modulus $Z = \dfrac{1000 \times 215^2}{6} = 7.70 \times 10^6 \text{ mm}^3$

Design moment of resistance $= \dfrac{f_{kx}}{\lambda_{mv}} + g_a Z$

$$= \frac{0.70}{2.80} + 0.067 \times 7.70 \times 10^6$$

$$= 2.44 \text{ kN m}$$

where f_{kx} = characteristic flexural strength in bending from Table 3 BS 5628 Part 1 = 0.70 for mortar designation 1; λ_{mv} = partial safety factor for materials from Table 4 BS 5628 Part 1 = 2.80 for normal manufacturing and special construction control.

Shear $V = \left(1.40 \times 0.27 \times 19 \times \dfrac{1.125^2}{2}\right)$

$$+ (1.60 \times 0.27 \times 1.50 \times 1.125) = 5.275 \text{ kN}$$

Design shear stress $v_h = \dfrac{5.275}{215} = 0.025 \text{ N/mm}^2$

f_v, the characteristic shear strength of masonry, Clause 25 BS 5628 Part 1 = $0.35 + 0.60\, g_a \text{N/mm}^2$ up to a maximum of 1.75 N/mm² for walls built using mortar designation (1). λ_{mv} = 2.50 for shear loads (Clause 27.4, BS 5268). Therefore v_h must be less than

$$\frac{f_v}{\lambda_{mv}} = \frac{0.35 + 0.60 \times 0.067}{2.50} = 0.156 \text{ N/mm}^2$$

Section 2

$h = 1.575$ m $\qquad t = 327$ mm

$$M_d = \left(1.40 \times 0.27 \times 19 \times \frac{1.575^3}{6}\right) + \left(1.60 \times 0.27 \times 1.50 \times \frac{1.575^2}{2}\right)$$

$$= 5.47 \text{ kN m}$$

$$g_a = \frac{14.58 + (0.9 \times 22.0 \times 0.45 \times 0.327)}{327}$$

$$+ \frac{(1.125 \times 0.112 \times 19) + (0.9 \times 1.50 \times 0.112)}{327}$$

$$= \frac{14.58 + 2.913 + 2.394 + 0.151}{327}$$

$$= \frac{20.03}{327} = 0.061 \text{ N/mm}^2$$

Section modulus $Z = \dfrac{1000 \times 327^2}{6} = 17.822 \times 10^6 \text{ mm}^3$

Moment of resistance $= \left(\dfrac{0.7}{2.8} + 0.061\right) \times 17.822 = 5.54 \text{ kN m}$

$$V = \left(1.40 \times 0.27 \times 19 \times \frac{1.575^2}{2}\right) + (1.60 \times 0.27 \times 1.50 \times 1.575)$$

$$= 9.93 \text{ kN}$$

$$v_h = \frac{9.930}{327} = 0.03 \text{ N/mm}^2$$

and is $\leqslant \dfrac{f_v}{\lambda_{mv}} = \dfrac{0.35 + 0.60 \times 0.061}{2.50} = 0.15 \text{ kN/mm}^2$

Section 3

$h = 1.875$ m $\qquad t = 440$ mm

$$M_d = \left(1.40 \times 0.27 \times 19 \times \frac{1.875^3}{6}\right) + \left(1.60 \times 0.27 \times 1.50 \times \frac{1.875^2}{2}\right)$$

$$= 9.03 \text{ kN m}$$

$$g_a = \frac{20.03 + (0.9 \times 22.0 \times 0.30 \times 0.44)}{440}$$

$$+ \frac{(0.9 \times 1.575 \times 0.112 \times 19) + (0.9 \times 1.50 \times 0.112)}{440}$$

$$= \frac{20.03 + 2.61 + 3.0 + 0.15}{440} = \frac{25.79}{440} = 0.058 \text{ N/mm}^2$$

Section modulus $Z = \dfrac{1000 \times 440^2}{6} = 32.27 \times 10^6 \text{ mm}^3$

Moment of resistance $= \left(\dfrac{0.7}{2.80} + 0.058\right) \times 32.27 = 9.96 \text{ kN m}$

$$V = \left(1.40 \times 0.27 \times 19 \times \frac{1.875^2}{2}\right) + (1.60 \times 0.27 \times 1.50 \times 1.875)$$

$$= 13.84 \text{ kN}$$

$$v_h = \frac{13.84}{440} = 0.031 \text{ N/mm}^2$$

and is $\leqslant \dfrac{f_v}{\lambda_{mv}} = \dfrac{0.35 + 0.60 \times 0.058}{2.50} = 0.153 \text{ N/mm}^2$

Section 4

$h = 2.175$ m $\qquad t = 552$ mm

$$M_d = \left(1.40 \times 0.27 \times 19 \times \frac{2.175^3}{6}\right) + \left(1.60 \times 0.27 \times 1.50 \times \frac{2.175^2}{2}\right)$$

$$= 13.85 \text{ kN m}$$

$$g_a = \frac{25.79 + (0.9 \times 22.0 \times 0.30 \times 0.552)}{552}$$

$$+ \frac{(0.9 \times 1.875 \times 0.112 \times 19) + (0.9 \times 1.50 \times 0.112)}{552}$$

$$= \frac{25.79 + 3.27 + 3.59 + 0.15}{552} = \frac{32.80}{552} = 0.059 \text{ N/mm}^2$$

Section modulus $Z = \dfrac{1000 \times 552^2}{6} = 50.78 \times 10^6 \text{ mm}^3$

Moment of resistance $= \left(\dfrac{0.7}{2.80} + 0.059\right) \times 50.78 = 15.69 \text{ kN m}$

$$V = \left(1.40 \times 0.27 \times 19 \times \frac{2.175^2}{2}\right) + (1.60 \times 0.27 \times 1.50 \times 2.175)$$

$$= 18.39 \text{ kN m}$$

$$v_h = \frac{18.39}{552} = 0.033 \text{ N/mm}^2$$

and is $\leqslant \dfrac{f_v}{\lambda_{mv}} = \dfrac{0.35 + 0.60 \times 0.059}{2.50} = 0.154 \text{ N/mm}^2$

Section 5

$h = 2.4$ m $t = 665$ mm

$$M_d = \left(\frac{1.40 \times 0.27 \times 19 \times 2.4^3}{6}\right) + \left(\frac{1.60 \times 0.27 \times 1.5 \times 2.4^2}{2}\right)$$

$$= 18.41 \text{ kN m}$$

$$g_a = \frac{32.80 + (0.9 \times 22.0 \times 0.225 \times 0.665)}{665}$$

$$+ \frac{(0.9 \times 19 \times 0.112 \times 2.175) + (0.9 \times 1.50 \times 0.112)}{665}$$

$$= \frac{32.80 + 2.96 + 4.16 + 0.15}{665} = \frac{40.07}{665} = 0.060 \text{ N/mm}^2$$

Section modulus $Z = \dfrac{1000 \times 665^2}{6} = 73.70 \times 10^6 \text{ mm}^3$

Moment of resistance $= \left(\dfrac{0.7}{2.80} + 0.060\right) \times 73.70 = 22.84$ kN m

$$V = \left(1.40 \times 0.27 \times 19 \times \frac{2.40^2}{2}\right) + (1.60 \times 0.27 \times 1.50 \times 2.4)$$

$$= 22.24 \text{ kN}$$

$$v_h = \frac{22.24}{665} = 0.033 \text{ N/mm}^2$$

and is $\le \dfrac{f_v}{\lambda_{mv}} = \dfrac{0.35 + 0.60 \times 0.060}{2.50} = 0.154 \text{ N/mm}^2$

OVERALL STABILITY

Overturning moment (about front toe of base)

$$\left(0.27 \times 19 \times \frac{2.7^3}{6}\right) + \left(0.27 \times 1.50 \times \frac{2.7^2}{2}\right) = 18.305 \text{ kN m}$$

Restoring moment

		(kN)	Lever arm, L_a (m)	(kN m)
Wall plus earth		46.33	0.767	35.48
Surcharge	1.50×0.75	1.125	0.89	1.0
Base	$24 \times 0.3 \times 1.265$	9.18	0.6325	5.66
Heel	$24 \times 0.3 \times 0.3$	2.125	1.116	2.36
		58.76		44.50
Inner leaf		18.15	0.20	3.63
Outer leaf		9.80	0.375	3.67
		86.71		51.80

Factor of safety against overturning

(temporary construction condition) $= \dfrac{44.50}{18.305} = 2.43 > 1.50$

therefore OK

Factor of safety (with superstructure complete) $= \dfrac{51.80}{18.305} = 2.82$

Eccentricity, $e = \dfrac{18.305}{86.71} + \dfrac{1.265}{2} - \dfrac{51.80}{86.71}$

$e = 0.211 + 0.632 - 0.597 = 0.246$ m

This is outside the middle third. Therefore

$$P_{max} = \frac{4W}{3(d - 2e)}$$

$$= \frac{4 \times 86.71}{3 \times (1.265 - 2 \times 0.246)} = 149.50 \text{ kN/m}^2$$

$$\text{Sliding force} = \frac{0.27 \times 19 \times 2.70^2}{2.0} + 0.27 \times 1.50 \times 2.70 = 19.79 \text{ kN}$$

RESISTANCE TO SLIDING

As bearing stratum is cohesive and $\o = 0$, the passive resistance $= \gamma h + 2c$. Adopt a maximum value for c of 50 kN/m². Therefore, at top of shear key:

$$P_p = + 2c = 2 \times 50 = 100 \text{ kN/m}^2$$

At bottom of shear key:

$$P_p = \gamma h + 2c = 0.3 \times 18 + 2 \times 50 = 105.40 \text{ kN/m}^2$$

Total passive resistance on shear key:

$$P_p = \frac{(100 + 105.40) \times 0.3}{2} = 30.81 \text{ kN}$$

Factor of safety against sliding $= \dfrac{30.81}{19.79} = 1.55$

With shear key 400 mm deep, $P_p = 41.44$ kN. Therefore

Factor of safety $= \dfrac{41.44}{19.79} = 2.09 > 2.0$

The passive resistance of the base depth has been ignored in these calculations as soil in front of similar bases can sometimes be removed.

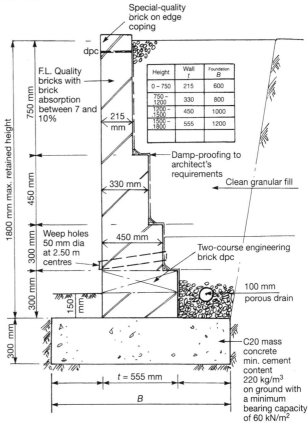

Fig. 7.29 Mass brick retaining wall for surcharge loads of 3.0 kN/m².

Developing on sloping sites

7.5.2 Reinforced cavity walls

Because of their low tensile strength, mass masonry walls are not very economical as retaining walls. However, masonry walls can be reinforced to increase their tensile resistance, resulting in substantial increases in bending strength. The reinforcement can be used within a concrete-filled cavity or within a cast-in concrete column in which pockets have been built by the bricklayer at specified design centres.

7.5.3 Pocket-type walls

Brickwork or blockwork is reinforced to resist lateral loading; the main reinforcing bars are concentrated into concrete pockets formed in the tension face of the retaining wall. The effective depth of the wall stem is measured from the compression face of the wall to the centroid of the tensile reinforcement. The design is based on BS 5628 Part 2, which is based on limit-state design criteria.

In pocket walls the pockets are usually 215 mm wide by 102 mm deep and are filled with a concrete mix using a 20 mm aggregate. A grade 35 concrete with a minimum cement content of 300 kg/m^3 is generally specified. The foundation base is designed in accordance with BS 8110 using a grade 35 concrete mix and high-yield steel reinforcement. The brickwork is laid in 1:4 sand:cement mortar with a minimum crushing strength of 20–35 N/mm^2, with an approved plasticizer. The backfill behind the retaining wall is DOT Type 1 sub-base. The ratio of wall height to effective depth should always be less than 18 to ensure that serviceability requirements are met. Pockets are usually spaced at 1.20–1.50 m centres.

Example 7.4 Pocket-type retaining wall

This example uses BS 5628: Part 2: 1985
A pocket-type retaining wall 3.0 m high retains a granular soil. The wall thickness is 330 mm and the pockets are 1.50 m apart.

$$\frac{\text{Height}}{\text{Effective depth}} = \frac{3 \times 10^3}{290} = 10.34 \quad < 18$$

Therefore serviceability OK

Assume angle of internal friction $\phi = 35°$.

$$K_a = \frac{1 - \sin 35}{1 + \sin 35} = \frac{1 - 0.57}{1 + 0.57} = 0.27$$

$$\gamma_w = 18 \text{ kN/m}^3$$

$$P_a = 0.27 \times 18 \times \frac{3.0^2}{2} = 21.87 \text{ kN/metre run}$$

Let $\tau_f = 1.40$.

Design shear force/pocket = $21.87 \times 1.50 \times 1.40 = 45.92$ kN

Design bending moment/pocket = $45.92 \times \dfrac{3.0}{3.0} = 45.92$ kN m

Using bricks with a crushing strength of 35 N/mm^2 and 1:4 mortar, $\tau_{mm} = 2.3$. Distance between pockets, $b = 1.50$ m. t_f = effective thickness between pockets = 0.5 d. Assume $d = 330 - 40 = 290$ mm.

Therefore $t_f = 0.5 \times 290 = 145$ mm. From BS 5628 Table 3, $f_k = 11.40$ N/mm^2 for mortar designation (1). Checking minimum wall strength:

$$f_k > \frac{M\tau_{mm}}{bt_f(d - 0.5t_f)} = \frac{45.92 \times 10^6 \times 2.30}{1500 \times 145 \times (290 - 145/2)} = 2.23 \text{ N/mm}^2$$

Required steel area/pocket:

Assume $z = 0.9d = 0.9 \times 290 = 261$ mm

$$A_s = \frac{M\tau_{ms}}{f_y z} = \frac{45.92 \times 10^6 \times 1.15}{460 \times 261} = 439 \text{ mm}^2$$

Substituting into

$$z = d\left(1 - 0.5 \frac{A_s f_y \tau_{mm}}{bdf_k \tau_{ms}}\right)$$
$$= 290\left(1 - \frac{0.5 \times 439}{1500 \times 290} \times \frac{460}{11.4} \times \frac{2.30}{1.15}\right) = 279 \text{ mm}$$

This is greater than 0.9 d, therefore assume $z = 0.9\, d = 0.95 \times 290 = 275.5$ mm.

$$A_s = \frac{M\tau_{ms}}{f_y z} = \frac{45.92 \times 10^6 \times 1.15}{460 \times 275.50} = 416 \text{ mm}^2$$

Therefore provide one T16 + one T20 bar in each pocket (515 mm^2). Using Table 10 (BS 5628: Part 2):

$$\frac{f_k}{\tau_{mm}} = \frac{11.4}{2.30} = 4.95 \text{ and } C = 0.95$$

Therefore $Q = 0.47$

Moment of resistance $M_d = \dfrac{0.47 \times 1500 \times 290^2}{10^6} = 59.30$ kN m

Minimum percentage $= \dfrac{0.10}{100} \times 1500 \times 290 = 435$ mm^2

Therefore provide one T16 + one T20 (515 mm^2). See Fig. 7.30.

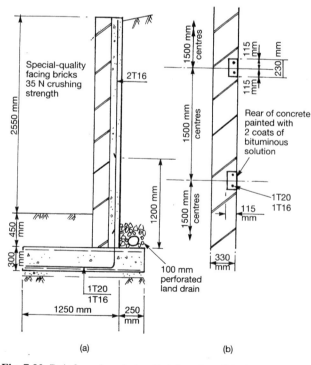

Fig. 7.30 Reinforced pocket wall: (a) section; (b) plan.

Check wall spanning between pockets (BS 5628 Part 1 applies, as unreinforced). Assume $f_k = 11.4/3.0$ for perforated bricks; $\tau_{mm} = 3.50$. Distance between pockets $= 1500 - 230 = 1270$ mm.

$$q_{lateral} = \frac{f_k}{\tau_{mm}}\left(\frac{t}{L}\right)^2 = \frac{11.4\times(330)^2}{3.5\times3.0(1270)} = 0.07 \text{ N/mm}^2 = 70 \text{ kN/m}^2$$

Maximum lateral pressure at base of wall $= 0.27 \times 18 \times 3.0 = 14.60$ kN/m². With $\tau_f = 1.40$ the maximum design pressure is $14.60 \times 1.40 = 20.4$ kN/m². This is well within the wall's capacity.

$$\text{Shear stress} = \frac{45.92\times10^3}{1500\times290} = 0.105 \text{ N/mm}^2$$

$$p = \frac{515}{1500\times290} = 0.00118 \quad \text{Shear span } a = \frac{45.92}{45.92} = 1.0$$

$$\frac{a}{d} = \frac{1\times10^3}{290} \qquad \text{Therefore } \frac{a}{d} = 3.44$$

The characteristic shear strength is given by $0.35 + 17.5p = 0.37$ N/mm², as indicated in BS 5628: Part 2, Clause 19.1.3.1.2. The shear span ratio multiplier is $2.50 - (0.25 \times 3.44) = 1.64$. Thus

$$f_v = 0.37 \times 1.64 = 0.60$$

and

$$\frac{f_v}{\tau_{mv}} = \frac{0.60}{2.0} = 0.30$$

Thus $f_v/\tau_{mv} > v$ (0.30 > 0.105), therefore wall stem is adequate for shear.

Example 7.5 Reinforced cavity retaining wall

See Fig. 7.31. A retaining wall 2.60 m high is to be constructed using a reinforced concrete-filled cavity 100 mm wide. Facing bricks with a minimum crushing strength of 35 N/mm² are to be used for the wall front face with engineering bricks on the rear face, all laid in a 1:4 sand:cement mortar. The two brick skins are to be tied with stainless steel fishtail ties at 450 mm centres vertically and 900 mm centres horizontally. The cavity is to be filled with Grade 35 concrete using a 10 mm aggregate. Design a suitable foundation base and stem reinforcement.

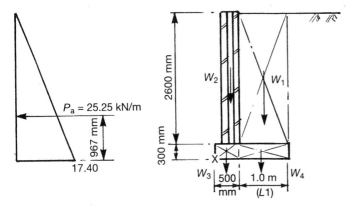

Fig. 7.31 Example 7.5: active pressure diagram.

For estimating purposes consider the base width to be 0.60 times the retained height which equals $0.60 \times 2.60 = 1.560$ m. Make the toe projection from the rear face of the wall equal to one third of the base width. Density of stone fill $= 18$ kN/m³. Density of concrete $= 24$ kN/m³. Density of wall stem $= 20$ kN/m³. Angle of Int. Friction $\phi = 30°$.

Using Rankine's formulae: Active pressure coefficient,

$$k_2 = \frac{1-\sin 30}{1+\sin 30} = \frac{0.50}{1.50} = 0.33$$

Design for a factor of safety of 2.0 against sliding. Coefficient of friction between base and ground $= \tan\phi = 0.57$.

Force causing sliding:

$$P_a = k_2\gamma_q \times \frac{H^2}{2} = 0.33\times18\times\frac{2.90^2}{2} = 25.25 \text{ kN/m}$$

Elements:

$W_1 = 18 \times 2.60 \times L_1 \quad = 46.80\,L_1$ kN/m

$W_2 = 20 \times 2.60 \times 0.30 \quad = 15.60$ kN/m

$W_3 = 24 \times 0.50 \times 0.30 \quad = 3.60$ kN/m

$W_4 = 24 \times 0.30 \times L_1 \quad = 7.20\,L_1$ kN/m

Therefore

$$25.25\times2.0 = (46.8L_1 + 7.20L_1 + 15.60 + 3.60)\times0.57$$
$$L_1 = \frac{50.50-10.94}{30.78} = 1.28 \text{ m}$$

therefore make $L_1 = 1.00$ m

Taking moments about X:

$$\text{Overturning moment} = 25.25\times\frac{2.90}{3.0} = 24.40 \text{ kN m}$$

Righting moments:

Element	Total load, W (kN)		Lever arm (m)	Moment (kN m)
W_1	$46.80\times1.0 =$	46.80	1.0	46.80
W_2		15.60	0.35	5.46
W_3		3.60	0.25	0.90
W_4	$7.20\times1.0 =$	7.20	1.0	7.20
	Total $=$	73.20		60.36

Therefore factor of safety against overturning

$$= \frac{60.36}{24.40} = 2.47 > 1.50, \text{ therefore OK.}$$

GROUND BEARING PRESSURE

Total vertical load $= 73.20$ kN. Net moment about X $= 60.36 - 24.40 = 35.96$. Therefore resultant acts at $35.96/73.20 = 0.490$ m from X. Therefore eccentricity on base $= 0.75 - 0.49 = 0.26$ m. This is just outside the middle third and some small tension will develop on the heel. The ground pressure contact length $= 3 \times 490 = 1470$ mm.

$$\text{Average pressure} = \frac{73.20}{1.0\times1.47} = 49.79 \text{ kN/m}^2$$

Maximum edge pressure (Fig. 7.32) $= 2 \times 49.79 = 99.59$ kN/m²

This is acceptable for a firm clay stratum.

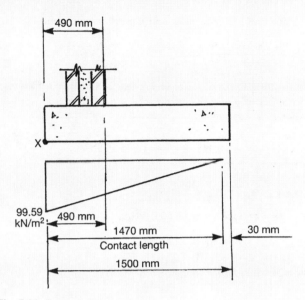

Fig. 7.32 Example 7.5: base pressures.

STEM REINFORCEMENT (FIG. 7.33)

Ultimate moment $= 0.33 \times 18 \times \dfrac{2.60^3}{6} \times 1.40 = 24.36 \text{ kN m}$

With mortar designation (ii), f_k for brick = 9.40; τ_{mm} = 2.30; τ_{ms} = 1.15; f_y = 460 N/mm². $M_d = Qbd^2$. Therefore

$$Q = \frac{24.36 \times 10^6}{10^3 \times 155^2} = 1.013$$

From Table 10:

$$\frac{f_k}{\tau_{mm}} = \frac{9.60}{2.30} = 4.0 \quad \text{Therefore } c = 0.85$$

Therefore

$$A_s = \frac{24.36 \times 10^6 \times 1.15}{460 \times 155 \times 0.84} = 467 \text{ mm}^2$$

Use T10 bars at 150 mm centres for up to 1.0 m above base with A252 fabric mesh lapped on for remainder.

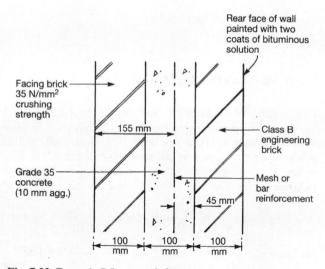

Fig. 7.33 Example 7.5: stem reinforcement.

158

Base reinforcement:

Downward moments $= (46.8 + 7.20) \times 1.40 \times \dfrac{1.0^2}{2.0} = 37.80 \text{ kN m}$

Upward moments $= 65 \times 1.40 \times \dfrac{0.97}{2} \times \dfrac{0.97}{3} = 14.30 \text{ kN m}$

Resultant moment = 37.80 − 14.30 = 23.50 kN m
Using BS 8110:

$$\frac{M}{bd^2 f_{cu}} = \frac{23.50 \times 10^6}{10^3 \times 255^2 \times 35} = 0.01$$

Therefore

$$A_s = \frac{M}{0.87 f_y l_a d} = \frac{23.50 \times 10^6}{0.87 \times 460 \times 0.95 \times 255} = 242 \text{ mm}^2$$

Minimum percentage $= \dfrac{0.13}{100} \times 10^3 \times 255 = 332 \text{ mm}^2$

Provide T 10 mm bars at 200 mm centres in top of base each way with T10 bars out of base at 150 mm centres into wall stem (Fig. 7.34). Provide A142 mesh in bottom of base.

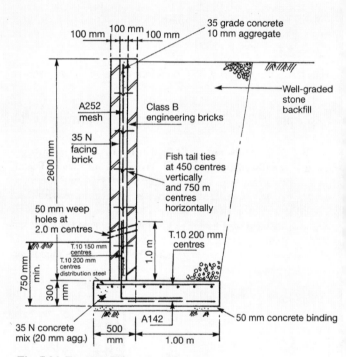

Fig. 7.34 Example 7.5: typical section.

Example 7.6 Concrete filled cavity retaining wall

A retaining wall is to be constructed up against the boundary of adjoining property. The wall will have to be designed without a heel slab. The most stable retaining wall is one which has the majority of its base supporting the retained soils. In these circumstances the major problem is coping with the sliding forces on the wall such that an adequate factor of safety of 2 is obtained. Generally, it is prudent to provide a substantial shear key to develop sufficient passive resistance. Shear keys are generally placed at the rear of walls since the bearing pressures in these positions are of a low order and excavating the key can sometimes cause localized ground disturbance.

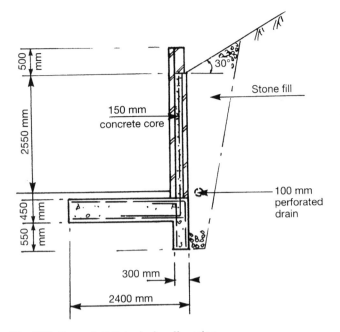

Fig. 7.35 Example 7.6: typical wall section.

The wall will be designed as a concrete-filled cavity wall in accordance with BS 5628 Parts 1 and 2 (1978) and BS 8110 Part 1 (1985). Figure 7.35 shows a typical wall section.

The retaining wall holds back a 30° slope. Angle of internal friction $\phi = 35°$; $\gamma_w = 20$ kN/m^3; coefficient of friction = 0.50; allowable ground bearing pressure = 150 kN/m^2.

$$k_a = \cos 30 \, \frac{\cos 30 - \sqrt{(\cos^2 30 - \cos^2 35)}}{\cos 30 + \sqrt{(\cos^2 30 - \cos^2 35)}}$$

$$= 0.866 \times \frac{0.866 - \sqrt{(0.866^2 - 0.819^2)}}{0.866 + \sqrt{(0.866^2 - 0.819^2)}} = \frac{0.866 \times 0.585}{1.147} = 0.44$$

Passive resistance coefficient $= \dfrac{1 + \sin 35}{1 - \sin 35} = \dfrac{1 + 0.57}{1 - 0.57} = 3.65$

OVERTURNING MOMENT

$$20 \times 0.44 \times \frac{3.0^3}{6.0} = 39.60 \text{ kN m}$$

Element	Weight (kN)	L_a (m)	Moment (kN/m)
Wall W_s	$8 \times 3.2 \times 1.0$ = 26	2.23	57.98
Base W_b	$2.4 \times 0.45 \times 24$ = 26	1.20	31.20
	Total W = 52		Total moment = 89.18

Factor of safety against overturning $= \dfrac{89.18}{39.60} = 2.25 > 1.50$

Resultant $W = \dfrac{89.18 - 39.60}{52.0} = 0.95$ m

Therefore eccentricity = 1.20 – 0.95 = 0.25 which is within the middle third

Ground pressure $= \dfrac{52.0}{1.0 \times 2.40}\left(1 \pm \dfrac{6 \times 0.25}{2.40}\right)$

$\qquad = 21.66 \pm 13.54 = 35.22$ kN/m^2 to 8.12 kN/m^2

(Fig. 7.36)

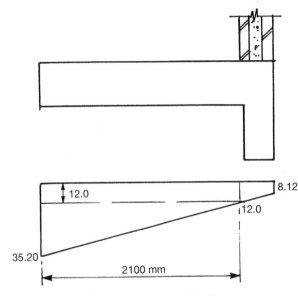

Fig. 7.36 Example 7.6: base pressures (kN/m^2).

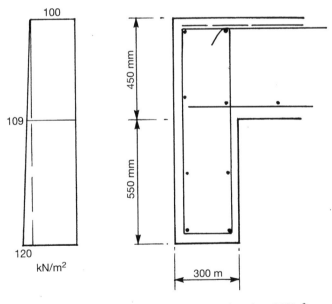

Fig. 7.37 Example 7.6: passive pressures on shear key (kN/m^2).

CHECK SLIDING

Ignore friction under base as bearing strata is cohesive. $P_p = \gamma H + 2c$ where c is limited to 50 kN/m^2. Therefore with 1.0 m deep shear key:

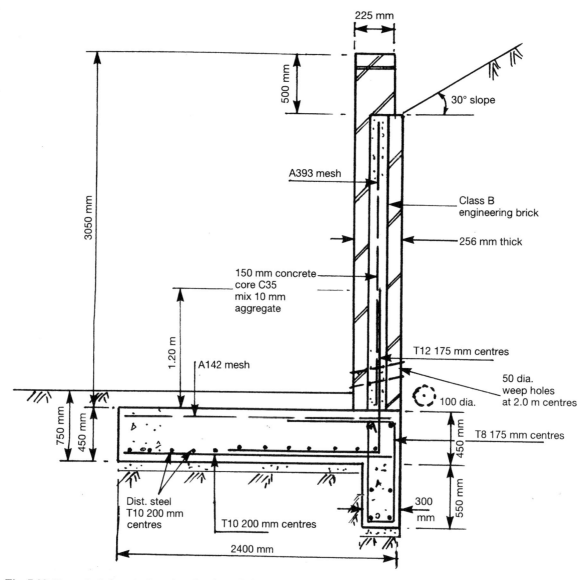

Fig. 7.38 Example 7.6: typical section showing reinforcement.

P_p top $= 2c = 100$ kN/m²

P_p bottom $= 20 \times 1.0 + 2 \times 50 = 120$ kN/m²

Total passive resistance $= \dfrac{120 + 100}{2.0} \times 1.0 = 110$ kN

Horizontal force $= 20 \times 0.44 \times \dfrac{3.0^2}{2.0} = 39.60$ kN

Factor of safety against sliding $= \dfrac{110}{39.60} = 2.77 > 2.0\,(\text{OK})$

BENDING MOMENT IN STEM

Use a partial safety factor of 1.40 for earth pressures.

Ultimate moment $= 20 \times 1.40 \times 0.44 \times \dfrac{2.55^3}{6.0} = 34$ kN m

f_k for brickwork $= 9.40$; $\tau_\text{mm} = 2.30$; $\tau_\text{mm} = 1.15$; $f_\text{y} = 460$ N/mm².

Moment of resistance $= Qbd^2 \dfrac{34.0 \times 10^6}{10^3 \times 175^2} = 1.10$

$\dfrac{f_\text{k}}{\tau_\text{mm}} = \dfrac{9.4}{2.3} = 4.0$

Therefore

$C = 0.83$ (BS 5628 Table 10)

$A_\text{s} = \dfrac{34.0 \times 10^6 \times 1.15}{460 \times 0.83 \times 175} = 585$ mm²

Provide T12 mm bars at 12 bars at 175 mm centres.
  Check 1.20 m above base.

Ultimate moment $= 20 \times 1.40 \times 0.44 \times \dfrac{1.35^3}{6.0} = 5.0$ kN m

$Q = \dfrac{5.0 \times 10^6}{10^3 \times 175^2} = 0.163$ Therefore $C = 0.95$

$A_\text{s} = \dfrac{5.0 \times 10^6 \times 1.15}{460 \times 0.95 \times 175^2} = 65$ mm²

Provide A395 fabric mesh in wall to provide minimum percentage of 0.05% distribution steel (see Fig. 7.33).

Base reinforcement

Upward moments:

$$12.0 \times \frac{2.046^2}{2.0} \qquad = +25.11$$

$$23.20 \times \frac{2.046^2 \times 2}{2.0 \times 3.0} \qquad = +32.37$$

$$\overline{\qquad\qquad +57.48\,\text{kN m}}$$

Downward moments:

$$0.45 \times 24 \times \frac{2.046^2}{2.0} \qquad = -22.60$$

$$\overline{\text{Unfactored base moment} \quad = +34.88\,\text{kN m}}$$

Ultimate moment $= 1.40 \times 34.88 = 48.83$ kN m

$$A_s = \frac{48.83 \times 10^6}{460 \times 0.87 \times 0.95 \times 400} = 321 \text{ mm}^2$$

Minimum percentage $= 0.13\% = \dfrac{0.13 \times 10^3 \times 400}{100} = 520 \text{ mm}^2$

Therefore provide T12 mm bars at 200 mm centres in bottom each way with A142 mesh in top.

SHEAR KEY REINFORCEMENT

See Fig. 7.37.

$$\text{Moment} = 109 \times \frac{0.55^2}{2.0} + 11 \times \frac{0.55^2 \times 2.0}{2.0 \times 3.0} = 17.58 \text{ kN m}$$

Ultimate moment $= 1.40 \times 17.58 = 24.60$ kN m

$$A_s = \frac{24.60 \times 10^6}{460 \times 0.87 \times 0.95 \times 250} = 258 \text{ mm}^2$$

Therefore provide T8 at 175 mm cts. in shear key in link form.

Example 7.7   Brick retaining wall

A retaining wall is to be designed to retain granular fill with a slope of 25° to the horizontal. The foundation bearing stratum is a firm shaley clay with an allowable bearing presure of 100 kN/m². The wall is to be constructed using engineering bricks and high-quality facing bricks with a minimum crushing strength of 35 N/mm². Mortar to be 1:4 cement : sand. Base concrete to be 35 N concrete using 20 mm aggregate. Cavity infill concrete to be 35 N strength using 10 mm aggregate. Wall ties to be galvanized solid fishtail type placed at 450 mm centres vertically and 900 mm centres horizontally on a staggered pitch. Movement joints to be provided at 12 m maximum centres. Allow for surcharge loading of 1.50 kN/m². $\beta = 25°$; $\phi = 35°$; $\gamma_w = 20$ kN/m³; Retained height of fill = 1.50 m.

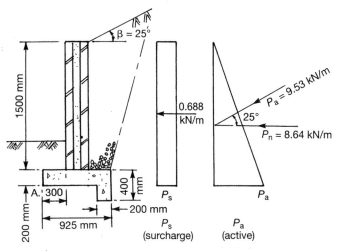

Fig. 7.39 Example 7.7: active pressure diagram.

OVERTURNING MOMENT ON BASE

$$k_2 = \cos 25 \; \frac{\cos 25 - \sqrt{(\cos^2 25 - \cos^2 35)}}{\cos 25 + \sqrt{(\cos^2 25 - \cos^2 35)}}$$

$$= 0.906 \times \frac{0.906 - \sqrt{(0.906^2 - 0.819^2)}}{0.906 + \sqrt{(0.906^2 - 0.819^2)}} = 0.33$$

Active pressure $P_a = 0.33 \times 20 \times \dfrac{1.70^2}{2} = 9.53$ kN/m

Horizontal component $= 9.53 \times \cos 25 = 9.53 \times 0.906 = 8.64$ kN/m.

Active pressure $P_s = 0.27 \times 1.50 \times 1.70 = 0.688$ kN/m

OVERTURNING MOMENT

$$\text{Overturning moment} = 8.64 \times \frac{1.70}{3} + 0.688 \times \frac{1.70}{2}$$

$$= 4.89 + 0.58 = 5.47 \text{ kN m}$$

Element	Weight, W (kN)	L_a (m)	Moment (kN m)
Wall	$22 \times 0.325 \times 1.50 = 10.275$	0.462	4.95
Base	$24 \times 0.20 \times 0.925 = 4.440$	0.462	2.05
Earth	$20 \times 0.30 \times 1.50 = 9.00$	0.775	6.98
Surcharge	$1.50 \times 0.30 = 0.45$	0.775	0.34
Total W	24.615 kN	Total M	14.32 kN m

Therefore factor of safety against overturning $= 14.32/5.47 = 2.61 > 1.50$ (OK)

$$\text{Resultant } W = \frac{14.32 - 5.47}{24.615} = 0.359 \text{ m from A}$$

$$\text{Therefore eccentricity} = \frac{925}{2} - 0.359 = 0.103 \text{ m}$$

BASE PRESSURES

$$P = \frac{24.615}{0 \times 0.925} \times \left(1 \pm \frac{6 \times 0.103}{0.925}\right)$$

$$= 26.61 \pm 17.77 = 44.38 \text{ kN/m}^2 \text{ max and } 8.84 \text{ kN/m}^2 \text{ min}$$

WALL STEM MOMENT

$$P_a = 0.33 \times 20\frac{1.50^2}{2.0} = 7.425 \times \cos 25 = 7.425 \times 0.906 = 6.72 \text{ kN}$$

$$P_s = 0.27 \times 1.50 \times 1.50 = 0.607 \text{ kN}$$

Let $\gamma_f = 1.40$ for earth loads and 1.60 for surcharge loads.

ULTIMATE MOMENT IN STEM

$$6.72 \times \frac{1.50}{3} \times 1.40 + 0.607 \times \frac{1.50}{2} \times 1.60 = 4.70 + 0.72 + 5.42 \text{ kN m}$$

$$\frac{f_k}{\tau_{mm}} = \frac{9.40}{2.30} = 4.10 \qquad\qquad Q = \frac{5.42 \times 10^6}{10^3 \times 150^2} = 0.24$$

Therefore

$$c = 0.95$$

$$A_s = \frac{5.42 \times 10^6 \times 1.15}{460 \times 0.95 \times 150} = 95 \text{ mm}^2/\text{m}$$

Minimum percentage $= \frac{0.13}{100} \times 150 \times 10^3 = 195 \text{ mm}^2/\text{m}$

Use A252 mesh reinforcement lapped on to T10 bars at 200 mm centres projecting from base.

ULTIMATE SHEAR

$$V = 6.72 \times 1.40 + 0.607 \times 1.60 = 9.40 + 0.97 = 10.37 \text{ kN,}$$

$$p = \frac{252}{10^3 \times 150} = 0.00168$$

$$v = \frac{10.37 \times 10^3}{10^3 \times 150} = 0.069 \text{ and } f_v = 0.35 + 17.5p = 0.37 > 0.069$$

Therefore no shear steel required.

MAXIMUM ULTIMATE BASE MOMENT

$$\frac{32.84 \times 0.30^2}{2.0} + \frac{11.54 \times 0.30 \times 2.0 \times 0.30}{2.0 \times 3.0} - \frac{0.20 \times 24 \times 0.30^2}{2.0}$$

$$= 1.47 + 0.34 - 0.216 = 1.59 \times 1.50 = 2.38 \text{ kN m}$$

$$A_s = \frac{2.38 \times 10^6}{460 \times 0.87 \times 0.95 \times 150} = 42 \text{ mm}^2/\text{m}$$

Minimum percentage $= \frac{0.13 \times 200 \times 10^3}{100} = 260 \text{ mm}^2$

Use A252 fabric reinforcement in top and bottom of base.

SHEAR KEY

Total sliding force $= \frac{0.33 \times 20 \times 1.70^2}{2.0} + 0.27 \times 1.50 \times 1.70$

$$= 9.53 + 0.68 = 10.21 \text{ kN}$$

PASSIVE RESISTANCE

$\gamma H + 2c$ where $c_{max} = 30 \text{ kN/m}^2$

Therefore

$$P_p = 0.60 \times 20 + 2 \times 30 = 72 \text{ kN/m}^2$$

Total passive resistance $= \frac{60 + 72}{2.0} \times 0.60 = 39.60 \text{ kN/m}$

Factor of safety against sliding $= \frac{39.60}{10.21} = 3.87$

Therefore reduce depth of shear key to 400 mm.

Total passive resistance $= \frac{60 + (0.40 \times 20 + 2 \times 30)}{2.0} \times 0.40$

$$= 25.60 \text{ kN}$$

Factor of safety $= \frac{25.60}{10.21} = 2.50 > 2.0 \text{ (OK)}$

Therefore use a 200 mm wide × 400 mm downstand at rear of base (Fig. 7.40).

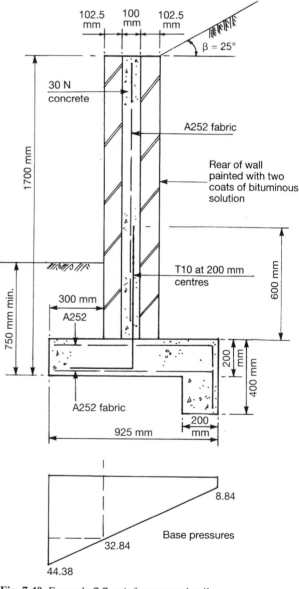

Fig. 7.40 Example 7.7: reinforcement details.

RETAINING WALLS DESIGNED USING BS 8002 (1994)

This next design example illustrates the use of the new Code of Practice BS 8002 for retaining structures. Some designers may prefer to adopt its philosophy, as it is compatible with the structural codes for concrete design BS 8110 and masonry design BS 5628 and the Euro-codes EC2 and EC6.

The basis of limit state design is to ensure that the probability of any limit state being reached is kept to an acceptably low level. With the limit state approach to design it is usual to adopt a two-stage process to ensure that:

1. The common **serviceability limit states**, i.e. excessive cracking, excessive deflection, etc. are not reached under the expected working loads, or
2. The **ultimate limit state** of collapse is not reached under the worst credible loading (which will be much greater than the normal working loads).

Therefore you must establish that your proposed design is strong, stable, robust, and safe to ensure that the ultimate limit state is kept to an acceptably low level. The next stage is to check that your design is going to be serviceable. This is generally achieved by adopting the good detailing guidelines in BS 8110, such as maximum bar spacing, minimum percentage of reinforcement, etc.

When using BS 8002 the partial safety factors of 1.4 for dead loads and 1.6 for imposed live loads are always used when calculating all design ultimate loads, **other than those due to earth pressures**.

With BS 8002 the worst credible earth pressures are determined in a different way: **they are not factored characteristic loads**.

Prior to 1994 retaining structures were designed using the Civil Engineering Code of Practice No. 2 (CP2) introduced in 1951, which was based on the permissible stress approach and adopted by many structural codes at that time. The main faults with using CP2 were:

- It failed to take into account the wall's interaction with the ground and did not allow for the fact that the mobilization of the maximum soil strength is dependent on the strain in the soil, which depends on the amount of lateral wall movement
- It did not recognize that effective stress analysis parameters should be used in the assessment of the earth pressures imposed behind and below retained structures
- It was not compatible with the modern limit state codes used in the design of the structural elements
- It failed to recognize the effect of heavy compaction plant used to place the backfill behind walls, which can result in high lateral earth pressures.

Before carrying out a detailed design of a retaining wall in accordance with BS 8002, the following points should be considered as a **minimum** requirement if one is to achieve a robust and stable retaining wall:

- An adequate ground investigation is **essential**
- Full consideration to be given to the choice of the materials to be used, the design life of the wall and its exposure conditions
- The choice of wall favours the use of less specialized labour and simpler construction procedures
- Ensure that **adequate** and **effective rear face drainage** is provided
- If the wall is part of a basement, ensure an adequate damp-proof system is installed, one that does not compromise the integrity of the retaining wall design, by making sure that the damp-proof membranes do not pass through the structural element of the wall
- Make sure the design is cost effective.

Example 7.8 Reinforced concrete wall, internal damp-proof membrane

A large housing scheme of three-storey flats is to be constructed on a sloping site. Because of the construction programme and the access road behind the flats, the backfill to the wall will have to be placed prior to the building works commencing above the ground-floor level. The wall will therefore need to be designed to cater for this temporary condition. The form of construction will adopt a reinforced concrete wall with an internal damp-proof membrane to the basement storey. The concrete stem will have external water-stops at construction joints and the rear face of the wall will be given two coats of an approved bituminous water-proofing solution.

Surcharge loading due to access vehicles will be based on the 10 kN per sq. metre required by BS 8002 and this loading will also cater for the load imposed by the construction plant placing the backfill materials.

The ground investigation has shown the natural bearing strata are medium density sands and approximately 1.0 metre of over-burden will be removed to create the wall base formation level. Backfill to the wall will be a graded coarse crushed rock placed in 150 mm layers.

Figure 7.41 shows a typical section through the wall where,

γ_m = Moist bulk density of the backfill material = 18.0 kN/m³

γ_s = Bulk density of the natural soils = 20.0 kN/m³

The existing soil strata are medium density sands with SPT values $N = 20$.

For natural soil strata

As there is a 1.0 metre depth of excavation to the formation, the effective overburden pressure is:

$20 \times 1.0 = 20$ kN/m².

From Fig. 2 BS 8002 the correction factor N^1 is 2.75 as the site is free draining, i.e. no water table, therefore

$N^1 = 2.75 \times 20 = 55$.

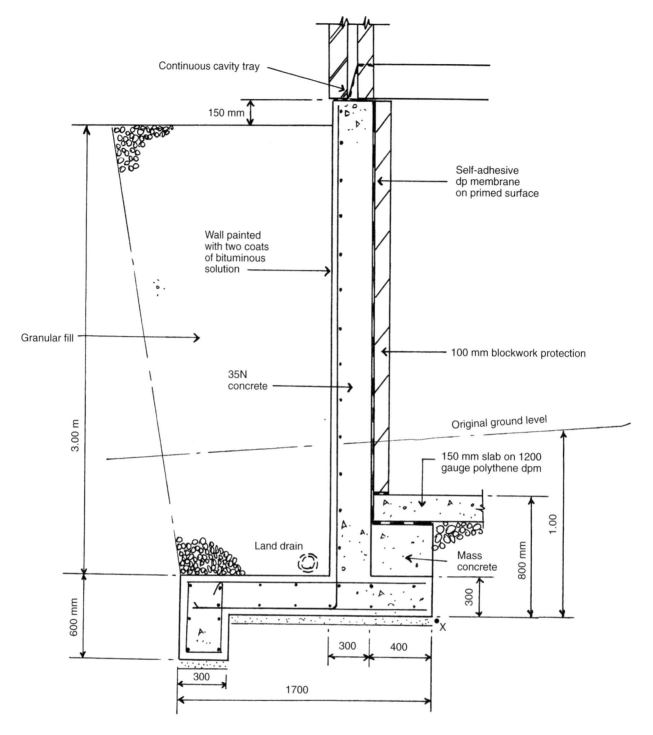

Fig. 7.41 Example 7.8: typical section through wall.

From Table 3 BS 8002, for natural soils:

$A = 0°$ for rounded particles

$B = 0°$ for uniform grading

$C = 8°$ (for $N^1 = 55$)

$\varnothing_{max} = \varnothing_{critical} = 30° + 8° = 38°$ (peak value)
Clause 2.2.4 BS 8002

$\varnothing_{critical} = 30°$

Therefore using effective stress parameters,

Mobilization factor $M = 1.20$, from Clause 3.2.4.

This will ensure that for soils that are medium dense, the wall displacements in service will be limited to 0.5% of the wall height.

Design $\varnothing$ is lesser of $30°$ or $\tan^{-1} \dfrac{[\tan 38]}{[1.20]} = 31.18$

Therefore $\varnothing_{design} = 30$.

For resistance to sliding

Design $\tan \delta = 0.75 \times \tan 30 = 0.38$

Backfill soils

γ_m = 18.0 kN/m² for angular material: $A = 4°$
For well graded rock $B = 4°$,
Assume $C = 0$ (conservative)

Therefore $\emptyset_{max} = \emptyset_{critical} = 30 + 4 + 4 = 38$

$\emptyset_{design} = \tan^{-1}\dfrac{[\tan 38]}{[1.20]} = 33°$ $\delta_{wall} = \dfrac{2}{3} \times 33 = 22°.$

BS 8002 Clause 3.2.6 gives the friction on the rear wall as

$\delta = \tan^{-1}(0.75 \times \tan 33) = 26°$

but in practice the bituminous paint on the rear face will reduce the frictional effect.

$\delta = 0$ and
β (inclination of the retained soil) = 0

Active pressure co-efficient $K_a = \dfrac{1 - \sin \emptyset^1}{1 + \sin \emptyset^1} = \dfrac{1 - \sin 33}{1 + \sin 33} = 0.337.$

Minimum unplanned excavation = 0.50 metres.

Ignore fill over the front toe as contributing to passive restraint because this could be removed.

CASE (1) TEMPORARY CONSTRUCTION CONDITION (Fig. 7.42)

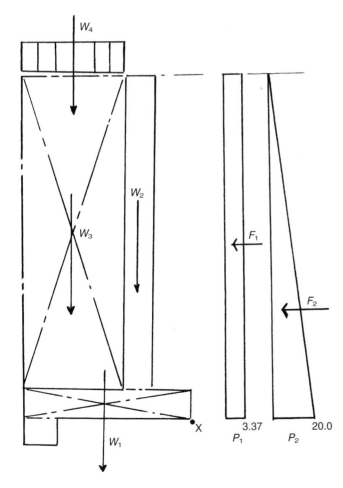

Fig. 7.42 Example 7.8: free body force diagram.

Serviceability limit state

Surcharge pressure $P_1 = 0.337 \times 10$ = 3.37 kN/m²
Active earth pressure $P_2 = 0.337 \times 18 \times 3.30 = 20.00$ kN/m²

Horizontal forces

$F_1 = 3.37 \times 3.30 = 11.12$ kN and $F_2 = \dfrac{20.0 \times 3.30}{2} = 33.00$ kN

F_3 (base friction) = $(W_1 + W_2 + W_3 + W_4) \times 0.38$ (angle of friction)

Vertical forces

W_1 (base)	= 0.30 × 1.70 × 24.0	= 12.24 kN
W_2 (stem)	= 0.30 × 3.00 × 24.0	= 21.60 kN
W_3 (backfill)	= 1.0 × 3.00 × 18.0	= 54.00 kN
W_4 (surcharge)	= 10 × 1.00	= $\dfrac{10.00 \text{ kN}}{97.84 \text{ kN}}$

Therefore $F_3 = 97.84 \times 0.38 = 37.17$ kN
< (11.12 + 33.0 = 44.21 kN)

As the base resistance F_3 is only marginally less than the active lateral forces, a nominal shear key 300 mm deep and 300 mm wide will be provided. This shear key is always best placed at the area of minimum ground pressure.

Overturning moment about X

$M_{F1} = 11.12 \times \dfrac{3.30}{2} = 18.34$ kN

$M_{F2} = 33.00 \times \dfrac{3.30}{3} = \underline{36.30}$ kN

Total = $\underline{54.64}$ kN

Restoring moment about X

$M_{W1} = W_1 \times \dfrac{1.70}{2}$ = 10.40 kNm

$M_{W2} = W_2 \times 0.55$ = 11.88 kNm

$M_{W3} = W_3 \times 1.20$ = 64.80 kNm

$M_{W4} = W_4 \times 1.20$ = $\underline{12.00}$ kNm

Total restoring moment = $\underline{99.08}$ kNm

Check stability and sliding (Case 1)

Sliding: Total lateral force = $F_1 + F_2 = 11.12 + 33.00$
= 44.21 kN

The base friction F_3 is 37.17 kN, therefore the wall is unstable laterally under the maximum horizontal pressures. It would be prudent to provide a small shear key at the rear end of the base where the earth pressures are lower to allow for some ground loosening during excavation for the base. This key will develop sufficient passive resistance. This is a temporary condition that will become adequate when the building over is constructed, adding additional dead loading, so increasing the frictional resistance.

Overturning: The restoring moment of 99.08 kN exceeds the overturning moment of 54.64 kN. This type of wall, with the majority of the base behind the stem, is the most stable type as the combined weight of the backfill and base counterbalances the horizontal forces so reducing the eccentricity of the vertical loading resultant.

Check ground pressures on base

R = Total W = 97.840

Therefore $x = \dfrac{99.080 - 54.64}{97.840} = 0.454$ m

Eccentricity $= \dfrac{1.70}{2} - 0.454 = 0.396$ m

This is outside the middle third and some tension will develop on the heel.

The ground pressure contact length is

$3 \times 0.454 = 1.360$ metres

Average pressure $= \dfrac{97.840}{1.0 \times 1.360} = 71.94$ kN/m²

Maximum edge pressure (Fig. 7.43) $= 2 \times 71.94 = 143.88$ kN/m²

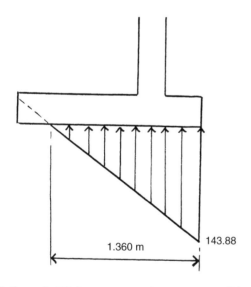

Fig. 7.43 Example 7.8: base pressures (temporary condition).

For sands with an N value of 20 blows an allowable bearing pressure would be 200 kN/m². Therefore base settlements will be low for this construction condition.

CASE (2)   FULL LOAD FROM BUILDING ABOVE WALL

The total serviceability line load from the walls above is 40 kN/metre run. The external leaf carries 13 kN and the internal leaf of blockwork carries 27 kN.

Additional moment about point $x = (27 \times 0.65) + (8 \times 0.45)$
$\qquad\qquad = 21.15$ kNm/metre

Final restoring moment = 99.08 + 21.15 = 120.23 kNm/m

Final vertical loading   = 97.84 + 40   = 137.84 kN/m

$x = \dfrac{120.23 - 54.64}{137.84} = 0.475$

Therefore eccentricity $e = \dfrac{1.70}{2} - 0.475 = 0.375$ m

Ground pressure contact length = $3 \times 0.475 = 1.425$ metres

Maximum average pressure $\qquad = \dfrac{137.84}{1.0 \times 1.425} = 96.72$ kN/m

Maximum edge pressure = $2 \times 96.72 = 193.45$ kN/m²

This is < 200 and therefore OK.

Ultimate limit state – structural design

For dead loads and imposed loads apply load factors of 1.40 and 1.60 respectively.

For earth pressures and forces use the Serviceability Limit State values, i.e. worst credible values.

Maximum overturning moment $= (F_1 \times 1.60) \times \dfrac{3.3}{2} + F_2 \times \dfrac{3.30}{3}$

as only imposed loading is factored

$\qquad = (11.12 \times 1.60 \times 1.65)$
$\qquad\quad + (33.0 \times 1.10) = 65.65$ kNm

Maximum dead loads

$(W_1 \times 1.4) + (W_2 \times 1.4) + (W_3 \times 1.4)$ + building line loads (13.0) + (20.0)

$= (12.24 \times 1.4) + (21.60 \times 1.4) + (54.00 \times 1.4) + 33$

$= 17.13 + 30.24 + 105.00 + 33 = \underline{185.37 \text{ kN}}$

Maximum imposed loads

Surcharge (10×1.6) + (building line load 7.0) = 23 kN/m

Total (dead load + imposed load) = 185.37 + 23 = 208.37 kNm

Maximum restoring moment about x – (kNm)

Factored dead loads

$\dfrac{(17.13 \times 1.7)}{2} + (30.24 \times 0.55) + (105.0 \times 1.20) + (13.0 \times 0.45)$

$+ (20.0 \times 0.65) = 176.04$ kNm

Factored imposed loads 16×1.20 (surcharge) $+ 7.0 \times 0.65$ (wall above) $= \underline{23.75}$

Total = 176.04 + 23.75 = 199.79 kNm

Therefore under design ultimate conditions:

$x = \dfrac{199.79 - 65.65}{208.37} = 0.643$ m

Therefore $e = \dfrac{1.70}{2} - 0.643 = 0.207$ m

which is within the middle third

$$P_{max} = \dfrac{208.37}{1.0 \times 1.70} + \dfrac{6 \times 208.37 \times 0.207}{1.70^2}$$

$$= 122.57 + 89.54 = \underline{212.11 \text{ kN/m}^2}$$

$$P_{min} = 122.57 - 89.54 = \underline{33.00 \text{ kN/m}^2}$$

Toe design

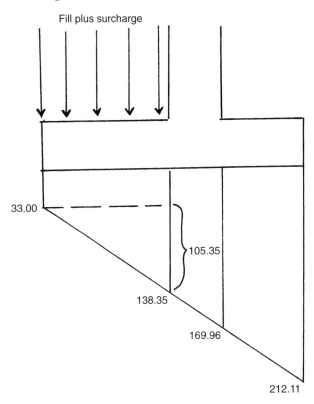

Fig. 7.44 Example 7.8: base pressures (on completion of building over).

At front of wall pressure $= \dfrac{(212.11 - 33.0)}{1.70} \times 1.30$

$$= 136.96 + 33.00 = 169.96 \text{ kN/m}^2$$

At rear of stem pressure $= \dfrac{(212.11 - 33.0)}{1.70} \times 1.0$

$$= 105.35 + 33.00 = 138.35 \text{ kN/m}^2.$$

Base design

Exposure conditions are assumed as moderate, the concrete is in contact with non-aggressive soils.

Use Grade 35 concrete with 40 mm minimum cover with $F_{cu} = 35$ N/mm².

Toe

$V_{pp} = \dfrac{212.11 + 169.96}{2} \times 0.400 - 0.3 \times 0.40 \times 24 \times 1.40$

$$= 76.414 - 4.032 = 72.382 \text{ kN}$$

$M_{pp} = 169.96 \times \dfrac{0.40^2}{2} + 42.50 \times \dfrac{0.40}{3} - 4.032 \times \dfrac{0.40}{2}$

$$= 13.59 + 5.622 - 0.806 = 18.40 \text{ kNm}$$

Check minimum steel percentage

$0.13\% = \dfrac{0.13}{100} \times 1000 \times 300 = 390$ mm²/m

T10 at 200 c/c provides 393 mm²/m

Lever arm factor $k = \dfrac{18.40 \times 10^6}{1000 \times 254^2 \times 35} = 0.0083$.

Use factor of 0.95

$A_s = \dfrac{18.40 \times 10^6}{0.87 \times 460 \times 254 \times 0.95} = 190.53$ mm²/metre

Use minimum percentage T10 @ 200 mm c/c.

Check shear

$V_c = \dfrac{72.38 \times 1000}{1000 \times 254} = 0.28$ N/mm² and $\dfrac{100\,A_s}{b_v \times d} = \dfrac{100 \times 393}{1000 \times 254}$

$$= 0.154$$

From BS 8110 Table 3.9 (V_c) allowable $= 0.38$ N/mm²

Heel design

$V_{yy} = 1.60 \times 10 \times 1.20 + 1.40 \times 3.0 \times 1.20 \times 18$

$$+ 0.30 \times 1.20 \times 24.0 - 33.0 \times 1.20 - 105.35 \times \dfrac{1.2}{2}$$

$$= (19.20 + 90.72 + 12.086) - 39.60 - 63.21$$

$$= 122.00 - 102.81 = 19.20 \text{ kN}$$

This is < 72.38 kN so OK.

$M_{yy} = 122.00 \times \dfrac{1.20}{2} - 33.0 \times \dfrac{1.20^2}{2} - 105.35 \times \dfrac{1.20}{3} \times \dfrac{1.20}{2}$

$$= 73.20 - 23.76 - 25.28 = 24.16 \text{ kNm}$$

By inspection use T10 @ 200 c/c.

Wall stem

Fill is 3.0 metres high, so apply $\gamma = 1.60$ to surcharge loads.

$V_{ss} = (0.337 \times 10.0 \times 1.60) \times 3.0 + (0.337 \times 3.0 \times 18.0) \times \dfrac{3.0}{2}$

$$= 16.176 + 27.29 = 43.47 \text{ kN}.$$

$M_{ss} = \left(16.176 \times \dfrac{3.0}{2}\right) + \left(27.29 \times \dfrac{3.0}{3}\right) = 51.55 \text{ kNm}$

Lever arm factor $= \dfrac{51.55 \times 10^6}{1000 \times 254 \times 254 \times 35} = 0.02$

Apply $k = 0.97$

$A_s = \dfrac{51.55 \times 10^6}{0.87 \times 460 \times 254 \times 0.97} = \underline{523 \text{ mm}^2}$

T10 @ 150 c/c provides 523 mm²

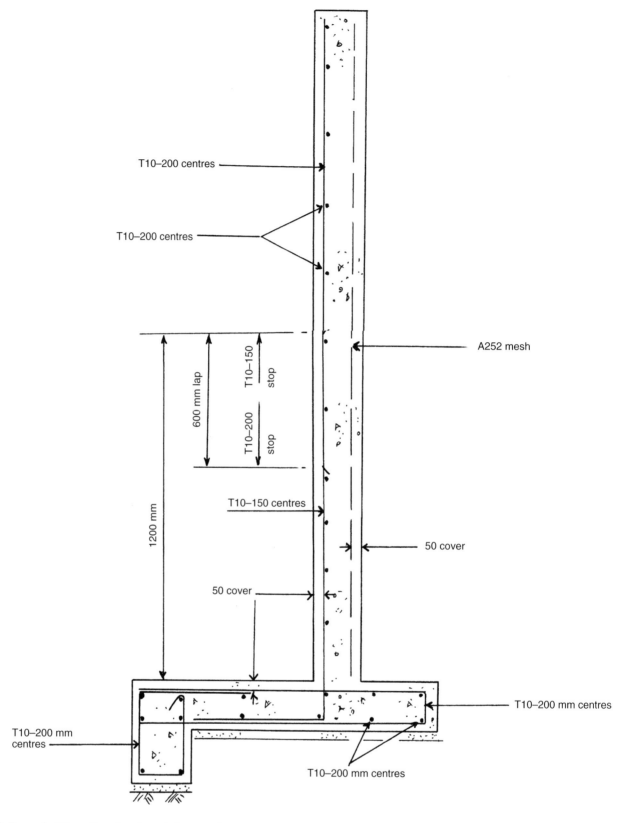

T10–200 centres

T10–200 centres

600 mm lap

T10–150 stop

T10–200 stop

A252 mesh

T10–150 centres

50 cover

1200 mm

50 cover

T10–200 mm centres

T10–200 mm centres

T10–200 mm centres

Fig. 7.45 Example 7.8: section showing reinforcement.

Check cracking

BS 8110 clause 3.12.11.2.7

$$\frac{A_s \times 100}{B \times d} = \frac{523 \times 100}{1000 \times 254} = 0.20\%$$

This is < 0.30% so OK.

Distribution steel: T10 @ 200 c/c longitudinally in wall stem and base both faces.

7.6 DAMP-PROOFING TO RETAINING WALLS

All domestic basements have to be damproofed in accordance with BS 8102 1990 *Protection against water of structures below ground level.* No matter how carefully a basement is designed and detailed, and even with good standards of workmanship, it cannot be made absolutely waterproof. Attention to the details outlined in BS 8102 should result in a basement which is relatively dry. For this guide we shall only deal with Type A and Type C structures: **tanked protection** and **drained cavity construction**. This involves wrapping the basement walls and floors in a continuous membrane which excludes water and vapour from the internal rooms.

When designing the basement walls the method of damp-proofing should always be considered to be of paramount importance. The damp-proof membrane must not weaken or pass through the structural walls and the construction should be kept simple to enable the damp-proof membrane to be installed easily. Emphasis should be on simplicity, build-ability and durability. Where possible, keep the basement as a simple box without complicated wall sections and level changes (Fig. 7.46).

7.6.1 Type A structures: tanked protection

For plain reinforced concrete or masonry basements, the continuous damp-proof membrane can be installed externally or internally.

(a) External tanking

External tanking (Fig. 7.47) is considered to be the best method.

Advantages:

1. Any groundwater pressures force an external membrane against the walls, whereas an internal membrane can be pushed off if it is not sandwiched.
2. The membrane is unlikely to be damaged or punctured from internal fixings or services.
3. It also gives protection to any reinforcement in the wall.

Disadvantages:

1. Access at a later date can be difficult and expensive.
2. Adhesive membrane require good, smooth walls for good bonding.
3. Workmanship is dependent on weather conditions and damage arising from other trades.
4. Incoming services are a potential problem.
5. Backfilling can damage the membrane if it is not suitably protected.

The two main types of tanking used in domestic basements are hot applied mastic asphalt and self-adhesive bituminous and polythene sandwich sheet membranes used

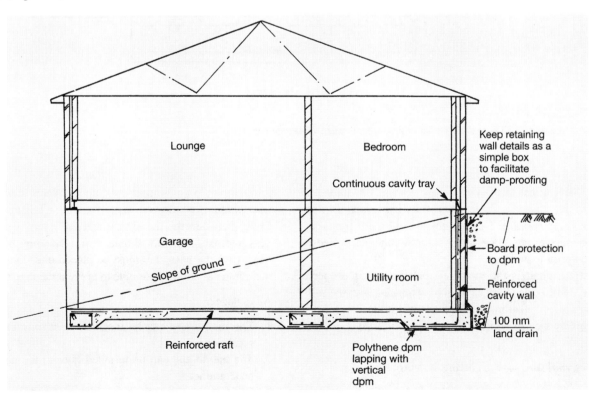

Fig. 7.46 Damp-proofing a domestic basement.

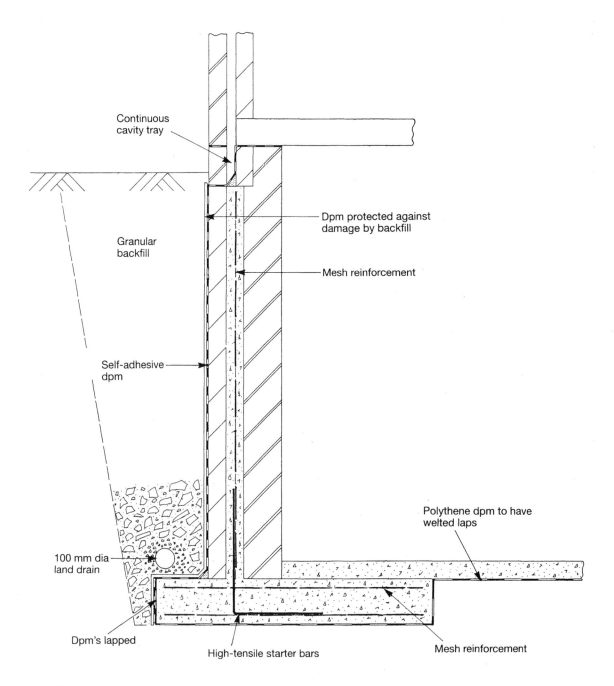

Fig. 7.47 Retaining wall with external damp-proof membrane (reinforced concrete).

mainly on vertical faces. Under base slabs, 1200 gauge or 2000 gauge polythene sheet is generally used, lapping with the vertical membrane. This polythene can be substituted by a self-adhesive membrane but a smooth concrete screed is required prior to laying.

Mastic asphalt is the most reliable tanking if properly applied but it is also very expensive compared with the self-adhesive sheet membranes. Most domestic basements are protected by the self-adhesive type.

(b)   Internal sandwiched tanking systems

These systems (Fig. 7.48) should only be considered when external tanking has been ruled out because of poor access

problems or poor ground conditions. When internal sandwich tanking is adopted there must be a full mortar-filled cavity between the membrane and the internal brickwork. This internal brickwork should be free-standing: i.e. no wall ties. Also, the base slab must be placed over the horizontal membrane as soon as possible to prevent damage occurring.

Advantages:

1. The membrane can be installed in favourable weather conditions.
2. The membrane can be installed later in the construction programme.
3. Easier access is possible in the event of repairs being required.

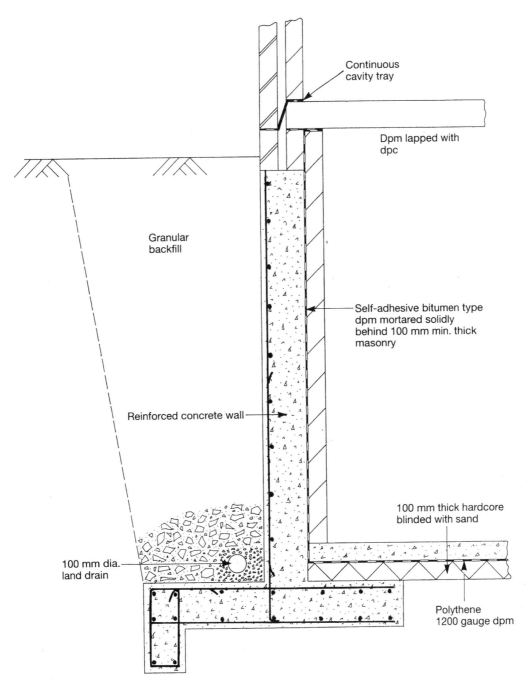

Fig. 7.48 Retaining wall with internal damp-proof membrane (reinforced concrete).

Disadvantages:

1. If the membrane gets punctured, water will travel to other defects in the structure some distance from the perforation.
2. The membrane below the slab may be subject to damage. To avoid this, the floor should be laid as soon as possible. If this cannot be done then the membrane must be well protected.
3. It requires the structural walls to be so designed so that no wall ties are required to pass through the membrane.
4. This membrane does not give protection to any re-

inforcement in the external face of the basement retaining wall.

Polythene or brush-applied liquid bituminous membranes are generally not considered suitable for domestic basements for the following reasons.

1. Workmanship in regard to masonry construction below ground level is often poor, resulting in badly filled perpendicular joints and gaps in bed joints.
2. Polythene is difficult to lap properly in a vertical situation and is easily punctured.

171

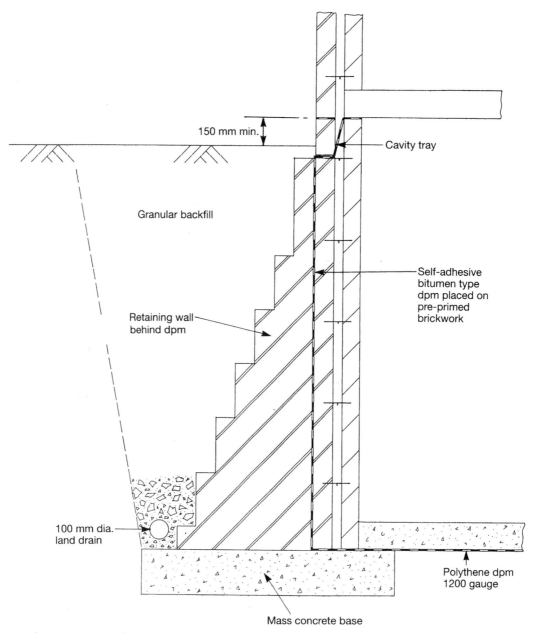

150 mm min.

Cavity tray

Granular backfill

Self-adhesive
bitumen type
dpm placed on
pre-primed
brickwork

Retaining wall
behind dpm

100 mm dia.
land drain

Polythene dpm
1200 gauge

Mass concrete base

Fig. 7.49 Drained cavity construction.

(c) Installation of tanking membranes

When installing a mastic asphalt or self-adhesive sheet membrane particular attention must be paid to the following points.

1. All walls should be finished with a relatively smooth finish free from mortar snots with no open joints.
2. Walls should be dry before application of the priming coat or asphalt. This increases the bonding to the wall surface.
3. The manufacturer's installation instructions must be followed. These generally specify a priming coat to be applied to the wall before applying the membrane.
4. Suitable protective boarding must be placed over the face of the membrane to prevent damage from backfill materials. There are many proprietary boards now available which are rot-proof; some also contain built-in vertical fin drainage.
5. Provide clean granular fill behind walls and a perforated drain at the bottom of the wall stem, linking if possible

into the site-drainage system.
6. Where services must pass through a wall, a caulking groove should be formed to be sealed with a bitumastic sealant. In certain situations, where large pipes are being installed, it may be necessary to provide a puddle flanged pipe sleeve to prevent water ingress.

7.6.2 Type C structures: drained cavity construction

This form of construction (Fig. 7.49) requires that the cavity be built within the thickness of the retaining walls. This cavity is to collect and remove to a sump or site drainage system any water which passes through the principal protective layer. The cavity will need to be ventilated to prevent the transmission of water vapour across to the inner skin. It is recommended that inner walls be built free-standing to avoid bridging the cavity.

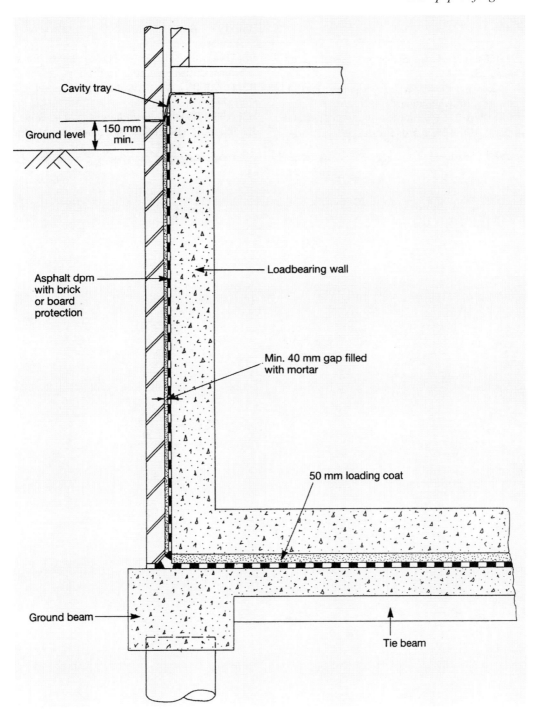

Fig. 7.50 Pile cap/retaining wall. Horizontal asphalt 30 mm (three 10 mm coats); vertical asphalt 20 mm (three coats).

Advantages:

1. Well suited to traditional construction tolerances, and a reliable system.
2. Water ingress may be examined prior to final installation of internal wall.

Disadvantages:

1. Pumps may be required to remove accumulated water.
2. Floor space is reduced.
3. No protection is given to any reinforcement in the retaining wall.
4. Pumping costs, maintenance and cavity cleaning have to be considered.

BIBLIOGRAPHY

Bishop, A.W. (1955) The use of slip circles on the stability of slopes. *Geotechnique*, **5**, 7–17.

BSI (1978) BS 5628: *Structural use of masonry*. Part 1. Unreinforced masonry, British Standards Institution.

BSI (2000) BS 5628: Part 2. Structure use of reinforced and prestressed masonry, British Standards Institution.

BSI (1997) BS 8110: Part 1. Structure use of concrete, British Standards Institution.

BSI (1985) BS 8110: Part 3. Design charts for singly reinforced beams, British Standards Institution.

BSI (1990) BS 8102: *Protection against water of structures below ground*.

BSI (1994) BS 8002: *Earth Retaining Structures*.

Jumikis, A.R. (1962) *Soil Mechanics*, Van Nostrand, New York.

Chapter 8
Building on filled ground

Houses with foundations placed on made ground, i.e. fills, may be at risk from differential settlement because of the nature of the fill, the condition of the fill, or for causes connected with the reason why the fill was placed in the first instance.

Where the fill is placed as part of the construction works, and this is done to a high standard with good quality control and proper engineering supervision, the risks can be assessed, and may be quite small. Where, however, the site is covered with existing fills, the risks are more difficult to assess and, short of complete excavation of the fills, the risks cannot usually be fully quantified.

Sites used as dumps for household or industrial refuse may contain such things as old steel drums, tins and car bodies, which may crush when loaded or which may collapse from corrosion in time. Generally such fills can be very localized and may not be encountered in a site investigation (Fig. 8.1).

Any site on an abandoned colliery tip or mining complex should be investigated for the presence of old lagoons. These are rarely encountered in tips more than 35 years old, but are present in more recent tips. British Coal records are not always reliable. Another consideration on old colliery tips is the acidity of the groundwater. Chemical analysis should be carried out on the fills and groundwater to enable the correct grade of concrete to be specified for foundations below ground.

Most existing tips have been placed without regard to any compaction of the materials and where excessively large-sized materials have been deposited they are likely to be surrounded by loose voided materials. This situation is particularly hazardous at the edges of quarries or sites bounded by large retaining walls. A study of old Ordnance Survey maps may show up such situations (Fig. 8.2).

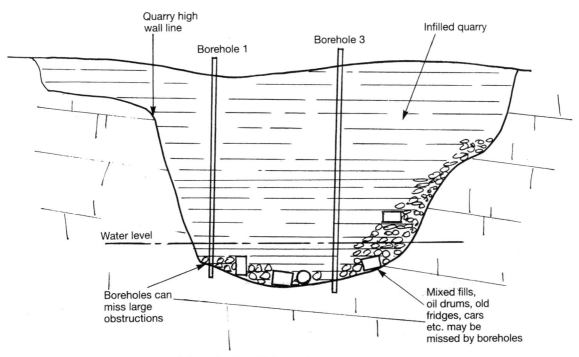

Fig. 8.1 Investigating filled quarries.

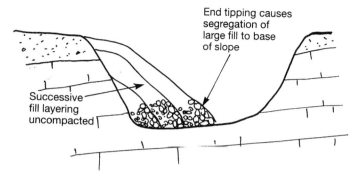

Fig. 8.2 Uncompacted opencast or quarry fills.

Where the history of the filling can be established it is usual to find that sites filled over a long period are more likely to be variable in their material composition and compaction density than those filled quickly. Those sites formed by end tipping over a quarry edge are also likely to be more variable than those sites where the fills are placed in discrete layers and spread and rolled (Fig. 8.3).

Fills placed in water will generally be looser than fills placed in the dry as the binders or clay matrix separate out.

The reasons for placing fill on a site, if they can be established, may also give cause for concern. There may have been surface hollows arising out of solution features such as swallow holes in limestone strata, or crown holes due to shallow old mine workings which have been filled in. These underlying hazards could still result in potential foundation problems on other parts of the site (Fig. 8.4).

Fills placed in thin layers and compacted by appropriate compaction plant can often be placed to give an equivalent of 95% dry density. The major problem with deep fills is the amount of additional settlement arising out of self-weight consolidation; these settlements can take up to seven years to finally settle out (Fig. 8.5). Most of the consolidation settlement should take place in the first two to three years. In these situations stiff raft foundations can be a suitable foundation solution.

The use of reinforced strip footings on such sites is not recommended because:

1. variations in the compaction can result arising from variations in the placed materials;
2. weather conditions during placement can result in strength variations;
3. if cohesive materials are being placed in the upper levels at low moisture contents they could swell giving rise to differential movements within short distances.

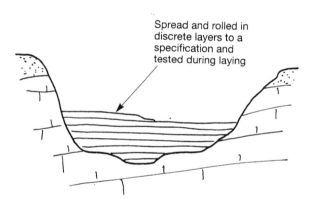

Fig. 8.3 Well-compacted clay, shale fills. Fully tested during placing and monitored on completion.

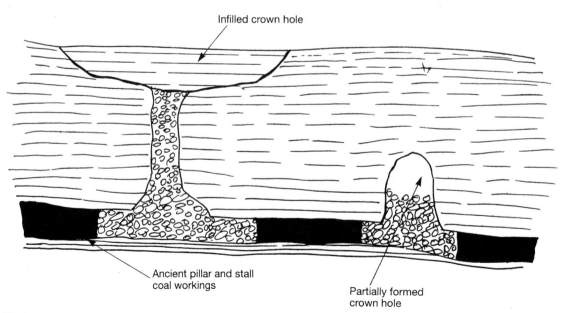

Fig. 8.4 Infilled crown holes.

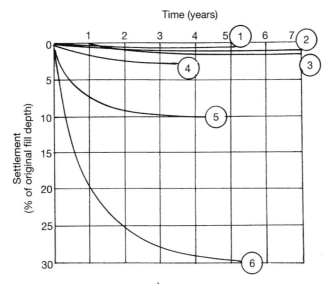

Fig. 8.5 Self-consolidation settlement in fills (Meyerhof, 1951). 1, well-graded soil, well compacted: 2, rock fill, medium compaction; 3, clay and chalk, lightly compacted; 4, sand, uncompacted; 5, clay, uncompacted; 6, mixed refuse, well compacted.

8.1 OPENCAST COAL WORKINGS

Generally, these have been backfilled following on from coal extraction. It is very rare for these fills to contain significant amounts of deleterious material as the materials being placed are the original overburden strata and drift deposits. The materials are generally deposited by dragline shovel operations with the materials being scattered from a great height, resulting in segregation of the larger particles.

This form of filling usually results in a large magnitude of consolidation settlements, and problems of high differential settlements can arise around the high wall batter planes which virtually become no-build zones (Fig. 8.6). Even the provision of stiff raft foundations may not be sufficient to prevent a dwelling from tilting excessively on areas where the fill depth increases in magnitude over a short horizontal distance.

A problem now being recognized with old opencast backfilled sites is that the original water table may reestablish itself, so softening the lower coal measure shales and mudstones as it rises. This can result in additional settlements, sometimes long after the normal time period over which consolidation settlements occur. Where fills are compacted to about 95% of their dry density, the effects of water table re-establishment are generally insignificant.

Some fill materials, such as colliery shales and coal washing plant residues, have a potential for swelling due to pyritic oxidation within the fill, activated by water and oxygen.

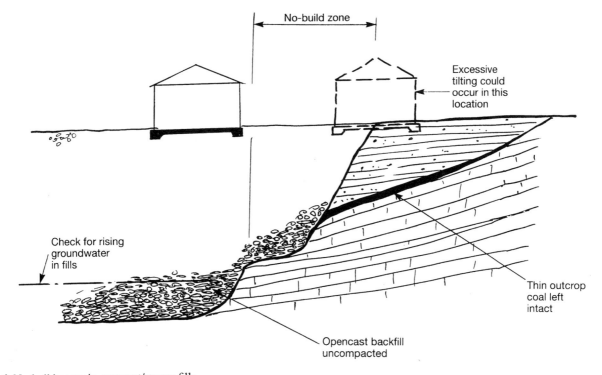

Fig. 8.6 No-build zones in opencast/quarry fills.

Such materials should be examined in detail and their chemical composition determined.

Where fills contain combustible matter such as coal and industrial wastes, tests should be taken to determine the calorific content of the materials, as combustion can occur in fills with calorific contents of 10 000 kJ/kg. There may be an external source for this to occur but some materials are known to burn because of spontaneous combustion within the material itself, which in turn can give off toxic gases.

The following action should be taken on old opencast fill sites.

1. Obtain the mine abandonment plans from the Opencast Executive or Mining Records Office.
2. Carry out a detailed borehole investigation and supplement it with trial holes excavated around the high wall batter planes. Determine fill densities by SPT and triaxial tests on undisturbed samples.
3. If the top of the high wall batter plane cannot be reached with a JCB machine because of the depth of replaced fills, then a series of closely spaced boreholes may be required at about 3 m centres to establish the angle of excavation.
4. Determine the chemical composition of the fills and check on their stability and soundness. Check for sulphates and acidity.
5. Establish whether there is a possibility of groundwater re-establishment. If water is already present there may be a requirement for installing piezometers to monitor the groundwater levels, especially if some form of ground improvement is being considered which could affect porewater pressures in the fills.

Once these investigations have been concluded, the foundation proposals can be evaluated in regard to the precise conditions of the fill materials. The site layout can be prepared to ensure that no dwellings are placed in the sterilized high wall batter planes. It is worthwhile designing the estate layout so that roads are placed over these areas.

8.2 FOUNDATIONS

The following types of foundations can be adopted on old infilled opencast coal or backfilled quarry sites.

8.2.1 Stiff raft foundations

These are suitable if field tests have shown that the fill materials are chemically stable and homogeneous, and long-term consolidation settlements have ceased.

The average bearing pressure below a pseudo edge beam raft foundation, if the total raft area is considered, is about 25 kN/m^2. However, with this type of raft the localized line loads at wall positions result in higher pressures below the edge beams and internal loadbearing walls. In order that settlements are kept within reasonable limits these line load pressures should be limited to about 50 kN/m^2.

The width of the edge beam or central thickenings in contact with the ground should be made wide enough to

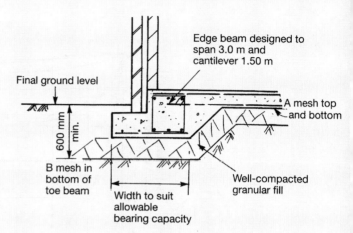

Fig. 8.7 Stiff edge beam rafts.

sustain the allowable ground pressures. The edge beams carrying the main loadbearing walls can be designed to span 3.0 m using $wL^2/10$ (the bending moment formula for semi-encastre beams) and cantilever at the corners a distance of 1.50 m. If clay fills are evident on the surface then the rafts should have an adequate depth of stone fill below the edge beams to prevent frost heave or drying out of the clays (Fig. 8.7).

Terraced blocks should be provided with full 25 mm wide settlement joints at every third narrow or second large unit. These joints should extend through the foundation and the units each side of the joint should have their own stability by the introduction of a block wall linking the front and rear walls to form a box (Fig. 8.8).

Consideration must be given to the effects of settlement on services entering or leaving the dwelling. Sufficient allowance for vertical movement at external wall positions can be provided by having sliding pipe collars. In addition, service pipes should have generous falls and be provided with flexible joints (Fig. 8.9).

8.2.2 Piled foundations

If the fill materials are weak but not excessively deep, then the use of piled foundations can be considered. The piles

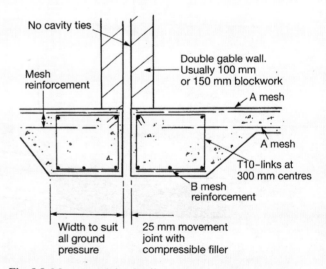

Fig. 8.8 Movement joint detail.

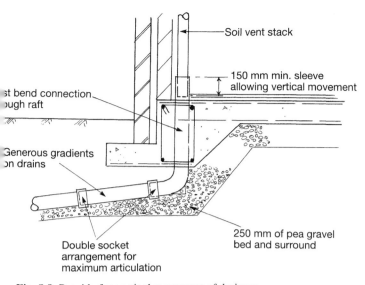

Fig. 8.9 Provide for vertical movement of drainage.

must be designed to carry the building loads plus the additional loads due to negative skin friction on the pile shaft as the fills consolidate. Any increase in ground levels will also result in additional loads on the pile shaft.

One advantage of piling is that dwellings can be constructed over 'no-build' zones, but careful consideration needs to be given to the type of pile to be used. Where an inclined rock stratum is present it is essential that the pile toe is adequately socketed into the rock. In these situations the

use of driven piles is not recommended unless they are tube piles which can be socketed in by down-the-hole drilling (Fig. 8.10).

If bored piles are used, taken through the fills into good bearing strata, the depth of penetration can be determined by calculating the skin friction and end bearing values using appropriate factors of safety. Again, account must be taken of the additional load due to the effects of negative skin friction. Table 8.1 gives some typical load capacities for bored piles in stiff and hard clays.

8.3 SUSPENDED GROUND-FLOOR CONSTRUCTION

Where sites are sloping, or there is existing fill present, it is often necessary to provide fully suspended ground floors. It is a requirement of the *NHBC Handbook* (Chapter 5.2) that, where make-up fill below ground slabs is in excess of 600 mm, then a fully suspended floor must be used. If fill is present, fully suspended construction is required irrespective of its depth.

Various construction methods can be used, such as:

- timber full-span joists with concrete oversite and ventilated air space – i.e. no sleeper walls off oversite;

Table 8.1. Load-carrying capacity of bored piles (kN)

Ground strength classification	Pile diameter (mm)	Pile lengths (m)			
		2.50	3.0	4.0	4.50
Stiff clays	250	40	50	60	70
(unconfined shear strength > 75 kN/m^2)	300	50	60	75	90
	350	65	80	95	110
Hard clays	250	55	65	80	90
(Unconfined shear strength > 140 kN/m^2)	300	70	85	100	115
	350	95	110	125	140

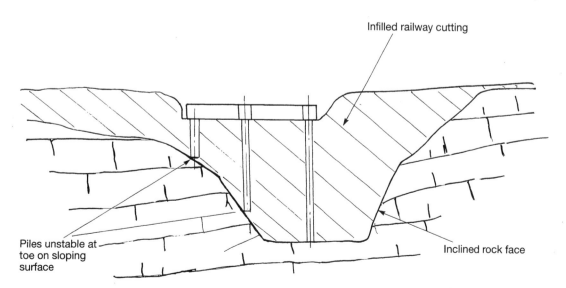

Fig. 8.10 Avoid driving piles on sloping rock strata. Driven piles are not recommended in localities where the pile toe cannot be socketed in or bedded down solid. Generally such piles drift out of plumb during driving.

Building on filled ground

- precast concrete beam and block flooring system with nominal ventilation;
- *in-situ* reinforced concrete slab cast on temporary earth shutter.

Where retaining walls have to be constructed to split-level designs the backfilled working space, though usually made up of consolidated granular fill, will be subject to consolidation settlements and the floors over such areas need to be fully suspended.

The provision of suspended ground floors for domestic buildings is outlined in the Chapter 5.2 standards. This requirement is for the following situations:

- on sites where the depth of upfill exceeds 600 mm;
- for sites with desiccated clays to cater for dumpling heave;
- on sites where vibrocompaction ground improvement is being used.

For clay sites which are affected by trees timber full-span joists or precast concrete beam and block are used, as a voided floor is a requirement of NHBC Chapter 4.2.

The imposed loads on domestic floors are 1.50 kN/m² and, for garages, 2.50 kN/m². Generally, lightweight block partitions are allowed for at 1.0 kN/m². Therefore the designs for a domestic floor slab can be used for a garage of equal span. Table 8.2 gives the area of reinforcement, slab thickness and minimum distribution steel for a 1.0 m width of slab designed as simply supported using a 30 N/mm² concrete mix and high-tensile reinforcement.

Example 8.1 Suspended floor slab over filled ground

A double garage is to have a fully suspended *in-situ* reinforced concrete floor. The internal width of the garage is 5.90 m. Concrete

mix is to be 30 N/mm² and high-tensile fabric mesh will be used. An internal spine wall is not being used because the depth of made ground is excessive.

Effective span = 5.900 + 0.100 = 6.00 m

Dead load, 225 mm slab = 0.225 × 24 = 5.40 kN/m²

Imposed load = 2.50 kN/m²

Factored dead load = 1.40 × 5.40 = 7.56 kN/m²

Factored imposed load = 1.60 × 2.50 = 4.00
Total = 11.56

Therefore ultimate moment = $\frac{11.56 \times 6.0^2}{8}$ = 52.02 kN m

With polythene dpm and 40 mm concrete cover the effective depth is

$225 - 40 - \frac{10}{2} = 180$ mm

Therefore

$\frac{M}{bd^2 f_{cu}} = \frac{52.02 \times 10^6}{10^3 \times 180^2 \times 30} = 0.053$

From Design Chart No. 1 (Fig. 2.11) the lever arm factor = 0.925. Therefore

Lever arm = 180 × 0.925 = 166.50 mm

Therefore

$A_{st} = \frac{M}{0.87 l_a p_{st}} = \frac{52.02 \times 10^6}{0.87 \times 166.50 \times 460} = 780$ mm²

Use B785 mesh in bottom of slab.

Distribution steel, 0.13% = $0.13 \times \frac{225}{100}$ = 293 mm²/metre

Transverse steel area with B785 = 252 (mm²)
plus 6 mm HT bars at 300 mm centres = 94
Distribution steel provision = 346

Table 8.2. Domestic floors and garage floors. Partitions: lightweight blockwork (1 kN/m²). Finishes: 18 mm chipboard on 35 mm polystyrene

Span (m)	Slab thickness (mm)	Finishes and partitions (kN/m²)	Unfactored dead load, g_k (kN/m²)	Unfactored imposed load, q_k (kN/m²)	Factored design load (kN/m²)	Design ultimate moment (kN m)	Minimum steel (mm²/m)	Area of steel (mm²/m)
2.50	125	1.40	3.40	2.50	8.76	6.84	193	B385
2.75	125	1.40	3.40	2.50	8.76	8.28	193	B385
3.00	150	1.40	4.00	2.50	9.60	10.80	193	B385
3.25	150	1.40	4.00	2.50	9.60	12.68	193	B385
3.50	150	1.40	4.00	2.50	9.60	14.70	252	B503
3.75	150	1.40	4.00	2.50	9.60	16.88	252	B503
4.00	175	1.40	4.60	2.50	10.44	20.88	252	B503
4.25	175	1.40	4.60	2.50	10.44	23.57	252	B503
4.50	200	1.40	5.20	2.50	11.28	28.55	260*	B503
4.75	200	1.40	5.20	2.50	11.28	31.81	260*	B785
5.00	200	1.40	5.20	2.50	11.28	35.25	260*	B785
5.25	200	1.40	5.20	2.50	11.28	38.86	260*	B785
5.50	225	1.40	5.80	2.50	12.12	45.83	293*	B785
5.75	225	1.40	5.80	2.50	12.12	50.09	293*	B785
6.00	225	1.40	5.80	2.50	12.12	54.54	293*	B1131

* Supplement mesh with loose bars. Concrete grade C30. Reinforcement high-tensile mesh. Concrete cover 40 mm min.

This will be acceptable for a garage floor slab but a B503 mesh plus a layer of A393 could also be used and this would give $896\,mm^2$ main steel with $645\,mm^2$ transverse steel.

8.4 COMPACTION OF FILLS TO AN ENGINEERED SPECIFICATION

Over recent years, derelict land has been reclaimed, usually with the aid of derelict land grants, by removing the fills or contaminated materials and replacing them with suitable reclaimed materials laid and compacted in discrete layers to an engineered specification. These methods have become widely established, especially in areas where coal can be mined by opencast methods.

In the past, opencast mines were backfilled loose using dragline plant, and such sites could not be developed for 25–30 years after filling. The object of placing fills to an engineered specification is to enable development to take place within 1–2 years after filling. This can make the coal extraction a viable and economic operation.

8.4.1 Procedure

The aim is to extract shallow coal deposits and restore the site for development by compacting the backfill materials, so that a restored surface is achieved with an adequate bearing capacity for surface structures. In addition, the long-term consolidation settlements should be minimized to 0.25–0.50% of the fill thickness, and areas of sterilized zones (areas with large variations in fill depth) caused by differential settlements over excavated slopes should be minimized.

Backfilling should be carried out to a specification based on Department of Transport Specification for Highway Works, Part 2, Series 600 Earthworks (1986). The backfilled materials should be spread in layers 300–400 mm thick and compacted with the appropriate number of passes of the compaction plant. Adjustments can be made to the natural moisture content of the placed materials by mixing and blending the different specifications of materials and, when necessary, by spraying water from bowsers.

Special attention must be paid to cutting back the looser fills on the outer slope of the compacted benches prior to placing subsequent benches in order that a good 'keying in' is achieved and the continuity of compacted densities are maintained between benches.

The type of plant used generally consists of sheepsfoot compactors, D8 bulldozers, Bomag vibratory towed rollers, water bowsers, dump trucks and spreading plant.

Any old mine galleries which may be encountered must be stopped off with a clay-type material to form a low-permeability plug. Any mine shafts encountered which extend to lower workings will need to be capped off at rockhead in accordance with British Coal procedure.

Materials which, when excavated, are too large for recompaction can either be placed in the sterilized zones or put through a primary crusher. Alternatively, they can be placed in proposed landscape areas.

The key to achieving good compaction results in such operations lies in the site testing carried out before and after treatment. Good site supervision and accurate surface monitoring are essential in order that the long-term consolidation settlements predicted by 'modelling' can be verified.

8.4.2 Site testing before backfilling

Representative samples of the coal measure strata and the superficial deposits should be taken and tested in the laboratory in accordance with BS 1377. These tests should include moisture content, plastic limits, liquid limits, sulphate and pH analysis, and determination of the optimum dry density/moisture content ranges of the various materials.

With these data the compaction parameters can be established based on the maximum dry density and optimum moisture contents. Tests can be carried out during the filling operations to enable comparisons to be made with the results obtained from compaction field trials.

(a) Compaction field trials

These field trials are invaluable and provide confirmations as to the correct type of plant to be used, the minimum number of passes per layer and the maximum thickness of the layers.

(b) Site testing

This is generally carried out by the resident engineer and his staff on a daily routine basis. The bulk density of the placed fills are checked by using nuclear density gauges which will have been calibrated from sand replacement tests in the compaction field trials. During the operations the use of regular sand replacement testing is recommended to confirm the accuracy and settings of the nuclear density gauges.

Where materials are found not to have been placed to the required density then the resident engineer may have the area recompacted. Alternatively he may have the materials replaced, or introduce different materials to achieve the required moisture content. In certain circumstances he may have to instruct the contractor to remove materials which do not meet the specification.

(c) Level monitoring

As sections of the site are filled up to the required levels a settlement-monitoring grid should be established as soon as possible. The levelling stations should be placed at a depth at which seasonal movements will not affect them. They should also be adequately protected as they are a very important element in establishing when the site can be developed.

The survey data collected are used to collate with the predicted model for consolidation settlements. This monitoring period should if possible extend over a period of at least 12 months for the initial phases of the operations with a minimum of 6 months for the final monitoring points.

(d) Site supervision

This is an absolute requirement on deep-fill compaction sites. There must be a resident engineer on site together with sufficient experienced technical staff to ensure that the works are carried out to the specification. The use of a small on-site laboratory can be very useful on a large fast-moving site.

The field staff must maintain detailed records of all field testing, daily weather readings, site progress records and photographs on a daily basis. Any changes to the specification must be agreed with the resident engineer.

8.4.3 Foundations

For low-rise buildings it is recommended that pseudo-raft foundations be used with appropriate measures in regard to movement joints. On sites where the fill materials are predominantly cohesive it can often be observed in the monitoring that the materials are swelling, and this is one of the reasons that rafts are recommended.

8.4.4 Roads and drainage

Precautions should be taken in the design of the highways and services where these cross the sterilized zones.

8.4.5 Groundwater

In loosely compacted fills the re-establishment of the groundwater levels has been known to cause large inundation settlements. When materials are placed to a very high dry density of 90–95% the·rise of the groundwater is slow and has very little effect on the strength of the fills.

8.5 GROUND IMPROVEMENT TECHNIQUES

Vibro-compaction is not always possible on opencast fill sites owing to difficulty in placing the vibrating poker through the fills. In addition, there is no point in treating fills which are still settling under their own weight.

One of the problems of using vibro-compaction on deep-fill sites is that the stone columns inserted into the upper fills act like vertical french drains transmitting surface or perched water down into the lower fills. This can result in inundation settlements of the fills and localized softening of cohesive strata. Claims that the foundation seals off the clays are not valid, and there are recorded case histories of such problems.

Vibro-compaction stone columns have been used successfully in speeding up the settlement of newly constructed road embankments by dispersal of porewaters as the clays are loaded. The process will affect filled sites in the same way.

8.5.1 Dynamic consolidation

This form of ground improvement has been used on opencast backfill sites. It requires specialized assessment. The effects of increased porewater pressures and the subsequent removal of the free water produced can give rise to problems.

8.5.2 Surcharge loading

Inducing settlements by preloading with soils has been successfully carried out on housing sites. The main problems with the method are the long timescale and monitoring periods required, and the costs incurred in providing heavy earth-moving plant.

The higher the surcharge, the shorter the timescale, but care must be taken when the monitoring stations are placed. If possible, the settlement indicators should be placed centrally in the mounds to indicate the maximum settlements. Settlement indicators placed around the edges and base of the mounds will be subjected to heave, and control indicators may be required.

Chapter 9 deals with ground improvement techniques in detail.

8.6 COMPACTION OF STRUCTURAL FILLS

Where raft foundations are to be constructed on imported granular fills it is important to ensure that the correct grade of stone is used and that proper compaction is carried out. Table 8.3 should be used as a guide for specifications. Compaction of fill materials increases the resistance of the fill to

Table 8.3. Compaction of earthwork materials

Mass/metre width of roll (kg)	Cohesive soils		Well-graded granular soils and dry cohesive soils		Uniformly graded materials	
	D (mm)	N	D (mm)	N	D (mm)	N
Smooth-wheeled rollers						
2100–2700	125	8	125	10	–	–
2700–5400	125	6	125	8	–	–
> 5400 kg	150	4	150	8	–	–
Grid rollers						
2700–5400	150	10	–	–	–	–
5400–8000	150	8	–	–	–	–
> 8000 kg	150	4	150	12	–	–
Tamping rollers						
> 4000 kg	225	4	150	12	250	4
Vibrating rollers						
1800–2300	150	4	150	4	225	12
2300–2900	175	4	175	4	250	10
2900–3600	200	4	200	4	275	8
3600–4300	225	4	225	4	300	8
4300–5000	250	4	250	4	300	6
> 5000	300	4	300	4	300	4

D = depth of layer; N = number of passes of roller. All works to be in accordance with the Specification for Highway Works to achieve 95% of optimum density.

deformation. It also reduces the permeability of the fill and prevents the possibility of inundation settlements by the action of water.

The most critical element in soil compaction is the moisture content at which the materials are placed. Maximum dry densities are obtained at an optimum moisture content for a given stone type. Where granular fill is being compacted using vibratory rolling plant, the dry density increases with the decrease in moisture content. The compaction of granular fills is more effective if the primary control is the depth of the compacted layer. Increasing the thickness of the layer results in a reduction in the average dry density.

Flat-plate vibrators are generally not considered suitable for compacting structural fills because of their shallow depth of compaction and lower power rating. They are more suitable for compacting soil in confined trenches. Self-propelled vibrating rollers are the most useful means of compacting structural fills; on large sites they can be towed by a separate tractor unit. Large vibrating rollers have a considerable impact effect on the fill layers for either cohesive or granular materials.

8.6.1 Materials specification

(a) Type 1 Sub-base materials

Type 1 granular fill shall be crushed rock, crushed slag, crushed concrete, or well-burnt non-plastic shales which are inert. The material shall be well graded and lie within the grading envelope of Table 8.4.

Table 8.4. Type 1 Sub-base range of grading

BS sieve size	% by mass passing
75 mm	100
37.50 mm	85–100
10 mm	40–70
5 mm	25–45
600 μm	8–22
75 μm	0–10

The material passing the 425 μm BS sieve shall be non-plastic as defined in BS 1377 and tested in compliance therewith.

(b) Type 2 sub-base materials

See Table 8.5.

Table 8.5. Type 2 Sub-base range of grading

BS sieve size	% by mass passing
75 mm	100
37.50 mm	85–100
10 mm	45–100
5 mm	25–85
600 μm	8–45
75 μm	0–10

8.6.2 Definitions

Chalk: Any porous material of natural origin composed essentially of calcium carbonate.

Argillaceous rock: Consists of shales, mudstones and slate composed of clay and silt particles.

Pulverized fuel ash: The resultant ash from pulverized coal burnt at power stations and having a maximum particle size of 3 mm.

Furnace bottom ash: Agglomerated pulverized fuel ash obtained from the bottom of the power station furnace and having a particle size no larger than 10 mm.

8.6.3 Suitable fill materials

- Cohesive soils including clays and marls with up to 20% gravel or rock and having a moisture content of not less than the plastic limit minus 4.
- Well-graded granular and dry cohesive soils including clays and marls containing more than 20% gravel or rock and/or a moisture content of less than the plastic limit minus 4.
- Sands and gravels.

8.6.4 Unsuitable materials

- Material from swamps or marsh land.
- All organic or part-organic materials.
- Material susceptible to spontaneous combustion.
- Colliery shales, ironstone shales and similar materials which have a potential for expansion due to oxidation of pyrites.
- Frozen materials or materials which are frost-susceptible.
- Any materials which have a higher moisture content than the maximum permitted for such materials as defined in the specification.
- Clays with high plasticity index exceeding 55%.

8.6.5 Compaction

The most-used method specification for compaction of structural fills is that used for highway construction and referred to as the Dept of Transport Specification for Roads and Bridge Works.

The number of passes and the thickness of the compacted layers are designed to achieve an adequate compacted dry density. On large sites where mixed fills may be compacted it is generally prudent to carry out compaction trials using the type of plant proposed for the works. The purposes of compaction trials are:

- to determine the thickness of layers and number of passes required to show that the compaction results meet the end result specification;
- to ensure that a method specification gives satisfactory results;

183

- to enable various types of compaction plant to be appraised for performance and end result.

It is most important to ensure that any compaction trial uses materials which are representative of the materials to be employed in the full site operations.

8.6.6 Testing on site

Structural fill should be regularly tested with sample testing of various layers. Tests to determine the *in-situ* dry density can be carried out using various methods such as:

- sand replacement test;
- nuclear density gauge meter (one test every $400\,m^2$).

Both these methods can be used for cohesive or granular soils. Generally, the nuclear density gauges are calibrated from sand replacement tests carried out during the field trials. The main advantages with the nuclear density gauge is its speed and economy.

BIBLIOGRAPHY

BSI (1990) BS1377: Methods of testing for Civil Engineering purposes, British Standards Institute.
Department of Transport (1986): *Specification for Highway Works*, Part 2, series 600: Earthworks, HMSO, London.
Meyerhof, G.G. (1951) Building on fill. *Structural Engineer*, **29** (2).
NHBC (1992) *NHBC Standards*, Chapter 5.2: Suspended Ground Floors.
Tomlinson, M.J. and Wilson, D.M. (1973) Pre-loading of foundations by surcharge on filled ground. *Geotechnique*, **23** (1), 117–120.

Chapter 9
Ground improvement

Significant increases in allowable bearing capacities can be achieved using various ground improvement techniques. These various techniques modify the *in-situ* soil properties generally by increasing the density of the soils.

It is not always necessary to use specialist contractors and there are often various measures that can be adopted which, if suitable, can improve soil densities at a relatively low cost. For example, loose ash fill material overlying soft clays or firm strata may be shallow enough to remove and replace in discrete layers using appropriate vibratory compaction plant. It should be possible to achieve a final density such that pseudo-raft foundations can be used. The main benefit of this method is that the high costs involved in removing fills to tips off site are avoided.

On sites where soft clays are present to depths in excess of 2–3 m it may be possible to remove some of the soft clays and replace them with well-compacted crushed stone fill, placing a pseudo-raft foundation on to the finished formation. If reinforced strip foundations are used the width of granular fill must be at least twice the footing width. The depth of consolidated stone below the footing must be at least 1.50 times the footing width, and this depth should be maintained under the whole dwelling to avoid differential settlements. This method is only recommended where the depth of stone replacement is limited to about 1.20–1.50 m so that the safety of site operatives is ensured. Once temporary shoring is required the method becomes impracticable and uneconomic (Fig. 9.1).

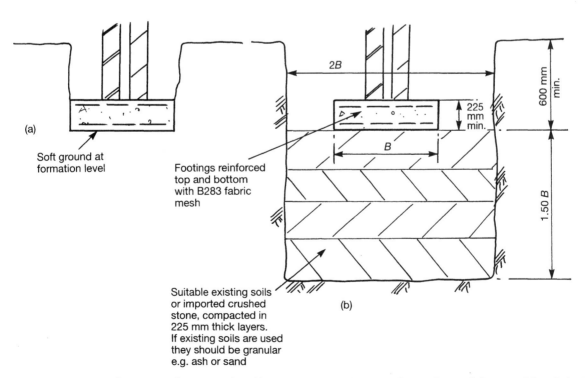

Fig. 9.1 Improving soils by replacement and recompaction. (a) If normal excavations reveal ground too soft for normal foundation loads, it may be possible to remove sufficient weak ground and replace with granular fill, as shown in (b).

On sites where weak natural sands or gravels are evident they can often be improved by recompaction using appropriate vibratory rolling plant. Where sands are weak as a result of a high water table, permanent dewatering systems using land drainage techniques may improve the bearing capacity; specialist advice should be sought on such sites. Quite often, marginal sites show a marked improvement in ground strength following the installation of the deep roads and sewer drainage systems.

On sites where the weak soils are deep there are several methods of ground improvement techniques which are available and which are usually carried out by specialized contractors. Such sites require the services of consulting engineers if the works are to be submitted to local authorities or the NHBC for building control approval.

9.1 VIBRO-COMPACTION TECHNIQUES

The main purpose of vibro-compaction is to stabilize and increase the natural loadbearing capacity of soils, so reducing settlements to within acceptable limits. This is achieved by using a large torpedo-like vibrating poker connected to a large crane rig. Air or water jetting is introduced through holes about the vibrating poker to aid compaction and assist in removing flotsam.

9.1.1 Types of treatment

The three common methods adopted for various ground conditions are:

- vibro-replacement;
- vibro-flotation;
- vibro-compaction.

(a) Vibro-replacement

This uses stone columns installed using the dry method of treatment which uses air jetting. It is suitable for treating soft clays, some silts and inorganic fills (Fig. 9.2).

(b) Vibro-flotation

Using stone columns installed using the wet method of treatment, this involves using water jetting which is more suitable for treating soft clays with undrained shear strengths less than 30 kN/m^2, some silts, and inorganic fills (Fig. 9.3). It is also more applicable for use on sites with a high water table. The process is similar to the dry method except that the soft clay soil is removed, and displaced by water pressure jets fitted to the vibrating poker. The soft soil is flushed out of the hole and the water pressure also keeps the sides of the hole stable.

If very weak soils are treated using the dry method, 'necking in' can occur when the poker is withdrawn, resulting in a stone column containing soft materials. The same situation can occur on sites with a high water table.

Some specialist firms have developed air-flush machines using a bottom-feed technique for placing the stone in ground conditions that would normally be dealt with using the wet method.

(c) Vibro-compaction

This uses the wet or dry method, depending on the water table depths. The existing loose sands or gravels are vibrated using the vibrating poker to densify the natural soils. If required, stone can be charged into the probe points to increase the ground stiffness further. The vibrating poker penetrates down through the soils under its own weight and compacts the surrounding ground, for a distance approxi-

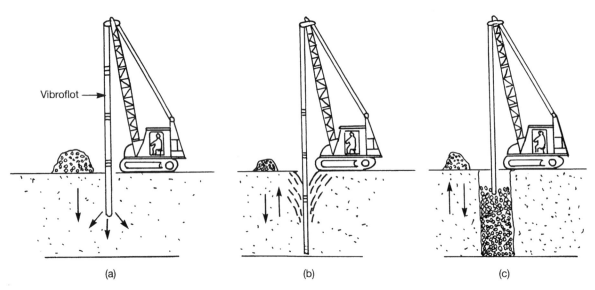

(a)   (b)   (c)

Fig. 9.2 Vibro-replacement: (a) void formed by air jetting from poker vibrator; (b) stone is placed and compacted; (c) compacted stone column formed at close centres.

mately equal to a 2.0–2.50 m radius from the centre of the poker. The probes are generally installed at about 1.50–3.0 m centres depending on the density required (Fig. 9.4).

The use of vibrated stone columns does not increase the strength of the clays themselves but the combined ground and stone column formation improves the overall strength of the clay soils by two or three times their original *in-situ* strength. Settlements of the treated areas will be reduced to between one half and one third of the magnitude of that which would have occurred without treatment.

It is important to recognize that the stone columns produced by vibro-compaction techniques cannot be considered as piles. When a rigid pile in a cohesive soil is loaded, it settles, developing end bearing pressures and skin frictional stresses along the pile shaft. A stone column made of compacted uncemented stone particles also develops end

bearing pressures and skin frictional stresses. However, the stone column also bulges and must therefore be supported by lateral stresses exerted by the soils.

9.1.2 Ground conditions

Seek specialist advice from an independent engineer if vibro-compaction is being considered. The independent engineer should arrange for a detailed site investigation and site appraisal to be carried out to determine:

- the nature, extent and depths of weak ground (Fig. 9.5) – avoid placing dwellings over major changes in strata;
- the suitability of vibro-compaction techniques for the site conditions and whether the wet method or dry method is applicable (Fig. 9.6);

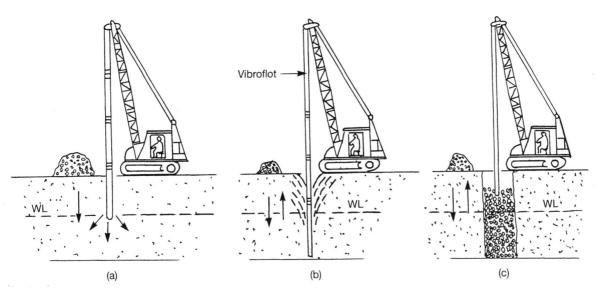

Fig. 9.3 Vibro-flotation (wet method): (a) void formed by water jetting; (b) stone is placed and compacted; (c) stone columns compacted up.

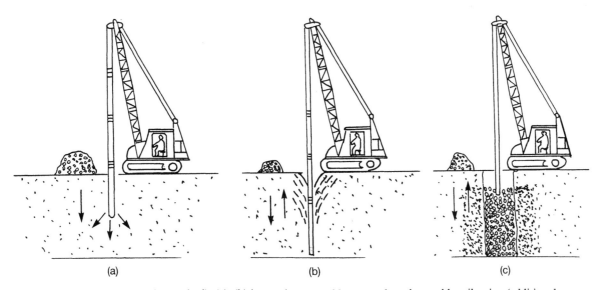

Fig. 9.4 Vibro-flotation (wet or dry method): (a), (b) improving natural loose sands and gravel by vibration (additional stone can be added if required); (c) stone columns formed with a densified zone round the probe point.

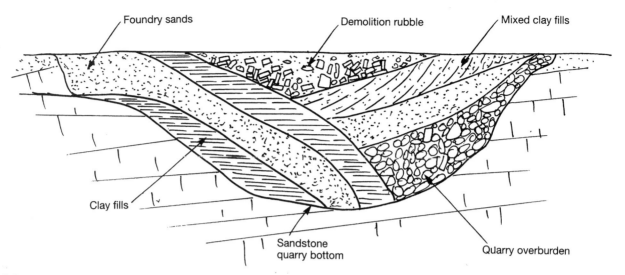

Fig. 9.5 Check ground conditions. Avoid siting houses over fills which show large variations in composition of strata.

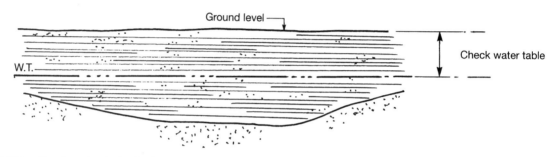

Fig. 9.6 Vibro-flotation: check depth of water table. With water table present, ensure wet process or bottom-feed methods are adopted.

- whether the site layout, house types, site drainage services and foundation designs are suitable for this type of treatment.

When specifying vibro-compaction techniques the following considerations should be applied.

1. Depth of fill or weak strata must be determined. The depth of treatment must be determined and if the fills are not to have full-depth treatment the specialist contractors must satisfy themselves as to the suitability of the untreated zones and ensure that suitable foundations for such conditions are provided.
2. Determine the nature of the soils to be treated with information as to the grading of the constituent materials (Fig. 9.7).
3. Determine whether there is a potential for gas migration or gas generation from any fills present (Fig. 9.8).
4. Determine the distribution of the various constituent materials in any fill materials which may be present.
5. Determine the groundwater levels and check the fills for contaminants. Decide whether to adopt the dry method or wet method (Fig. 9.9). If chemical waste materials are present these can often produce toxic gases which can migrate up the stone columns. In addition, water can pass down the columns and cause some chemical wastes to go into solution which could result in the stone columns bulging and settling (Fig. 9.9).

6. Determine the positions of existing and proposed services and their depths (Fig. 9.10). In situations where the proposed or existing drainage is within the 45° line of repose then the house foundation will need to be deepened such that the line of repose passes below the drain.

Sites generally unsuitable for vibro-compaction are as follows:

- Those sites where fills have been recently deposited and which are still undergoing consolidation settlements.
- Ground which contains thick bands of peat, organic materials and fills which are putrescent. Sites which contain timber, vegetation or topsoils in excess of 15% by volume should not be treated (Fig. 9.11).
- Fills which are poorly graded, containing voided areas, bottles and large objects such as car tyres or car bodies.
- Sites with very soft clays or silts with undrained shear strengths less than 30 kN/m². It may be possible to improve clay soils weaker than 30 kN/m² by using the wet method but clays with undrained shear strengths less than 15 kN/m² are not suitable for treatment.
- Sites where the plasticity index of the clays exceeds 40%.
- Filled sites which may be prone to solution features or inundation arising from a re-established water table, such as chemical fills, chalk fills etc.

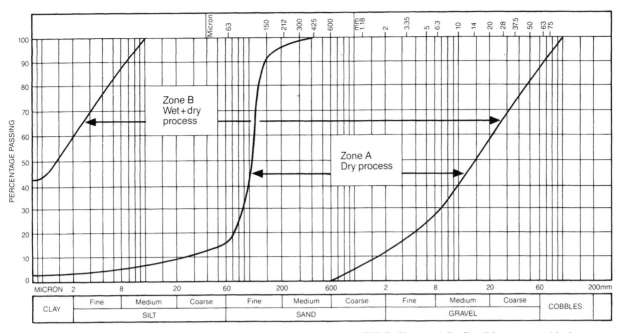

Fig. 9.7 Grading zones for granular strata when using vibro-compaction (from NHBC Chapter 4.6). Conditions acceptable for treatment are only those within zones A and B of the chart.

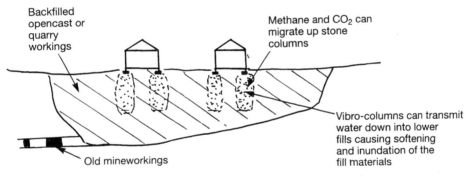

Fig. 9.8 Filled quarries/landfill sites.

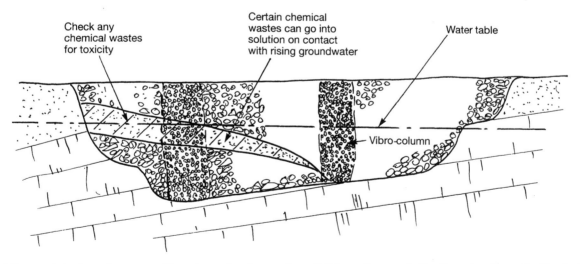

Fig. 9.9 Fills containing hazardous waste. Groundwater levels may rise or fall depending on whether the wet or dry method is adopted. This could affect adjacent buildings.

● Contaminated fills or toxic wastes where toxic or gas migration could result.

In deep-fill situations the specialist contractor may decide to carry out partial-depth replacement treatment. The independent engineer must ensure that the untreated zones of fill are stable and are likely to remain so. In these situations it is usual to provide a raft-type foundation to stiffen the foundation in the event of any residual settlement of the untreated fills.

189

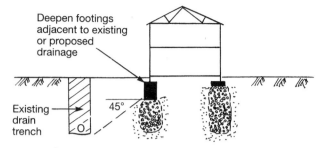

Fig. 9.10 Existing or proposed deep drainage.

When vibro-compaction is used on housing sites, there are factors relating to the housing layout which need to be taken into account:

- The location of existing or proposed **tree planting**. In clays with a plasticity index between 20% and 40% the depth of the foundations must be in accordance with *NHBC* Chapter 4.2 (Fig. 9.12).
- The presence of any existing or proposed **drainage** services.
- The location of any existing **buildings**. When existing buildings are close to the treated zones, consideration must be given to the short- and long-term effects of vibro-compaction treatment. Where large volumes of water are used in the wet process the groundwater levels may rise, resulting in ground heave or settlement of adjacent buildings (Fig. 9.13). Alternately, vibro-compaction stone columns can act like vertical french drains and cause a lowering of the groundwater levels. This may cause a loss of fines if the stratum is silty or granular, and could result in settlement of adjacent buildings.
- Consider the **housing layout** relative to the ground conditions and avoid having long terraces in locations where major variations in the ground are anticipated.

Provide movement joints at suitable locations in long terraces, especially at points where the storey heights change (Fig. 9.14).

The main problem and disadvantage with vibro-compaction ground improvement is that the treatment is carried out blind. It is not uncommon to find that areas have been treated which contain thick bands of topsoil fills or other organic matter and generally these are only revealed by accident during drainage excavations or if a stone column test fails. Random testing is most likely to miss such problems.

9.1.3 Engineering supervision

The most important consideration when using vibro-compaction on a site is to provide adequate engineering supervision and testing, independent of the specialist contractor. This supervision should ensure that the required depths of treatment have been achieved. It should ensure that the stone columns have been accurately positioned. Where excessive amounts of stone have been placed in certain probes this could be an indicator of very poor ground and such locations should be examined critically and subjected to more testing. It may be necessary to provide a stronger foundation, such as a raft.

(a) Testing following ground treatment

Testing should consist of quick plate bearing tests on individual stone columns selected at random as a control on workmanship, together with larger-scale dummy footing or zone tests to simulate the type of foundation loads likely to occur. The small-scale plate tests are usually carried out by jacking against heavy plant. It is *most* important that the

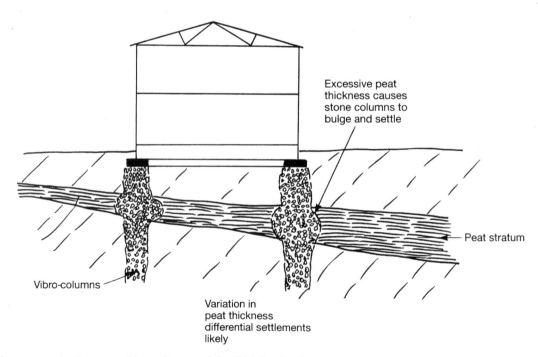

Fig. 9.11 Vibro-compaction is not suitable on sites containing thick bands of peat.

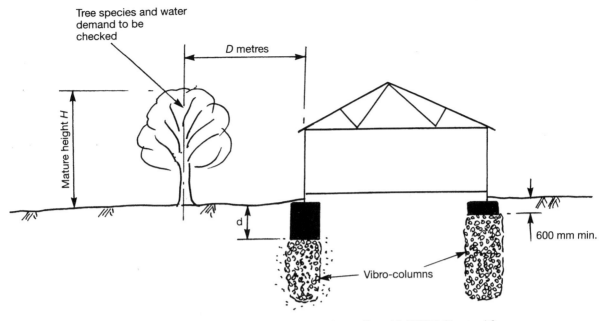

Fig. 9.12 Sites affected by trees: on clay sites check that foundation depth *d* complies with NHBC Chapter 4.2.

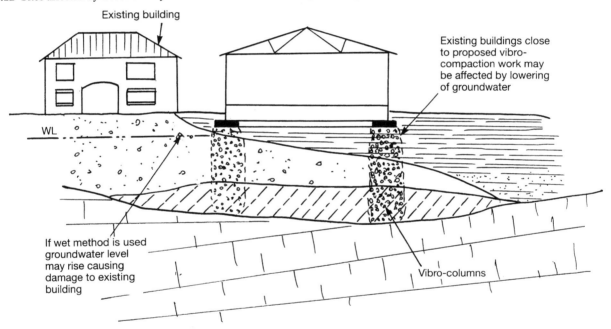

Fig. 9.13 Check ground conditions when close to existing buildings.

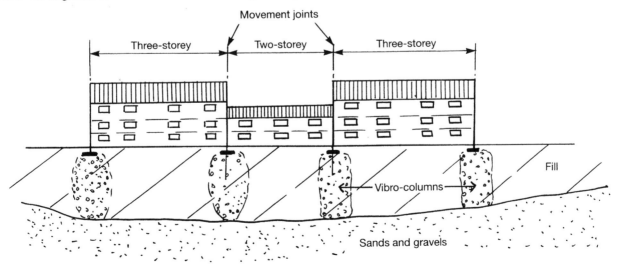

Fig. 9.14 Provide movement joints at critical locations in long blocks.

porewater pressures in the soils are allowed to dissipate before testing columns, otherwise unreliable results could arise.

Most specialist contractors like to be off site as soon as the site has been treated and it is useful therefore to have the majority of the testing carried out at the early stages in the ground treatment.

(b) Loaded skip tests

For a more meaningful test, to establish that the total ground mass has been brought up to a specified bearing capacity, it is recommended that a loaded skip test be carried out (Fig. 9.15).

A loaded skip positioned over a dummy footing or raft pad can be tested about 3 days after vibro-compaction has finished. By this time, any porewater pressures should have stabilized. This type of test is generally monitored for longer periods than the plate bearing tests on stone columns and if meaningful results are to be obtained this timescale should be about 2–3 weeks.

(c) Plate test procedure

Generally the testing should be in accordance with the Institution of Civil Engineers' *Specification for Ground Treatment* (ICE, 1987) or in accordance with BS 5930:1981. All such tests should be carried out under the direct supervision of competent experienced personnel.

The hydraulic jack and load-measuring gauges should be located so that they are stable when the maximum load required is applied. The dial gauges should be positioned clear of the loaded plate. The area under the plate should be cleared of any loose material and blinded with a layer of sand not exceeding 15 mm in thickness. The load applied should be measured using a proving ring or similar load-measuring device which should be calibrated before and after each series of tests. Three deflection gauges should be positioned symmetrically around the plate.

The maximum load to be applied should be three times the working load and it should be applied in six approximately equal increments. Following each application of loading increment the settlements should be measured at intervals of 1 min until no change is detected, and then read at intervals of 5 min. The load should be held for 10 min for each incremental stage. The maximum load should be kept on for 15 min and measured on release of the load and again after 5 min.

For a plate bearing test a report should be prepared giving the maximum load applied, the period of time it was held, the maximum recorded settlement and the recovery on unloading.

This type of test should be carried out over the duration of the contract period at a rate of one test for every 100 vibro-compaction probes. Generally, settlements on stone columns on low-rise housing should be up to a maximum of 10 mm. Any settlements in excess of 15 mm should be investigated.

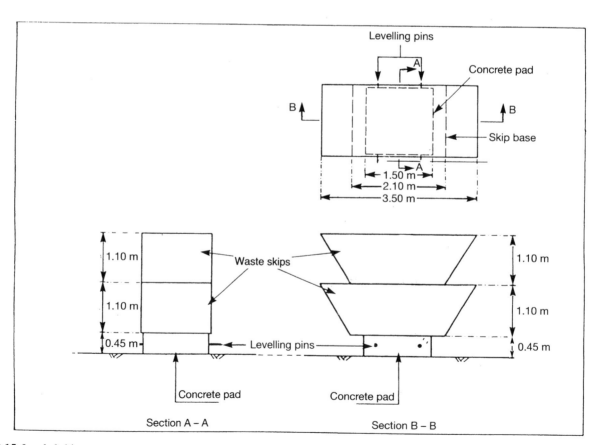

Fig. 9.15 Loaded skip test.

(d) Zone testing

Zone testing (Fig. 9.16) is usually carried out over groups of stone columns using suitable kentledge. This type of testing allows the interlocking ground matrix around the stone columns to be quantified. The loadings applied in the test should simulate the dwelling loads likely to be placed on the fills with an appropriate load factor of about three times the unit loading required.

(e) Trial holes post-treatment

It is often useful to carry out random trial pit investigations or boreholes following ground treatment, taking SPT tests, pressuremeter tests etc. to check on the improvement achieved.

(f) Zone test procedure

The zone test is a loading test carried out with a raft slab or a loaded skip situated over a strip, pad foundation or dummy footing. The ground pressures exerted over a wider and deeper zone of treated ground will be more relevant to the loads imposed by the dwelling.

The surface around the test area should be cleared of all loose material and blinded with a 50 mm thick layer of blinding concrete. The builder's skip can be loaded with a known weight of sand and, if required, skips can be placed one on top of the other or placed in tandem. The settlement gauges should be placed symmetrically around the loading platform and be mounted on a reference frame which will not be affected by seasonal ground movements.

The test load should be applied in increments of one quarter working load in three stages:

1. to working load;
2. to 2.0 times working load;
3. to 2.50 times working load.

The incremental loading should be applied until the rate of settlement under the preceding load is less than 0.5 mm per hour. The deflection gauges should be read every 5 min and the average of the values taken as the rate of settlement. The test load should be removed in similar stages and the final recovery should be measured.

9.1.4 Design of vibro-compaction stone columns

Experiments have shown that the ultimate load capacity of an isolated stone column is governed primarily by the maximum lateral reaction of the soil around the bulging zone, and that the extent of vertical movement is limited.

From pressuremeter theory:

$$S_{rl} = S_{ro} + 4c + u$$

where S_{rl} = limiting radial stress; S_{ro} = *in-situ* lateral stress around the stone column; u = porewater pressures. Generally, the value of $u = 0$, as the porewater pressures are dissipated fairly quickly by the installation of the stone column. Therefore

$$S_{rl} = S_{ro} + 4c$$

At failure of the stone column in the bulged section

$$S_v^i = K_p S_{ro}^i$$

where

K_p = coefficient of passive earth pressure for the stone used in the column,
= $(1 + \sin \phi) / (1 - \sin \phi)$ when ϕ = angle of internal friction of the stone column;
S_v^i = vertical effective stress;
S_{ro}^i = lateral effective stress.
Therefore

$$S_v^i = \frac{(1+\sin \phi)}{(1-\sin\phi)} \cdot (S_{ro} + 4c)$$

$$S_{ro}^i = K(\gamma h + p)$$

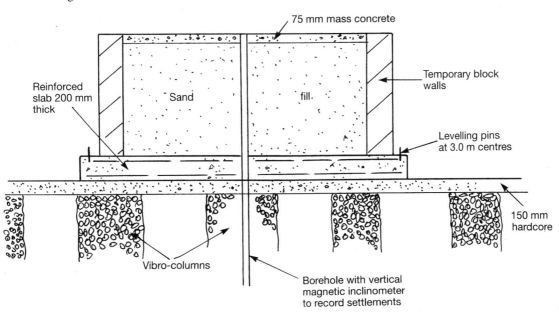

Fig. 9.16 Long-duration zone test.

193

where

K = coefficient of earth pressure;

h = the critical depth, i.e. that depth at which failure of the stone column by bulging occurs (this is generally one or two columns diameters measured below the foundation level);

γ = density of the surrounding soil mass;

p = the extra load imposed on the surrounding soil mass by the foundations.

Values for K vary, depending on the ground conditions, but in general an at-rest value of 1.0 can be used.

Therefore

$$Q_{ult} = \frac{(1+sin\phi)}{(1-sin\phi)}.(\gamma + 4c + p)$$

Example 9.1 Design of stone columns

Triaxial compression tests carried out on borehole samples give undrained cohesions in the range 45–70 kN/m². A value of C_u = 60 kN/m² can be used. ϕ = 44°; γ = 19 kN/m³. With stone columns at 2.05 m centres, calculate the safe working load per column and the reduction in settlements.

Failure of the stone columns is most likely to occur by bulging in the fills and will occur approximately 1.60 m below the ground surface. Using Hughes and Withers (1974), the ultimate vertical stress in a stone column is given by

$$Q_{ult} = \frac{1+sin\,44}{1-sin\,44} \times (1.60 \times 19 + 4 \times 60 + 25)$$

$$= \frac{1.69}{0.31} \times (30 + 4 \times 60 + 25) = 1608 \text{ kN}$$

Using a 650 mm stone column

$$\text{Area} = \frac{0.65^2\,\pi}{4} = 0.331 \text{ m}^2$$

Therefore ultimate load on column = $1608 \times 0.331 = 532$ kN

Assume 33% of the load is carried by the surrounding ground and 67% is carried by the stone columns. Therefore

$$\text{Effective ultimate load carried by each stone column} = \frac{532 \times 3}{2}$$

$$= 798 \text{ kN}$$

Applying a factor of safety of 2.50:

$$\text{Safe working load} = \frac{798}{2.50} = 320 \text{ kN}$$

Consider the worst loading case: A 900 mm wide strip footing loaded at 150 kN/m² with compaction points at 2.05 m centres.

Maximum load on each compaction = $150 \times 0.90 \times 2.05 = 278$ kN

Safe working load = 320

Therefore design OK for strength.

Check reduction in settlement. Using approximate method: maximum depth of compressible fill = 3.90 m; foundation depth = 800 mm. Without treatment, settlement in compressible layer:

p = $m_v\,h\,\sigma z$

m_v = 0.10 m²/MN (obtained from soil report)

h = 3.90 − 0.8 = 3.10 m

σz = 150×0.41

Using Boussinesq's pressure charts we have

$$\frac{z}{b} = \frac{0.8}{0.9} = 0.88$$

$$\frac{a}{b} = \frac{2.05}{0.90} = 2.27$$

Therefore

$\sigma = 0.10 \times 3.10 \times 150 \times 0.41 = 19$ mm

Reduction in settlement after treatment. Area of stone column = 0.332 m². Area of strip footing = $2.05 \times 0.90 = 1.845$ m². From Priebe charts

$$\frac{A}{A_c} = \frac{1.845}{0.332} = 5.60$$

Settlement reduction factor = 2.10

Hence approximate settlement after treatment $= \dfrac{19}{2.10} = 9.0$ mm

Therefore design for settlement OK.

This settlement is likely to be conservative since no account is taken of the improvement in the *in-situ* fills. In addition the stone columns will provide short drainage paths allowing dissipation of porewater pressures which will result in an increase in the ground strength and an improved factor of safety.

Example 9.2 Improving mixed clay fills

A site investigation has revealed mixed clay fills to a depth of 5.50 m and the report has recommended the fills be improved using vibro-compaction to bring the ground up to 100 kN/m² capacity. The depth of treatment will be 5.20 m with probe spacings at 2.0 m. Dwellings are to be placed on 600 mm wide strip footings reinforced top and bottom with B283 mesh. Ground floors are to be precast beam and block with polystyrene, polythene vapour barrier and moisture-resistant chipboard as the floor finishes.

BEARING CAPACITY

The ultimate bearing capacity of the stone columns is calculated using the standard formula developed by Hughes and Withers (1974):

$$\sigma_v = \frac{1+sin\,\phi}{1-sin\,\phi}.\gamma_b h + 4c$$

where

ϕ = angle of friction for stone column = 45°;

γ_b = bulk density of soil = 19 kN/m³;

h = depth at which lateral stress is considered = 1.30 m;

c = 40 kN/m² for clayey fills.

Therefore

$$\sigma_v = \frac{1+0.707}{1-0.707} \times 19 \times 1.30 + 4 \times 40$$

$$= 5.83 \times 24.70 + 160 = 1076.8 \text{ kN/m}^2$$

Therefore ultimate load $= \dfrac{\sigma_v \pi D^2}{4} = \dfrac{1076.8 \times \pi \times 0.70^2}{4} = 414$ kN

where D = stone column diameter = 700 mm.

Applying a factor of safety of 3.0

$$\text{Design load} = \frac{414}{3.0} = 138 \text{ kN}$$

LOAD ON STONE COLUMN AND SURROUNDING SOIL MASS

Total load = $100 \times 2.0 \times 0.60 = 120$ kN < 138 therefore OK.

This is less than the safe capacity of the stone column, therefore the design is safe. In practice, the load would be shared between the stone columns and the soil mass between the probes.

SETTLEMENT

With the strip footing fully loaded to a bearing capacity of 100 kN/m² over the soil profile shown, the mean vertical pressure can be obtained from Design Chart No. 5 (Fig. 9.17).

The depth of stressed ground is split into the various strata strengths and the total settlements are calculated as shown in Table 9.1, where

d_h = thickness of each layer;

z = depth from ground level to centre of strip being considered;

b = width of foundation;

p/q = mean pressures obtained from Design Chart No. 5 for values of z/b;

m_v = coefficient of consolidation.

Therefore

layer (a) s = $1.45 \times 30 \times 0.1 = 4.35$ mm

layer (b) s = $1.45 \times 13 \times 0.1 = 1.89$ mm etc.

Table 9.1. Example 9.2: calculation of total settlements

Layer	d_h	z	z/b	$\Delta p/q$	Δp	m_v	Δs
a	1.45	1.325	2.2	0.3	30	0.1	4.35
b	1.45	2.775	4.6	0.13	13	0.1	1.89
c	1.45	4.225	7.0	0.10	10	0.1	1.45
d	0.65	5.275	8.8	0.09	9	0.15	0.88
e	1.47	6.335	10.6	0.07	7	0.05	0.51
f	1.47	7.805	13.0	0.05	5	0.05	–
g	1.47	9.275	15.5	0.03	3	0.05	–

Total settlement = 9.10 mm

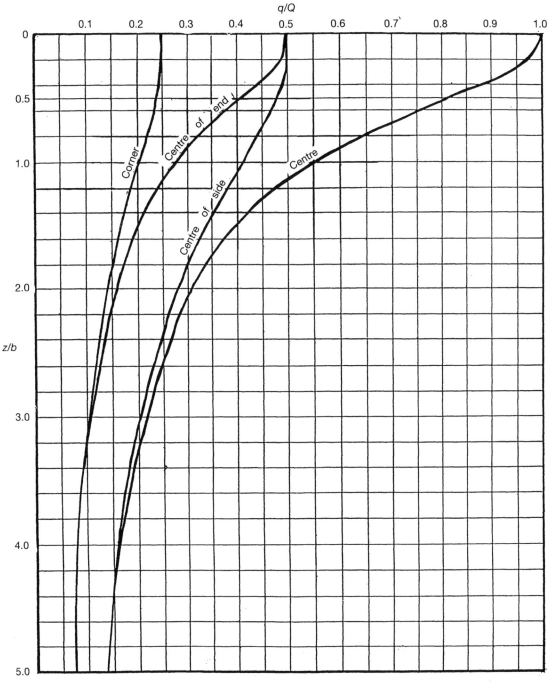

Fig. 9.17 Design Chart No. 5: distribution of vertical stress beneath a long strip footing.

Example 9.3   Partial depth treatment of filled ground

A site investigation report has revealed the presence of a backfilled quarry on a housing development. The upper fills in the quarry are described as soft to firm silty clays within a stony matrix for the top 4 m, consisting of end-tipped quarry overburden materials. These fills were deposited 30 years previously and contain no contaminants or putrescent materials. Below the 4 m depth the fills are in the medium dense to dense category, as established by dynamic probe testing.

The site is to be developed with bungalows and two storey houses constructed on pseudo-raft foundations placed on a 600 mm minimum thickness of recompacted fill. Design the vibro-compaction proposals and estimate the likely settlements if the allowable bearing capacities under the edge beams are to be brought up to 100 kN/m^2, using a 900 mm wide edge beam toe.

DESIGN CRITERIA (Hughes and Withers, 1974)

The shear strength and compressibility of the upper fills based on the SPT results place them into the soft to firm category. BS 5930: 1981 indicates that soft to firm clays have shear strengths in the 40–60 kN/m^2 range. Taking $C_u = 40$ kN/m^2, then

$$M_v = \frac{1}{100 C_u} = 0.25 \text{ m}^2/\text{MN}$$

Check for shear failure:

$$Q_{ult} = 5.70 \, C_u = 5.70 \times 40 = 228 \text{ kN/m}^2$$

With factor of safety = 2.50

$$Q_{all} = \frac{228}{2.50} = 91.20 \text{ kN/m}^2$$

Provided final settlements are of a low order then the effective factor of safety is equal to $5.70 \, C_u/100 = 2.28$.

SETTLEMENT CRITERIA

To determine the total settlements, the vertical stress on the soils around the stone columns must be determined, based on equal vertical strains in the fills and stone columns.

$$Q_c = R \, Q_s$$

where Q_c = vertical stress in stone columns; Q_s = vertical stress in fill materials; $R = E_{column}/E_{fills}$. Usually R is within a range of 10–25 (Hughes and Withers, 1974).
For vertical equilibrium

$$A \, Q = A_c \, Q_c + A_s \, Q_s$$

where Q = allowable bearing pressure; A = foundation area; A_c = stone column area; $A_s = A - A_c$.

Using a foundation width of 900 mm and stone column area of 0.30 m^2 at 2 m centres,

$$\frac{A}{A_c} = \frac{0.90 \times 2.0}{0.30} = 6$$

If we consider a single stone column the ultimate working stress is given, from Hughes and Withers (1974), by:

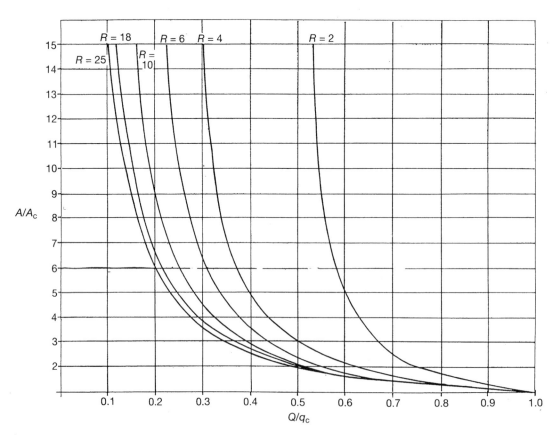

Fig. 9.18 R value graph: $R = E_{column}/E_{soil}$.

$Q_{cu} = K_c (Q_{ro} + 4 C_u)$

$Q_{ro} = K_s (\gamma h + \text{surcharge})$

$K_c = \tan^2 (45 + \phi_c/2) = 5.80$ for ø = 45

K_s and $K_o = 0$; $\gamma_w = 18$ kN/m³; depth $h = 1.20$ m (column diameter + foundation depth). Therefore

$Q_{ro} = 18 \times 1.20 = 21.60$ kN/m²

$Q_{cu} = 5.80 \times (21.60 + 4 \times 40) = 1050$ kN/m²

With a factor of safety of 2.50:

$$Q_c = \frac{1050}{2.5} = 420 \text{ kN/m}^2$$

Therefore

$$\frac{Q}{Q_c} = \frac{100}{420} = 0.24$$

From Fig. 9.18 $R = 15$.
  The vertical soil stress

$$Q_s = \frac{Q_c}{R} = \frac{420}{15} = 28 \text{ kN/m}^2$$

Using Fig. 9.19:

Settlement $S = M_v Q_s B R_o$

where B = breadth of footing; R_o = integral of stress function. At depths exceeding $z/B = 3.50$ the vertical stress increase will be insignificant and the analysis can therefore be limited to this depth. For z/B of 3.50, $R_o = 1.57$ (Fig. 9.19).
Therefore

$S = 0.25 \times 10^3 \times 28 \times 0.90 \times 1.57 = 0.01$ m = 10 mm

Therefore carry out vibro-compaction to a minimum depth of 4 m with compaction points at 2.0 m centres in accordance with NHBC Chapter 4.6, 1992.

9.1.5 Foundations on vibro-compaction sites

Where full-depth treatment is carried out on weak ground the use of reinforced strip footings together with a fully suspended ground floor should suffice in accordance with NHBC Chapter 4.6, 1992. Because the poker vibrator loosens the ground surface, these foundations must be placed

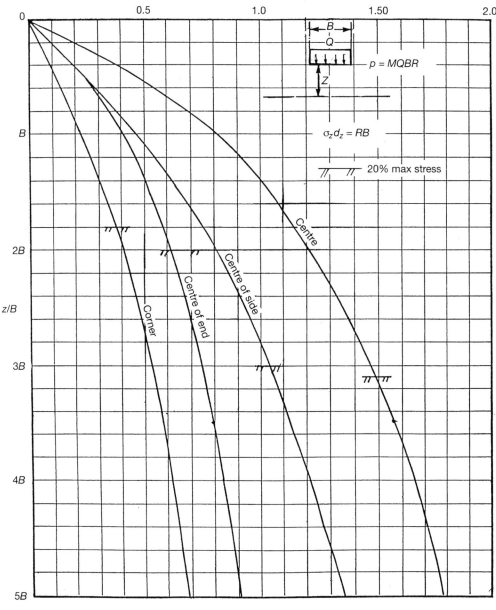

Fig. 9.19 Integral of stress beneath long strip footing.

at least 600 mm below the treated surface level. In clay soils the foundation should be placed at the appropriate depth to cater for seasonal movements (Fig. 9.20).

If, during the ground treatment works, some of the stone columns are placed out of position, some redesign of the foundation will be required. It may be necessary to place a raft foundation down in lieu of strip footings and in this event the top 600 mm of soils should be removed and recompacted using suitable vibratory plant.

Where only partial depth treatment is carried out a raft foundation is generally the most suitable foundation provided the soils below the treated depth have been assessed and judged to be adequate. Again, the top 600 mm of ground needs to be removed and recompacted in thin layers with a vibratory roller (Fig. 9.21). The raft itself should be designed to span 3.0 m and cantilever 1.50 m at the corners with the slab element designed to span between the centres of probes.

If, during excavations for foundations, weak or unsuitable strata are encountered, then the strip foundations should be taken below the unsuitable strata and stepped back once the length of unsuitable ground runs out (Fig 9.22).

Any hard spots such as old walls or foundations, should be removed and well-compacted replacement crushed stone should be provided to give a minimum cushion of 900 mm between the foundation and the obstruction (Fig 9.23).

9.2 DYNAMIC CONSOLIDATION

This system of ground improvement involves dropping a large heavy block or tamper about 5–20 tonnes in weight from various heights, usually between 2.0 m and 25 m, using free-fall crane jibs. The energy put into the ground packs the individual soil particles tighter, so densifying the strata.

The types of soil which can be treated successfully include granular fills, sands and gravel deposits which are loose, hydraulically placed sand fills and even loose silty sands, depending on the silt content. Marginal soils such as peats, soft saturated clays and silty alluvium can be treated by this method, but they require special consideration. The method is generally only economic for large open sites in excess of 12000–15000 m².

One of the major restrictions on its use is the effect of the shock waves which travel through the ground, and its use close to existing buildings or existing services may give cause for concern. In such situations it is prudent to carry out some prior testing to determine the magnitude of the

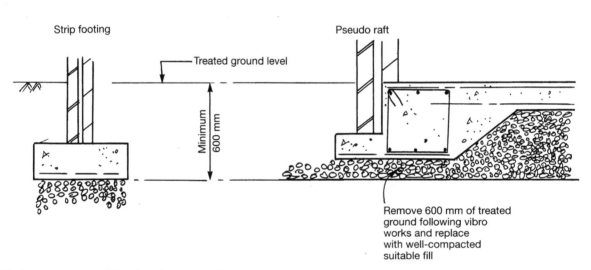

Fig. 9.20 Ground treatment following vibro-compaction.

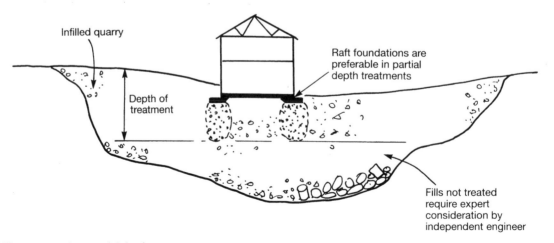

Fig. 9.21 Vibro-compaction: partial depth treatment.

198

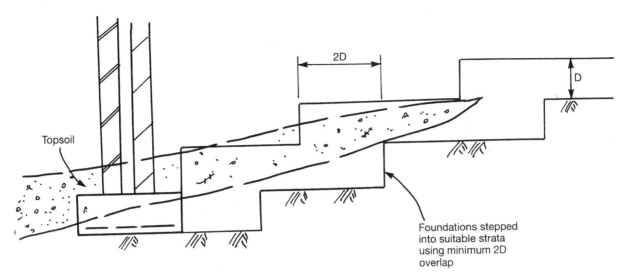

Fig. 9.22 Foundation treatment when unsuitable formation is encountered following vibro-compaction.

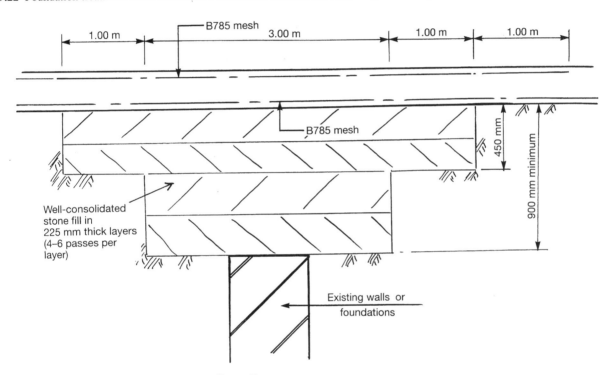

Fig. 9.23 Avoiding hard spots: foundation over cellar walls.

peak particle velocities transmitted along the ground (Fig. 9.24). Some existing buildings have fire sprinkler systems which can be activated by such ground disturbances. Distances of 30–50 m are generally considered a safe distance but in some cases these can be reduced by excavating deep cut-off trenches to break the path of the shallow ground vibrations.

The method of treatment consists of dropping the tamping weight over a grid pattern in a series of passes. Generally, the first pass is aimed at compacting the deepest loose layers of strata over a particular area. The second pass is generally on intermediate grid positions; several passes may be required until the grid is closed up and the whole of the site has been treated. Between passes all the compaction craters are backfilled with suitable material and levelled off before the next pass.

A final pass, known as the **ironing pass**, completes the treatment and is aimed at compacting the upper surface materials. To achieve this, the tamping weight is dropped from a height of 2–3 m at overlapping spaces. The site can be finished off by using a large vibrating roller in preparation for construction of foundations.

During such ground treatment there is likely to be a build-up in excess porewater pressures. These need to have dissipated before the next series of tampings begins. On certain strata the dissipation of porewater pressures can be accelerated by the formation of drainage paths in the soil mass.

Prior to carrying out any dynamic consolidation a full site investigation should be carried out to obtain samples for laboratory testing, *in-situ* testing and information on the underlying soils to assess their likely behaviour during treatment. The levels of any groundwater should be determined.

199

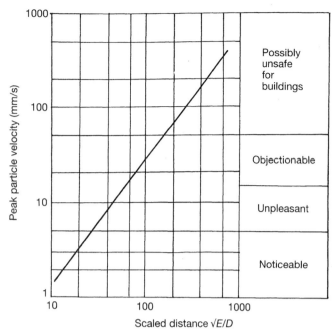

Fig. 9.24 Chart of peak particle velocity versus single-impact energy and distance from impact point: *E* in joules, *D* in metres.

Once the site investigation is available the initial treatment plan can be worked out and the following critical parameters can be evolved:

1. the size of weight to be used, *W* (tonnes);
2. the free-fall height, *H*(m);
3. total energy input;
4. the tamping pattern and the number of passes required to enforce the settlements required;
5. with the information on porewater or water levels the time interval between successive passes can be established.

The treatment depth, *D* (metres) is given by

$$D = 0.50\sqrt{WH}$$

for silty soils, fills and rubble (modified Menard formula). The number of drops per imprint can then be determined relative to the spacing of the drops, the size of the weight and the drop height.

On certain sites the fill materials could be lowered by up to 10–15% in depth following compaction. Because of the size of the plant required for this work it is often necessary to provide a thick crushed stone blanket for the machines to travel on. On weak ground this is generally about 1.0 m thick.

It is important to provide the correct amount of imported fill so that after compaction the site levels can be brought up to final levels by importing the required amount. If the fills are too high before treatment, and the fills do not compact down as much as intended, then surplus materials will have to be excavated and removed off site at considerable cost.

It is therefore sensible to leave the site lower than the final levels following treatment.

Example 9.4 Dynamic consolidation

A backfilled stone quarry has been filled with quarry overburden, mixed stoney clays and building rubble. The maximum depth of fill

is 8 m. It is proposed to improve the ground bearing capacity up to a minimum of 50 kN/m² and reduce settlements to within acceptable limits. Stiff edge beam pseudo-rafts will be constructed for the two-storey dwellings on completion of the site testing.

PRELIMINARY DESIGN

Using the modified Menard formula for mixed granular fills:

$$D = 0.50\sqrt{WH}$$

where *D* = depth of treatment (m); *H* = drop height of tamper (m); *W* = weight of tamper (tonnes).
It is considered acceptable to consolidate the top 6 m of fill. Therefore

$$6 = 0.50\sqrt{8.0\,H}$$

$$36 = 0.25 \times 8\,H$$

$$H = \frac{36}{8 \times 0.25} = 18.0\text{ m}$$

Check depth using

$$D = \frac{WHK}{B^2}$$

where
D = depth of treatment;
W = weight of tamper;
H = drop height of tamper;
B = plan size of tamper;
K = dynamic resistance of soil skeleton.
From the results of previous work,
K lies between 0.10 and 0.16, depending on the soil type and density.
From site investigation take a value of 0.13 for *K*.
Therefore

$$H = \frac{DB^2}{KW} = \frac{6.0 \times 1.80^2}{0.13 \times 8.0} = 18.60\text{ m}$$

Therefore drop an 8 t weight from 20 m height.

Energy required = 0.60 *W H* t m/m²

Consider a 5.0 m square grid pattern which is to be given three passes using various drop heights.

Pass No. 1   Five drops from a height of 20 m at each 5.0 m grid intersection.

Pass No. 2   Three drops from a height of 15 m at each 5.0 m offset square grid position.

Pass No. 3   A double continuous ironing pass: the 8 tonne weight to be dropped twice at overlapping positions from a height of 5 m.

ENERGY INPUT

Pass	*W* (tonnes)	Drop height, *H* (m)	No. of drops	Grid	Energy/m²
1	8	20	5	5.0 m	$\frac{5 \times 20 \times 8}{5 \times 5} = 32$
2	8	15	3	5.0 m	$\frac{3 \times 15 \times 8}{5 \times 5} = 14.4$
3	8	5	2	Continuous	$\frac{2 \times 5 \times 8}{1.8 \times 1.8}$ $= 24.7$

Total energy = 71.1

For a 6 m design depth of treatment this is equivalent to

$$\frac{71.10}{6.0} = 11.80 \,\text{t m/m}^3$$

Using an 8.0 t tamping block of the solid steel type. Average enforced settlements over the treated area will be in the order of 200–300 mm. Porewater pressures to be monitored during and between the various passes.

9.2.1 Testing

On large sites a trial area can be compacted and testing equipment can be calibrated during this operation. Standard tests on completion, such as Dutch cone penetrometer, SPT Menard pressuremeter, and loading, can be carried out to verify the consolidation performance.

One of the most useful ways of assessing the improved density of filled ground is to excavate deep trial pits at specific locations around the site. In certain cases it may be necessary to provide safety shoring in accordance with statutory requirements but the benefits obtained in being able to examine the strata at depth justify the cost.

During operations, rapid-response piezometers are usually installed to measure any changes in porewater pressures, and these can be used to determine the time interval between finishing the consolidation and starting the testing programme.

Large-scale zone tests or dummy footing tests can be carried out similar to those described for vibro-compaction sites.

9.3 PRELOADING USING SURCHARGE MATERIALS

Preloading with temporary surcharge to enforce settlements and speed up the consolidation settlements can be a cheap and effective way of improving filled ground or weak and highly compressible strata. Generally, the additional weight consists of imported materials such as crushed stone, sand and clays, and if the site is large enough for these materials to be used in other areas following preloading then the economies will be even greater.

When a soil mass is subjected to additional loading from house foundations, the ground compression causes a reduction in the pore spaces of the soil matrix. If the ground is fully saturated there will be an initial excess of porewater. For the soil to compress, the porewater has to drain out of the soils. This has the effect of gradually transferring the new compressive stresses to the soil particles, increasing the effective pressure until all the excess porewater pressures have dissipated. This process is known as **consolidation**.

For soils with low permeability this process can take several years. Highly compressible peats, soft silty alluvium etc. will settle significantly over long time periods. In such soils the drainage paths are long and contorted. By introducing sand wicks or proprietary vertical drains the drainage paths can be shortened and the consolidation process can be speeded up (Fig. 9.25). As the surcharge loads are applied, the fills or weak strata consolidate. Water from the soil matrix is squeezed out, reducing the void ratio of the soil mass.

The process can be lengthy, because until the porewater pressures have stabilized, settlements are still occurring. The timescale is very dependent on the soil structure, its permeability and the efficiency of its lateral drainage paths. If time is the major constraint on development the process can be quickened by installing sand wicks or vertical drains. However, it requires a detailed geotechnical study to determine the spacing and installation of such drainage.

In determining the required height of the applied surcharge fill, the surcharge should result in a larger increment

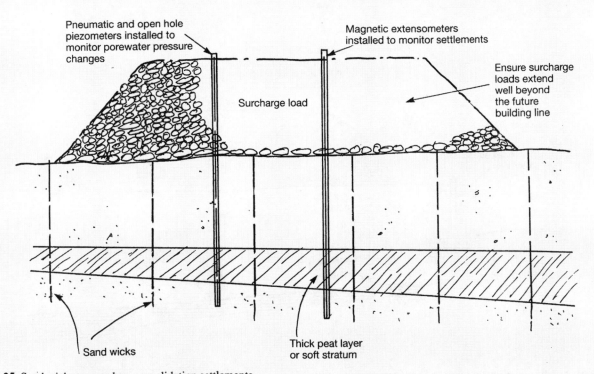

Fig. 9.25 Sand wicks to speed up consolidation settlements.

of effective vertical stress than that which will be applied because of the weight of the houses to be built on the site. A load factor of about 1.20 times the anticipated bearing capacity resulting from the foundation loads generally produces an adequate margin of loading.

The major points to be determined when considering this method of ground improvement are as follows.

1. Is there a ready supply of suitable material for surcharging the ground at an economic rate? If the material can be re-utilized on completion of loading as sub-base for roads, houses etc. then it can be an economic method.
2. Will the preloading be effective?
3. Check the area around the proposed surcharge mounds to ensure that no structures, services, or retaining walls will be damaged by the lateral movement and heave effects of the soil mass under and away from the surcharged mounds (Fig. 9.26).
4. What height of surcharge is required and how long will it need to be left in place? If time is a key element the height of surcharge may need to be increased.
5. At what depth will the surcharge load improve the properties of the loose fills or weak soils?

During the placing of the surcharge materials, surface and below-ground levelling stations can be installed for monitoring by precise levelling techniques. In addition, magnetic extensometers can be installed in boreholes to record level changes. A full range of piezometers needs to be installed to monitor the porewater pressure variations.

9.4 IMPROVING SOILS BY CHEMICAL OR GROUT INJECTION

There are various injection methods presently used for improving weak strata:

- cement-based;
- chemical (sodium silicates);
- organic resins (phenplasts);
- high-pressure cement–bentonite injections.

The various methods have limitations, depending on the particular soil types. The first three methods are not very suitable for fine-grained soils such as silts. In addition, these methods can be very expensive, in terms of plant and material costs. Another disadvantage is the effects of the chemicals on any groundwater present and other environmental factors.

The high-pressure cement–bentonite injection technique can be used on most soils, especially those with a high silt content. The grout suspension contains cement only with a resultant high compressive strength.

The aim of injection grouting is to consolidate loose deposits of sands and gravels, fill materials and alluvium. A detailed site investigation with field trials is generally advisable if a grout injection method is to be adopted for improving weak ground. The use of grout injection is a complex technique, best carried out by specialists who are skilled in the operation and control of the grouting and have vast knowledge of special grout mixes for various soil conditions.

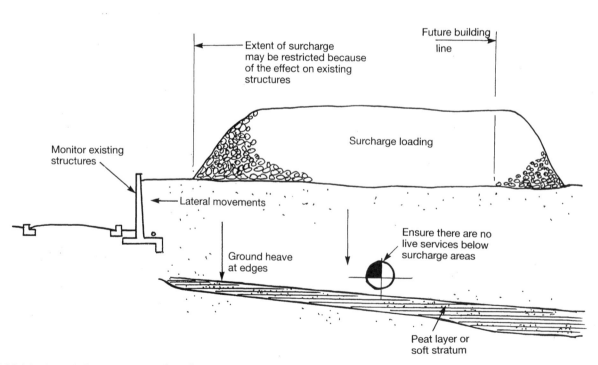

Fig. 9.26 Monitor existing structures and services.

The grout is generally placed via boreholes using traditional drilling equipment. Once the drilling depths have been reached the flushing operation for clearing the borehole is replaced by a high-pressure valve in the jet housing located at the base of the drilling rod. The suspension of cement–bentonite mixture flows out under pressure from the jets in the lower section of the lance at high velocity. Rotation of the lance during withdrawal results in a cement-stabilized column of soil.

The size and compressive strength of these cementitious columns can be varied by adjusting the rate of withdrawal, pumping pressures, and the use of various jet diameters. The columns can be used in a contiguous operation, enabling horizontal walls to be formed. Alternatively, separate columns on a closed spaced grid can be used for general ground improvement. The system can be used for general ground improvements, underpinning, and reducing settlements to existing foundations.

When grouting sands and gravel strata, a casing pipe is generally required to prevent collapse of the hole during drilling. Grouting is then carried out from the base of the casing as it is raised in small increments. Alternatively, the 'tube-a-manchettes' arrangement using perforated pipes with rubber sleeves can be used.

Organic resins have been developed to create special grout mixes which are very viscous during injection. The setting time of these suspensions can be varied from a few seconds to hours. Unfortunately these resins are very expensive materials.

BIBLIOGRAPHY

BSI (1999) BS 5930: *Code of practice for site investigations*, British Standards Institute.

Greenwood, D.A. (1970) Mechanical improvement of soils below ground surface. *Proc Symposium on Ground Engineering*, Institution of Civil Engineers, London, pp. 11–22.

Hughes, J.M.O. and Withers. N.J. (1974) Reinforcing of soft cohesive soils with stone columns. *Ground Engineering*, 3, 42–49.

Institution of Civil Engineers (1987) *Specification for Ground Treatment*, Thomas Telford, London.

Menard, L. (1972) The dynamic consolidation of recently placed fills and compressible soils. *Travaux*, 4S2.

Menard, L. and Broise, Y. (1975) Theoretical and practical aspects of dynamic consolidation. Extract from *Geotechnique*, Thomas Telford, London.

NHBC (1992) *NHBC Standards*, Chapter 4.6: Vibratory ground improvement techniques, National House-Building Council, London.

Tomlinson, M.J. and Wilson, D.M. (1973) Pre-loading of foundations by surcharge on filled ground. *Geotechnique*, 23 (1), 117–120.

West, J.M. (1976) The role of ground improvement in foundation engineering, in *Ground Treatment by Deep Compaction*, Institution of Civil Engineers, London, pp. 71–82.

Chapter 10
Building up to existing buildings

If a new dwelling is to be constructed directly against an existing building, special foundations are usually required if the existing building is to remain stable during and after construction of the proposed new building.

If the soils are granular or cohesive, the two buildings should be kept separate if possible, so that the new settlements can take place without affecting the existing building fabric. If the buildings are on rock strata there is no reason why they cannot be physically linked if it is desirable. However, consideration should be given to the age of the existing building and the likelihood of it being demolished in the future. Obviously, if the new dwelling has its own built-in stability this would not present a problem.

If the existing building is very old, and the foundations consist of shallow belled-out brickwork, it may be necessary to underpin the existing walls prior to construction of the new foundations (Fig. 10.1). This underpinning should be carried out in short lengths to suit the soil conditions present and with full regard being given to the condition of the existing building.

The general length of underpin segments is 0.9–1.0 m, but if an existing building is in a poor condition or the soil conditions are poor then the maximum segment may need to be reduced to 500 mm. The maximum length of underpinning excavation at any one period should not be more than one sixth of the length of the wall being pinned. To ensure that a 48 hour lapse between adjacent pours is maintained the sequence of operations should be based on a 1–4–2–5–3–6 pattern.

When building up to existing buildings, it is always prudent to have the building surveyed and to have record photographs prior to work commencing. Discussions should be held with the owners or occupiers of the existing property as to the new proposals; it may be necessary to satisfy the owner's own professional advisors on stability grounds. The more information that is known about an existing building, the less is the risk of its being damaged. Many new buildings have been constructed alongside buildings which contained basements or part basements and which generally necessitated major design and construction modifications.

10.1 SITE INVESTIGATION

This is the most essential element when building up against an existing building, especially if the new works have to be placed on piles. During construction for a new retail store in York in 1972 the old listed buildings of historical importance adjoining the new site were damaged as a result of ground subsidence. A site investigation had been carried out which indicated bands of wet loose sands at various depths, typical of the area close to the River Ouse. Bored piles were specified on the external walls with driven piles on the internal core. During construction of the bored piles, the sands migrated from below the existing foundations, causing severe subsidence. To enable the piling to be completed the type of pile was changed to a continuous flight auger type which was relatively new in the UK in 1972.

10.2 FOUNDATION TYPES

These can be divided into three main categories:

- wide mass concrete or reinforced footings;
- raft foundations;
- piled foundations.

In Example 10.1 these different methods have been developed, the only constant being the existing foundation. These methods can be adapted to suit the ground conditions encountered in the site investigation.

Example 10.1 Foundations beside existing building

A two-storey house is to be constructed up to an existing warehouse building. Surveys have revealed that the warehouse was built in 1920 using engineering bricks. Trial pits excavated alongside the gable end of the warehouse showed the brick walls were stepped

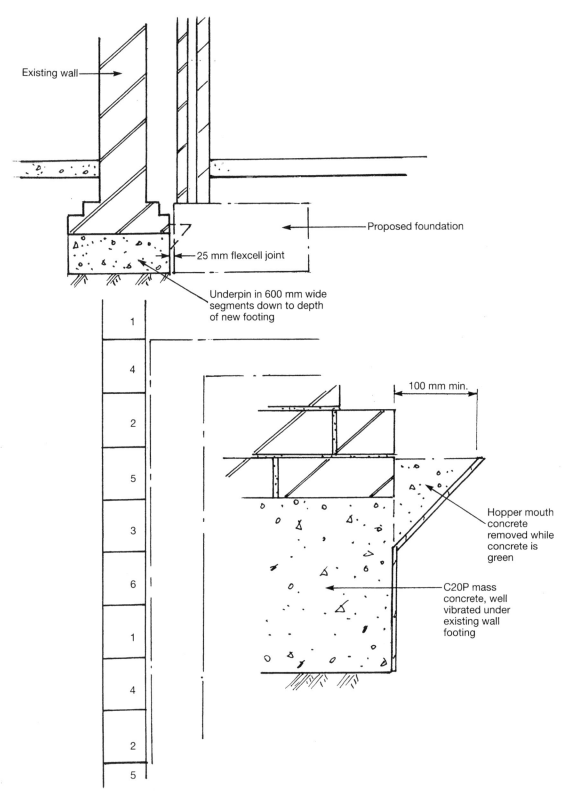

Fig. 10.1 Underpinning an existing wall.

out and terminated in a firm brown sandy clay at a depth of 600 mm below existing ground level. The ground floor levels of the new house and the existence of mature trees require the new foundations to be taken a minimum depth of 1.0 m below existing ground level. The ground bearing capacity of the firm clay has been put at 90 kN/m². Design a suitable foundation for the new dwelling.

MAJOR CONSIDERATIONS IN DESIGN

The existing warehouse gable wall will have to be underpinned to the same depth as the proposed foundations. In view of the type of construction and lack of a proper foundation it is recommended that this be carried out in 600 mm wide segments as indicated in Fig. 10.2.

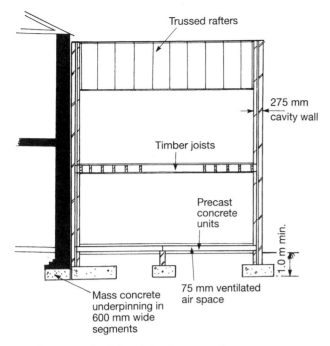

Trussed rafters

275 mm
cavity wall

Timber joists

Precast
concrete
units

1.0 m min.

Mass concrete
underpinning in
600 mm wide
segments

75 mm ventilated
air space

Fig. 10.2 Example 10.1: existing factory wall.

As it will be difficult to achieve adequate subfloor ventilation it is recommended that a precast ground floor or an *in-situ* reinforced concrete on a dpm is used. If the precast floor option is used then nominal ventilation of a 75 mm air space will be required.

Design codes:

- BS 6399 Part 1 Loading
- BS 648: 1964 Schedule of weights of building materials
- BS 8110 Structural use of concrete

Concrete grade 30 to be used with high-yield reinforcement.

LOADING TO MAIN WALL

Roof

	(kN/m^2)
Concrete interlocking tiles	0.55
Battens, felt	0.15
Self-weight of trusses	0.20
Plasterboard, insulation	0.17
Dead load =	1.07
Imposed load =	1.0	say 2.0 kN/m^2

First floor

This spans front to rear and will not be considered.

Ground floor

	(kN/m^2)
150 mm dp. beam and block floor	2.25

50 mm cement sand screed	1.20
Partitions	0.55
Dead load =	4.0
Imposed load =	1.50	say 5.50 kN/m^2

External walls

	(kN/m^2)
100 mm blockwork	2.0
102 mm brickwork	2.0
Plaster	0.25
Dead loads =	4.25	say 4.25 kN/m^2

Side-wall line loading

			(kN/m)	W_2	W_1
Roof	$2.0 \times 1.0 \times 0.50$	=	1.0	1.0	–
Ground floor	$5.50 \times 3.25 \times 0.50$	=	9.625	9.625	–
Wall	6.50×4.25	=	27.625	14.625	13.0
	Total	=	38.250	25.25	13.0

DESIGN PHILOSOPHY

In Fig. 10.3 the new foundation is shown to butt up to the existing foundation. If a mass footing is to be adopted there are two solutions which can be considered, as follows.

(a) Design the new foundation to take account of the eccentric loading but not of load spread through the base. With this method there will be a high edge pressure which should not exceed the allowable bearing pressure of 90 kN/m^2. This usually results in an impractical and wasteful solution.

(b) Assume that the wall loads will spread through the base by virtue of its inherent stiffness and make the base wide enough to keep the average ground pressure below 90 kN/m^2. Though no account has been taken of the frictional or adhesion forces on the sides of the foundation, the ground pressures using a smaller foundation are actually less than the theoretical method.

For this example both methods have been used; it is left to the judgement of the individual engineer to draw his own conclusions.

Method (a)

Wall load $W_1 = 2 \times 6.50 = 13.00$ kN/m

Wall load $W_2 = 9.625 + 1.0 + 14.625 = 25.25$ kN/m

Resultant $W = \dfrac{(13.00 \times 0.05) + (25.25 \times 0.20)}{13.00 + 25.25} = 0.149$ kN/m from A

where A is the cross-sectional area on the plan.
Maximum edge pressure is to be restricted to 90 kNg/m^2; Average ground pressure is to be 45.00 kN/m^2. Therefore

$$\frac{P}{A} = 45.00$$

$$P = 25.25 + 13.0 + 24B = 38.25 + 24B$$

$$\frac{38.25 + 24B}{B} = 45.00$$

Foundation width, $B = 1.82$ m

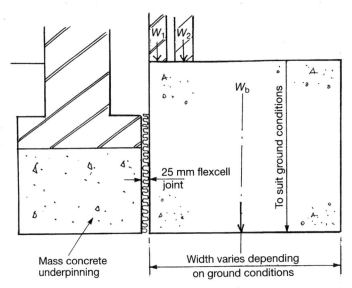

Fig. 10.3 Example 10.1: foundation eccentrically loaded.

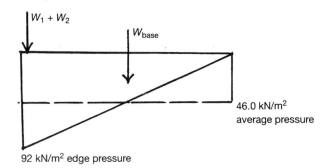

Fig. 10.4 Example 10.1: base pressures.

Therefore using base 1.85 m wide:

			kN/m
Weight of base	=	1.85 × 24 =	44.40
Wall load		=	38.25
		Total =	82.65

The distance x of the centroid of vertical loads from the edge of the foundation is given by:

$$\bar{X} = \frac{0.149 \times 38.25 + 0.925 \times 1.85 \times 24}{82.65} = 0.56 \text{ m}$$

Therefore eccentricity, $e = 0.925 - 0.56 = 0.365$ m

$$\text{Pressure} = \frac{P}{A} \pm \left(1 + \frac{6e}{d}\right)$$

$$= \frac{82.65}{1.0 \times 1.85}\left(1 + \frac{6 \times 0.365}{1.85}\right) = 44.67 + 52 = 96.67 \text{ kN/m}^2$$

Try base at 2.00 m wide.

			(kN/m)
Weight of base	=	2.00 x 24 =	48.00
Wall load		=	38.25
		Total =	86.25

$$\bar{X} = \frac{0.149 \times 38.25 + 1.0 \times 2.0 \times 24}{86.25} = 0.62 \text{ m}$$

Therefore

$e = 1.0 - 0.62 = 0.38$ m

Therefore ground pressure $= \frac{86.25}{1.0 \times 2}\left(1 + \frac{6 \times 0.38}{2}\right)$

$$= 43 + 49 = 92 \text{ kN/m}^2$$

Make base 2.00 m wide × 1.0 metre and provide nominal reinforcement top and bottom.

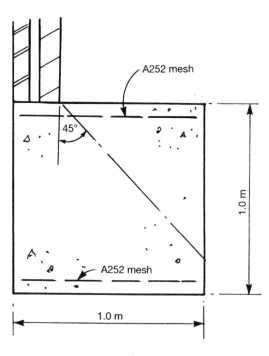

Fig. 10.5 Example 10.1: foundation section, method (b).

Method (b)

Using 1.0 metre deep foundation:

		(kN/m)
Wall load	=	38.25
Weight of base	=	24.0
Total	=	62.25

$$\text{Minimum base required} = \frac{62.25}{90.0} = 0.69 \text{ m}$$

Using 1.0 metre wide base the ground pressure is:

$$\frac{62.25}{1.0 \times 1.0} = 62.25 \text{ kN/m}^2$$

Use 1.0 m × 1.0 m strip footing with nominal reinforcement (Fig. 10.5).

RAFT SOLUTION

Instead of using a strip footing with a precast floor, the use of a raft foundation could have been considered. Where ground conditions are poor this would be the most suitable foundation solution. There may be a situation where the soils are made-up ground which is too deep to remove without affecting the stability of the existing foundations.

From previous design make raft downstand 750 mm wide (Fig. 10.6).

$$\text{Maximum ground pressure} = \frac{62.25}{0.750} = 83 \text{ kN/m}^2$$

The moment from the eccentric loading will be resisted by the raft foundation slab which should be reinforced with A393 mesh.

Raft design

Using a combined load factor of 1.50:

Wall load = (38.25 + 0.60 × 24) × 1.50 = 79.00 kN/m

With a 750 mm wide foundation the eccentricity = 375 − 149 = 226 mm.

Ultimate moment = 79.00 × 0.226 = 17.85 kN m

Using 30N concrete:

$$\frac{M}{bd^2} = \frac{17.85 \times 10^6}{10^3 \times 160^2} = 0.69 \qquad \text{From BS 8110 Chart 2}$$

$$\frac{100A_s}{bd} = 0.18$$

$$\text{Therefore } A_s = \frac{0.18 \times 10^3 \times 160}{100} = 288 \text{ mm}^2/\text{m}$$

Therefore provide A393 mesh in the top of the raft slab. For raft slab use A393 high-tensile mesh top and bottom.

During the construction of the new wall, great care must be taken to keep mortar droppings from building up at the ground-floor level. The use of cavity battens and access holes for cleaning should be insisted upon.

GROUND BEAMS AND PILES

Where the depth of the existing foundations are excessive or there is a cellar on the other side of the wall it will be necessary to use either a pad and beam solution or a piled solution using ground beams at the higher levels. The most suitable piling system to adopt is either a bored pile or a continuous flight auger pile taken down into the firmer strata. For estimating purposes a safe working load of 250 kN should be adopted for the piles.

Make ground beams 350 mm wide by 450 mm deep. To reduce the load on the foundations adjacent to the existing wall the roof, first floor and ground floor will span front to back. The overall plan dimensions of the new dwelling are 10 m × 8.0 m.

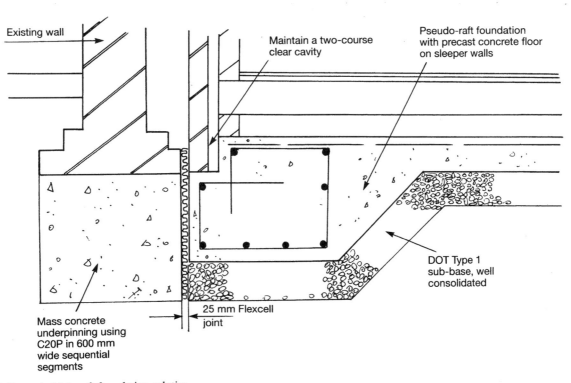

Fig. 10.6 Example 10.1: raft foundation solution.

Gable wall

Loading:

			(kN/m)
Roof nominal	0.50×2.0	=	1.0
Wall	6.50×4.25	=	27.625
Ground beam	$0.35 \times 0.45 \times 24$	=	3.78
	Total	=	32.405

Front and rear walls

			(kN/m)
Roof	$\dfrac{10.0}{2.0} \times 2.0$	=	10.0
Walls	6×4.25	=	25.50
First floor	$4 \times 0.50 \times 2.0$	=	4.0
Ground floor	$4 \times 0.50 \times 5.50$	=	11.0
Ground beam		=	3.50
	Total	=	51.0

Internal walls

			(kN/m)
Walls	2.50×2.50	=	6.25
First floor	$6 \times 0.50 \times 2.0$	=	6.0
Ground floor	$6 \times 0.50 \times 5.50$	=	16.50
Ground beam		=	3.50
	Total	=	32.25

Pile loadings

Design ground beams as continuous using $w_1^2/10$ and using 1.20 times elastic shears for pile loads.

Corner piles 1, 4, 9 and 12:   **Piles 5 and 8:**

(kN)

$$\frac{4.0}{2.0} \times 51 \quad = \quad 102$$

$$\frac{8.0}{2.0} \times 51 \times 1.20 \quad = \quad 244 \text{ kN}$$

$$\frac{4.0}{2.0} \times 32.4 \quad = \quad 64.80$$

Total   166.80

Piles 2, 3, 10 and 11:   **Piles 6 and 7:**

(kN)

$$\frac{6.0}{2.0} \times 32.40 \times 1.20 = 116.60$$

$$\frac{8.0}{2.0} \times 32.35 \times 1.20 \ = \ 154.80 \text{ kN}$$

$$\frac{4.0}{2.0} \times 32.25 \quad = \quad 64.50$$

Total   181.10

Cantilever

Ultimate moment $= 116.60 \times 1.50 \times 0.35 = 61.20$ kN m

$$\frac{M}{bd^2 f_{cu}} = \frac{61.20 \times 10^6}{350 \times 400^2 \times 30} = 0.036$$

Lever arm factor $= 0.95$

$$A_s = \frac{61.20 \times 10^6}{0.87 \times 460 \times 400 \times 0.95} = 402 \text{ mm}^2$$

Use two T16 in top of beams.

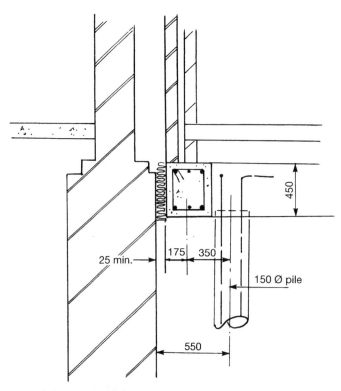

Fig. 10.7 Example 10.1: ground beam detail.

Beam E

Ultimate moment $= 32.40 \times 1.50 \times \dfrac{4^2}{10} = 77.60$ kN m

$$\frac{M}{bd^2 f_{cu}} = \frac{77.60 \times 10^6}{350 \times 400^2 \times 30} = 0.046$$

Lever arm factor $= 0.94$

$$A_s = \frac{77.60 \times 10^6}{0.87 \times 460 \times 400 \times 0.94} = 516 \text{ mm}^2$$

Use three T16 top and bottom.

Beam A

Ultimate moment $= 51 \times 1.50 \times \dfrac{4.0^2}{10} = 122.40$ kN m

$$\frac{M}{b^2 d f_{cu}} = \frac{122.40 \times 10^6}{350 \times 400^2 \times 30} = 0.072$$

From Design Chart No. 1 (Fig. 2.11) the lever arm factor $= 0.90$.

Therefore

$$A_s = \frac{122.40 \times 10^6}{0.87 \times 460 \times 400 \times 0.90} = 850 \text{ mm}^2$$

Use three T20 bars top and bottom.

Shear

Maximum shear $= 116.60$ kN.

Therefore

$$v = \frac{116.60 \times 10^3}{350 \times 400} = 0.83 \text{ N/mm}^2$$

$$\frac{100 A_s}{350 \times 400} = 0.28$$

Therefore $v_c = 0.42$

Therefore $v_c + 0.4 = 0.82$

Minimum links required using T8 bars.

Therefore

$$S_v = \frac{100.50 \times 0.87 \times 460}{0.40 \times 350} = 287 \text{ mm}$$

Therefore use T8 links at 250 mm centres.

Beams B and C

Ultimate moment $= 32.25 \times 1.5 \times \dfrac{4^2}{10} = 77.4$ kN m

$$\frac{M}{bd^2 f_{cu}} = \frac{77.4 \times 10^6}{350 \times 400^2 \times 30} = 0.046$$

Lever arm factor $= 0.94$

Therefore

$$A_s = \frac{77.40 \times 10^6}{0.87 \times 460 \times 400 \times 0.94} = 514 \text{ mm}^2$$

Use three T16 bars top and bottom. The piling layout and reinforcement details are as detailed in Figs 10.8 and 10.9.

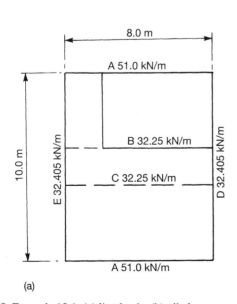

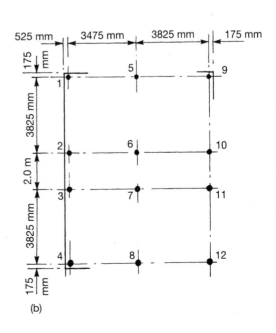

Fig. 10.8 Example 10.1: (a) line loads; (b) pile layout.

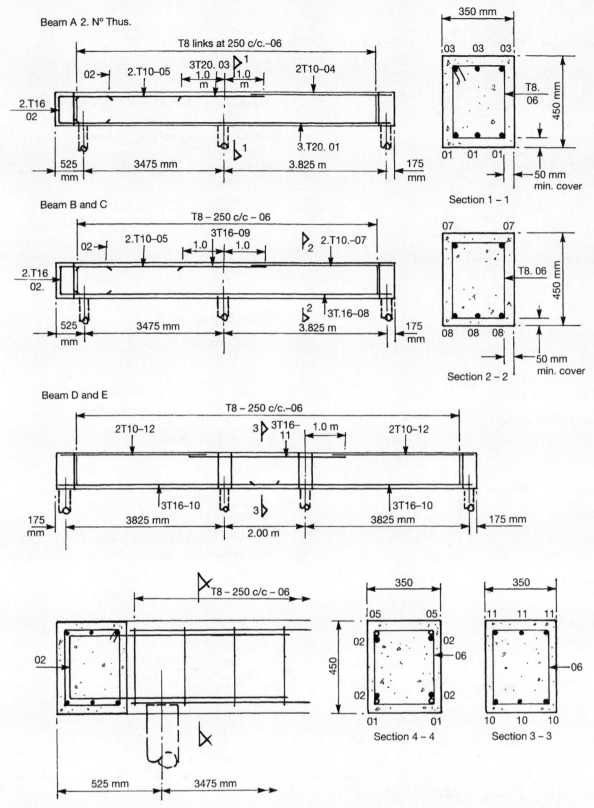

Fig. 10.9 Example 10.1: ground beam reinforcement.

10.3 UNDERPINNING

When building up against existing buildings, it may be necessary to underpin the existing walls or foundations for the following reasons.

- The existing foundations are too shallow and the depth of the proposed foundations, being deeper, could create unstable conditions if underpinning is not carried out.
- The existing building may be old and foundations may simply consist of the brickwork belled out at a shallow

depth. There may be external factors, such as mature trees, which on clay soils may require foundations to be deepened in accordance with the guidance laid down in NHBC Standards, Chapter 4.2.

- The existing foundations may have suffered settlement as a result of a fluctuation in the groundwater levels and it may be necessary to underpin down to a firmer stratum.

Underpinning is often used to stabilize foundation movements arising from the results of clay shrinkage, foundations placed on unsuitable formations or to remedy situations where deep drainage, placed parallel to the foundation, has been constructed without concrete backfill up to the foundation level.

The traditional method of underpinning is carried out using mass concrete placed in narrow segments, so transferring the foundation loads down to a stable and adequate formation (Fig. 10.10). Where existing buildings are found to have no proper foundation, or the brickwork is of a poor quality, the width of the segments may have to be reduced to 450–600 mm. In addition, the number of segments which can be removed at any one time may need to be reduced.

For reasonable-quality brickwork on a standard concrete strip footing a segmental width of 0.9–1.0 m is generally considered to be suitable, provided that the sum total of segments excavated at any one time does not exceed one sixth of the length of the wall being underpinned. To achieve this requirement, the segments should be excavated using the

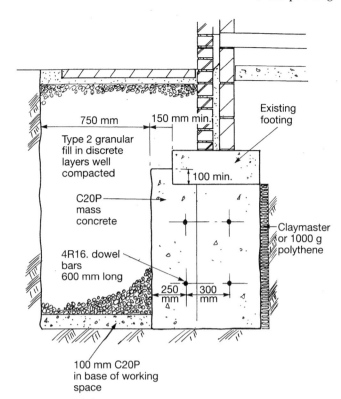

Fig. 10.10 Typical traditional underpinning.

1–4–2–5–3–6 sequential order for the total works. Using this sequence it is generally possible to maintain a minimum of 48 hours between concreting one bay and excavating its adjoining section (Fig. 10.11).

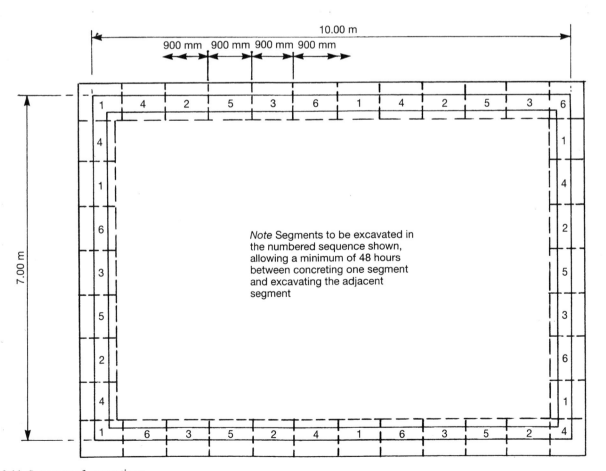

Note Segments to be excavated in the numbered sequence shown, allowing a minimum of 48 hours between concreting one segment and excavating the adjacent segment

Fig. 10.11 Sequence of excavations.

When the underpinning is localized over part of a building it will be necessary to step back up to the unaffected foundations, in 225 mm steps. A property may have been affected by a large tree and it may only be necessary to underpin the length of wall within the influence of the tree.

The traditional mass concrete method is limited to a maximum depth of about 2.0 m: any deeper and the questions of ground stability, practicality and economics have to be given careful consideration. Even for depths less than 2.0 m it will be necessary to comply with the requirements of health and safety legislation in ensuring that the trenches are adequately shored.

There are two ways of transferring the existing foundation loads on to the new foundation.

- Leave a 75 mm gap, which is packed up later using a strong dry-pack concrete. This method is not popular, as it delays the excavation process even longer, and it is another operation which can go wrong. Not all underpinning works are fully supervised, and the installation of the dry-pack concrete is critical if the underpinning is to be successful.
- Finish the mass concrete about 100 mm above the underside of the existing footing and use a vibrating

poker to expel any air pockets. This is the generally favoured method, using ready-mixed concrete. The use of a vibrating poker is essential and the amount of concrete shrinkage is negligible.

With both methods, it is essential that any clay on the underside of the existing footing is thoroughly cleaned off, prior to concreting. The individual concreted segments can be linked together using 600 mm long 16 mm diameter dowel bars. These are generally driven into the sides of the open excavation.

When underpinning properties that have been affected by trees or drought it may be necessary to place a polythene membrane or low-density polystyrene such as Claymaster at the rear face of the excavated segment prior to concreting, in order that 'locking in' is reduced. Should the clays rehydrate at a later stage, the potential effects of ground heave will have been mitigated.

The backfilling of the working space following completion requires careful consideration. Before backfilling it is good practice to ensure that the base of the working space has a concrete blinding at least 100 mm thick. This prevents the base of the formation from being softened by water, which tends to collect because of the sump effect. This blinding is

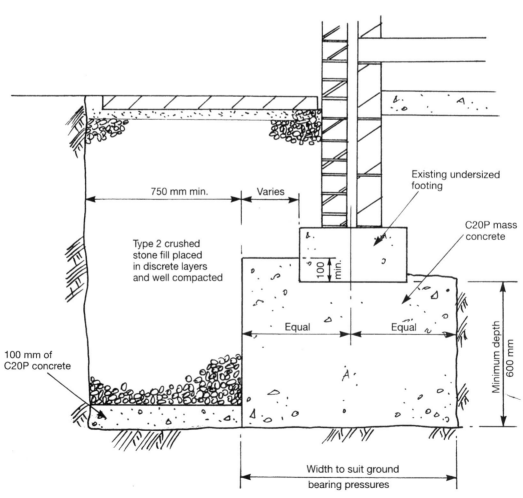

Fig. 10.12 Widening an existing footing by underpinning.

214

usually placed during the pouring of segments, using concrete which is surplus.

If the stratum removed is non-cohesive and the area around the property is not paved, there is no reason why the excavated material cannot be put back. However, in situations where residential buildings are being underpinned, there is often little room available to store the spoil on site. If the excavated materials are cohesive, and footpaths have to be reinstated around the property, it is unlikely that good compaction will be achieved, especially if the clays have dried out. In addition, the clays could rehydrate and affect the new foundations.

When working around residential buildings it is essential that the backfilled materials are well compacted to minimize the amount of long-term consolidation. In view of these requirements it is advisable to specify a clean well-graded crushed stone as the backfill medium.

Method statement

When carrying out underpinning the following methods should be adopted.

1. Remove existing pavings and stack for re-use.
2. Excavate all round lengths to be underpinned down to the level of the underside of the existing foundation. Cart off site or store for re-use if suitable.
3. Excavate in the required sequential order for individual segments down to required depths. Ensure that all clay is removed from the underside of the existing foundation.
4. Insert 16 mm dowel bars into each side of the excavation. Provide slip membrane at rear of segment if required.
5. Erect formwork and strutting. Ensure that formwork has been well oiled.
6. Concrete opened segments using C25 mass concrete or, if in ground containing sulphates, use a suitable mix as recommended in BRE Digest 250.
7. Repeat this process until all underpinning is completed.
8. Place a 100 mm thick layer of mass concrete at base of working space.
9. Backfill in 150 mm thick layers using a well-graded crushed stone, each layer being given four to six passes of a suitable trench compactor.
10. Replace footpaths on a 50 mm sand bed.
11. Tidy up and make good garden areas.

In certain circumstances the existing foundations may have been too narrow, resulting in excessive settlements. In this situation it may simply be a case of providing a wider foundation, and if this is done in the traditional manner then the minimum depth of pinning below the existing foundation should be 600 mm (Fig 10.12).

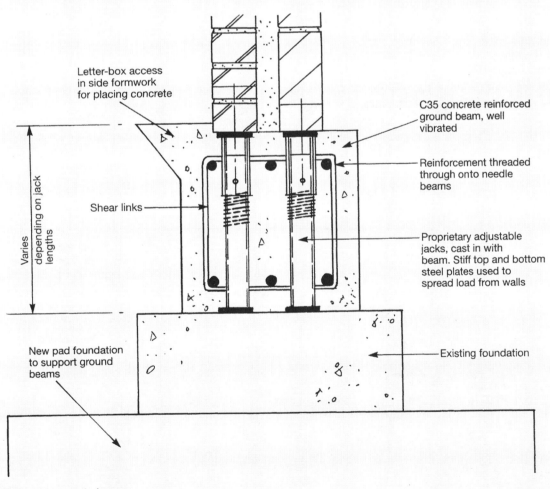

Letter-box access in side formwork for placing concrete

C35 concrete reinforced ground beam, well vibrated

Reinforcement threaded through onto needle beams

Varies depending on jack lengths

Shear links

Proprietary adjustable jacks, cast in with beam. Stiff top and bottom steel plates used to spread load from walls

New pad foundation to support ground beams

Existing foundation

Fig. 10.13 Pad and beam solution.

10.3.1 Beam and pad solution

This is generally carried out by specialist contractors with experience in major underpinning works. It is normally used when the depth to a suitable formation exceeds 2.0 m.

This method uses a system of deep pads placed at 3–4 m centres with reinforced ring beams spanning between. These beams are formed above the existing footing by removing brickwork and using sacrificial jacks or special precast concrete stools through which the reinforcing rods are threaded. The beams are poured *in situ* using a letterbox formwork arrangement (Fig. 10.13).

The installation of the pads will generally require the use of temporary shoring or the use of proprietary shore brace systems to ensure the safety of the site operatives. In addition, consideration must to be given to the eccentric forces which can develop from lateral pressures. These eccentricities could give rise to tension under the base and the size of pad should be designed to allow for these forces. The number of pads which can be excavated at any one time will depend on the extent of the works and the stability of the structure.

Once the pads and ground beams are in place, the existing foundations become redundant.

10.3.2 Pile and needle beam solution

Where the depth of excavation for pads is considered to be too great for practical purposes, or where the ground conditions are unfavourable (i.e. wet loose sands or very soft clays), then it is generally more economic to use a pile and needle beam solution (Fig. 10.14). This system can provide support either above or below the foundations and generally involves the use of mini-piling or stitch piling. It is possible to use larger piles for larger, heavier buildings but for two-storey residential properties piles are usually 150 mm diameter with a load-carrying capacity of about 100 kN.

This type of piling is generally carried out by specialist firms using piling equipment that can work in restricted headroom and close to walls. To avoid problems of vibration and noise most of these piles are bottom-driven using a Grundematt hammer, or are machine-augered.

Where internal access is not possible or desirable, the system can be adapted by introducing cantilever needle beams and a twin pile or counterbalance arrangement (Fig. 10.15).

This should be designed to ensure that the rear piles do not develop any tension. This is sometimes achieved by forming the external works in conjunction with the external footpath to create extra dead load. The major drawback to this method is in ensuring and being certain that the footpath will not require to be lifted to gain access to service pipes in the future. This method should only be adopted in areas where the ground is unlikely to be disturbed.

Where access is available from both sides of the wall the most favoured solution is the double ring beam with *in-situ* needle beams at 1.0 m centres as shown in Fig. 10.16. Generally, the main longitudinal beams span onto the piled needle beams which are spaced at 4–5 m centres.

As underpinning works constitute a significant alteration to existing foundations the works are subject to the Building Regulations (1985) and such works should be submitted to the appropriate local authority for approval.

Method statement

When underpinning by pile and beam method the following methods should be adopted.

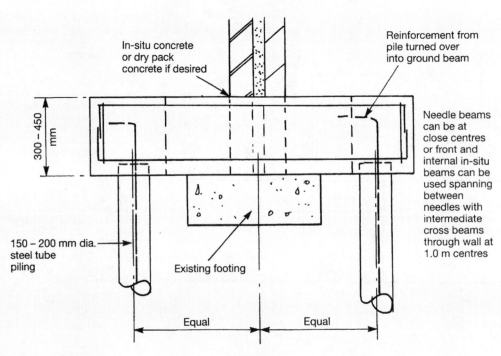

In-situ concrete
or dry pack
concrete if desired

Reinforcement from
pile turned over
into ground beam

300 – 450 mm

Needle beams
can be at
close centres
or front and
internal in-situ
beams can be
used spanning
between
needles with
intermediate
cross beams
through wall at
1.0 m centres

150 – 200 mm dia.
steel tube
piling

Existing footing

Equal   Equal

Fig. 10.14 Piling and needle beam solution.

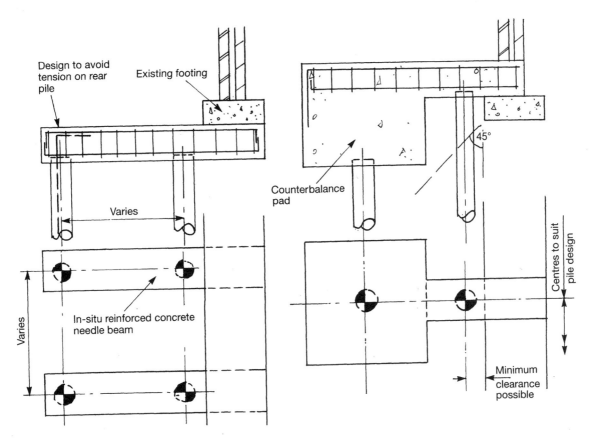

Fig. 10.15 Cantilevered needle system.

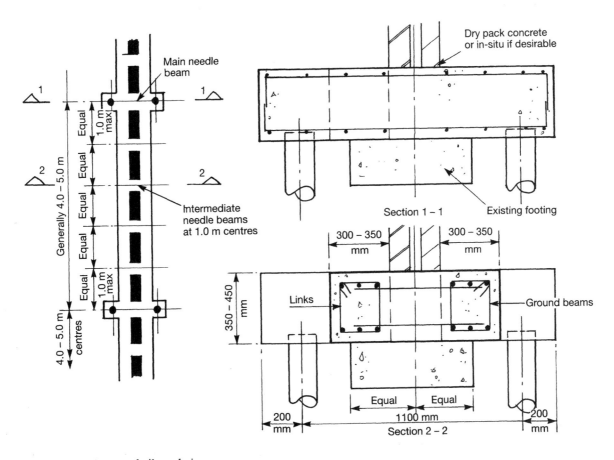

Fig. 10.16 Double ground beam and piles solution.

Building up to existing buildings

1. Remove existing pavings and stack for re-use.
2. Excavate down at each corner of the property to establish the foundation widths. This will enable the piles to be positioned clear of the foundations.
3. Drive piles to the required set and place any reinforcement and spacers down the pile. Place the high-strength concrete down the piles using a pipe inserted to the base of the pile. The concrete should be placed, maintaining a head of concrete in the tube at all times. Any water in the base of the pile will be displaced using this technique.
4. Excavate down to the underside of the proposed ground beam levels and reduce piles down to the cut-off level, i.e. 50–75 mm into the ground beam.
5. Break through the external walls at the required positions for the transverse needle beams.
6. Erect ground beam formwork and place reinforcement into position, ensuring that the steel is adequately spaced off the formwork. Ensure that the formwork is well oiled to assist in achieving a good strip-out.
7. Concrete ground beams, ensuring good compaction with the use of a 25 mm vibrating poker. Protect top of concrete with wet hessian or spray with a proprietary curing agent.
8. Backfill with clean stone and re-lay footpaths.
9. Make good cracking in walls. Remove any cracked masonry and replace.
10. Make good gardens and tidy up.

Case study 10.1   Investigation and underpinning of detached house on made ground, York

A large detached dwelling on the outskirts of York has suffered damage to the main external walls and there is evidence of internal damage in the form of cracking and doors binding.

INVESTIGATION

Trial pits excavated at each corner of the house revealed the following information.

- The southern, eastern and western foundations were placed on made ground. In addition, the foundations were not concentric with the walls, and further investigation established that there must have been a setting-out error in the foundation excavation which the bricklayers had overcome by building off centre.
- Along the southern boundary several mature trees existed and the clay soils below the made ground were desiccated down to 1.50 m as a result of the previous dry summers.

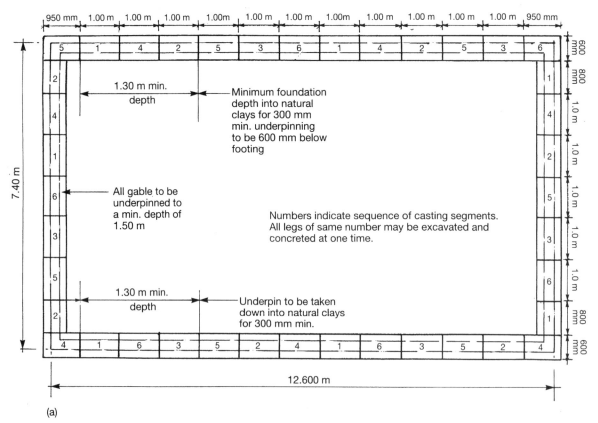

(a)

Fig. 10.17 Case study 10.1: Rufforth, York. (a) Plan (scale 1:50). Underpinning to be carried out in accordance with sequence on plan. A minimum of 48 h must elapse between excavating and concreting adjacent segments. All segments to have four R 16 mm dowel bars, pushed in for 300 mm.

Case Study 10.1 (*contd.*)

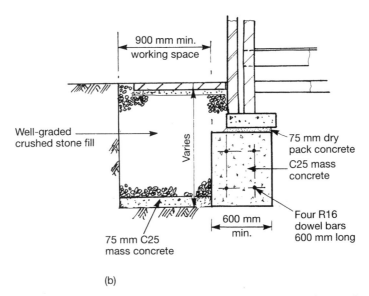

(b)

Fig. 10.17 (b) Typical section. Underpinning to have minimum dimensions noted on section subject to adequate formation on excavation. 75 mm dry pack concrete to be well rammed in under full width of foundation.

- The chimney breast on the southern elevation was not supported wholly on the foundation.
- Cracking was noted above the dampcourse on the northeast corner and investigation below the internal concrete oversite revealed made ground which contained significant amounts of ash and gypsum plaster. The oversite concrete was undergoing sulphate attack, which was considered to be the cause of the cracking on the north-east elevation. Examination of old Ordnance Survey plans showed that the dwelling had been constructed over former farm buildings.

RECOMMENDATIONS

The following remedial works were required to make good the deficiencies in the foundations and building works.

1. Underpin all foundations on made ground down to the firm clay strata to a minimum depth of 750 mm below path level; minimum depth of underpin to be 600 mm below the underside of foundation.

2. Underpin the southern elevations and return walls to a depth that is in accordance with the recommendations contained in NHBC Chapter 4.2. Tests have shown the clays to be in the medium shrinkability category with a plasticity index of 32%.

3. Remove the timber ground floor, remove oversite concrete and deleterious fills and replace with clean crushed stone, 100 mm concrete oversite placed on a 1200 gauge polythene dpm.

4. Build up the chimney breast off the new underpinned foundation using good quality commons which are to be pinned with slate and mortar packings.

The works were carried out as indicated in Fig. 10.17 and firm clays were confirmed in all external foundations.

During the removal of the internal oversite concrete it was found that internal load bearing walls had been constructed off this oversite and it was therefore necessary to carry out localized underpinning to those walls.

Case study 10.2   Differential settlement, Leyburn

A large detached dwelling in Leyburn, North Yorkshire, had developed cracking both externally and internally and there was sufficient evidence following a long monitoring period to show that the movements were progressive and likely to increase.

INVESTIGATION

The major damage was around the garage gable, and three trial pits were excavated in the positions shown in Fig. 10.18.

219

Case Study 10.2 (*contd.*)

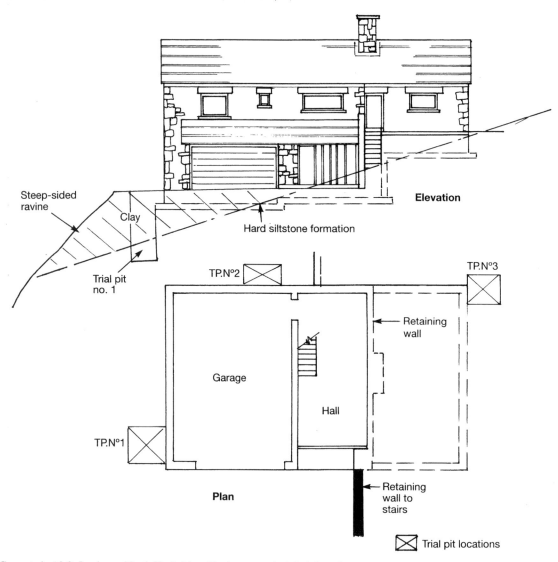

Fig. 10.18 Case study 10.2: Leyburn, North Yorkshire. Site layout and trial pit locations.

- Trial pit no. 1 at the front gable revealed foundation depth of 1.50 m. Adjacent to the trial pit was a large mature ash tree and during the excavation for the trial pit it was necessary to remove a 100 mm diameter tree root.
- Trial pit no. 2 in the rear conservatory revealed similar deep foundations but no evidence of any tree roots or desiccated clays.

  This trial hole was extended by hand auger to a depth of 3.50 m below path level and stopped on hard siltstones.
- Trial pit no. 3 excavated in the front drive, alongside the garage right-hand pier, revealed the foundations to be extremely shallow on to a hard siltstone formation.

There was a steep-sided ravine on the garage side and the dwelling had been constructed to a split-level design with a retaining wall formed at the junction of the garage and house.

When the siltstone levels were plotted on a section it was established that the foundations to the garage section, which also supported the floor and roof over, had been placed on the soft to firm clay stratum. Because of the split-level design and the severe dip on the siltstone, the remaining foundations had been placed on the hard siltstones. This had given rise to differential settlements which had been further exacerbated by the large ash tree at the front corner during a long period of very dry weather.

RECOMMENDATIONS

It was recommended that the garage and return wall foundations be taken down to the underlying siltstones to ensure that the dwelling was *wholly* on similar compotent strata. Because of the depth to the siltstones and the poor access to

Case Study 10.2 (*contd.*)

deep foundations it was decided to use a specialized piling system with external and internal ground beams and transverse *in-situ* needle beams at 1.0 m centres. Because of the steep ravine the front and rear ground beams were tied into a pad foundation well anchored into the siltstones at the high side. The remedial works are indicated in Fig. 10.19.

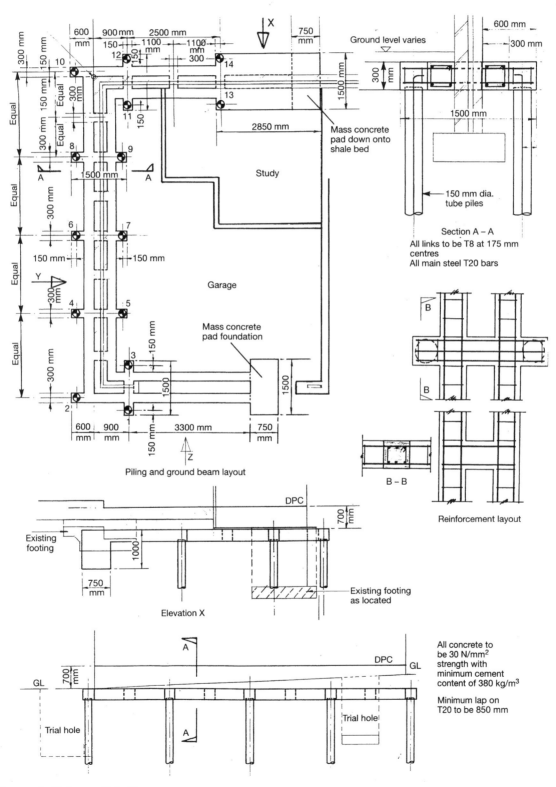

Fig. 10.19 Case study 10.2: remedial piling.

Case study 10.3 House on made ground, Beverley

An investigation on a semi-detached dwelling in Beverley had revealed severe cracking to the front and rear walls close to the party wall. The cracking was also evident internally, and examination of the ground floor revealed that the gable wall was moving laterally.

INVESTIGATION

Two trial holes were excavated close to the front and rear party wall and these revealed the foundations to be in what appeared to be firm natural orange-brown clays.

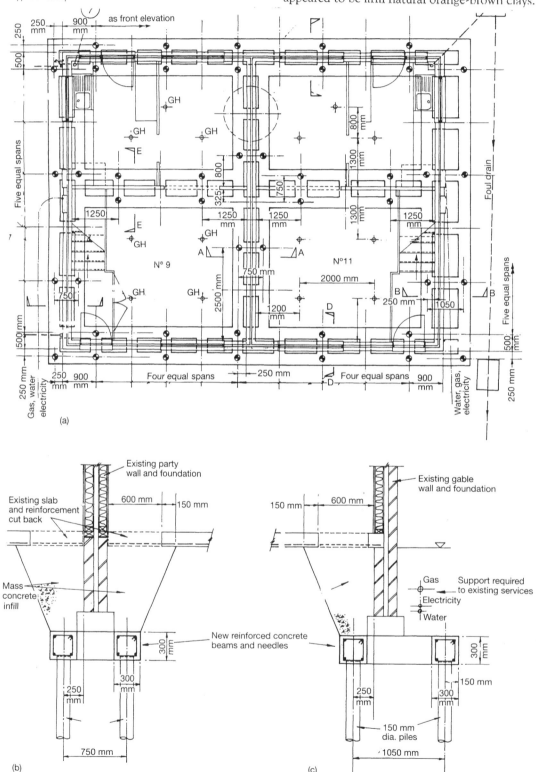

Fig. 10.20 Case study 10.3: remedial piling at Beverley, Yorkshire. (a) Pile and ground beam layout, scale 1:50; GH, grout hole; (b) section A–A; (c) section B–B.

Case Study 10.3 (*contd.*)

A further trial hole was opened up at the front gable and revealed the foundation to be on clay fill. This trial hole was extended to a depth of 2 m and a soft wet organic peaty stratum was encountered. Examination of old Ordnance Survey plans revealed the presence of a drainage dyke and an old railway embankment at the rear of the dwellings. It was obvious from the original survey drawings that this area of the site had been made up with site spoil.

A borehole investigation was carried out which revealed that the soft organic deposits extended too deep to enable traditional mass concrete underpinning to be used in the remedial works.

RECOMMENDATIONS

The borehole investigation and survey plans cast doubt on what had been considered to be firm natural clays encoun-

tered in the first trial pits. It was therefore decided to carry out a full pile and ring beam remedial scheme covering both houses.

The dwellings were of timber-frame construction and the ground floor slab was thought to be an *in-situ* reinforced suspended type.

During the installation of the piles and ring beams it was found that the foundations to both dwellings were on made up clays which had come from earlier phases on the development. In addition, the ground floor was found to be non-suspended and a new reinforced slab was laid on insulation with a polythene dpm spanning on to the new ring beams.

The remedial scheme is indicated in Fig. 10.20.

Case study 10.4 House founded on sloping rock formation, Scarborough

A residential property in a small village 4 miles west of Scarborough in Yorkshire was badly cracked as a result of movement of the gable wall. A plumb and level survey revealed that the gable wall had settled approximately 65 mm at the front gable corner. Cracking of the external masonry was also evident at the rear gable corner on the same gable elevation. The cracking evident on the front elevation was 1.20 m in from the gable wall and 900 mm at the rear elevation.

Two trial pits were excavated at the front and rear of the property in the position of the cracking. These revealed a weathered limestone rock stratum which gave way to a mixed clay fill. It was noticeable that the rock face sloped at about 45°. The ground floor to the lounge had a severe fall towards the gable wall. The weathered rock could be removed by pick, and it was therefore considered suitable to underpin the gable wall using 150 mm diameter steel tube piles driven to virtual refusal with a Grundamatt machine.

A remedial piling scheme was prepared incorporating an internal and external ring beam and transverse needle beams at pile positions. On completion of the piling and ground beams the external skin of brickwork was to be removed all around the dwelling and rebuilt to level. The ground floor was also to be rebuilt to level. This scheme is as detailed in Fig. 10.21.

The eight external piles were driven to depths of approximately 2.50 m and all the piles were plumb and driven to virtual refusal.

The piling internally was aborted after the first pile was driven as the pile started to drift off line and the maximum depth driven was only 1.0 m.

Further trial holes were excavated internally and it was discovered that the rock face commenced to slope off just inside the inside face of the existing foundation. The weathered rock was proven to be intact at shallow depth throughout the inside of the property.

A decision was made to place the internal needles on to a continuous reinforced foundation on the weathered rock, with mass concrete placed above up to the oversite level to provide an element of counterbalance purely as a precaution in case any of the external piles moved laterally. These modifications are detailed in Fig. 10.22.

Once excavation was carried out internally the walls at the front of the dwelling were seen to be severely cracked through both leaves. It was therefore decided to provide some temporary shoring along the gable and rebuild the front corner section completely from the bay window to the external chimney position along the gable end. This would allow the front window lintel to be jacked back up to level.

Case Study 10.4 (contd.)

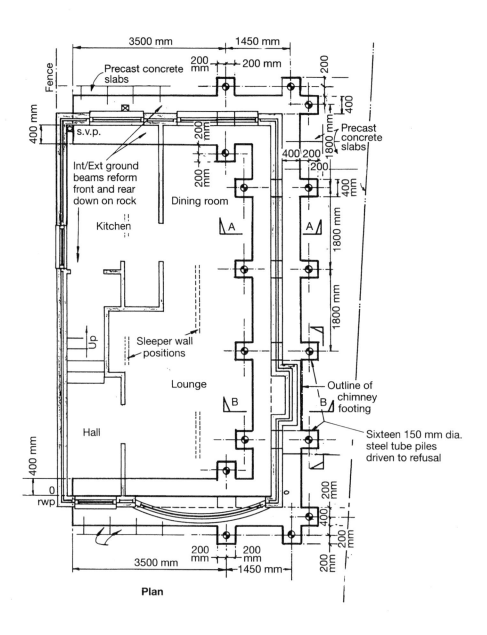

Plan

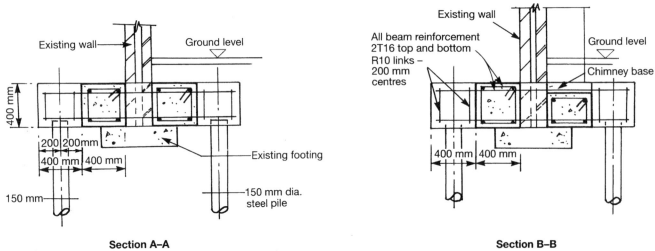

Section A–A

Section B–B

Fig. 10.21 Case study 10.4: initial pile and needle beam remedial works at Scarborough.

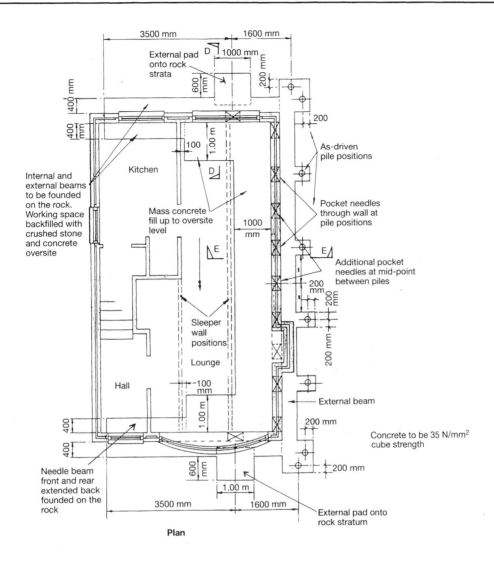

Plan

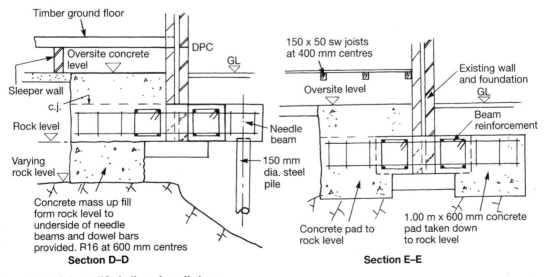

Fig. 10.22 Case study 10.4: modified pile and needle beam.

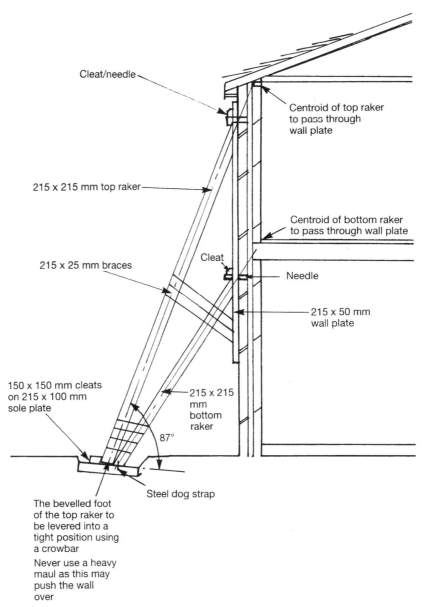

Fig. 10.23 Raking shore to defective building.

10.4   SHORING

If a building has been severely affected by subsidence or settlement of the foundations, and is considered to be potentially dangerous, then it is always a prudent measure to shore the building up. Shoring will help to stop further movements and will also increase the safety threshold for specialist contractors engaged to carry out remedial works to the foundations.

Generally, raking shores are used to prop walls which are showing signs of failure, such as bulging, out of plumb and cracking. The shores can be constructed from 215 mm × 215 mm timber sections with a 215 mm × 50 mm wall piece. The shore should be provided with a suitable sole plate support at ground level and a timber needle fixing at the head, secured with a shear cleat.

The angle of the shoring depends on the space available at the site. The recommended angle of inclination is between 60° and 75° to the horizontal. Where possible, the centre line of the shore should pass through the centre of the floor or roof wallplate. The angle between the sole plate and the angled shore should be slightly less than 90° to ensure a tight fit when the shore is crowbarred into its final position.

The wall plate should be attached to the wall using steel wall hooks or bolts at the top and bottom and at 2.50 m intervals. The top needle is fashioned from 100 mm × 100 mm timber and is usually 300–400 mm long. This needle is placed through the wall plate into a pocket in the brickwork. The needle is strengthened against the upthrust from the shore by a timber cleat nailed to the wall plate. The top of the angled shore is notched to fit into the needle. This assists in the erection and also prevents the shore from falling out if it becomes loose. The sole plate is usually timber and made the same width as the shore. If the ground bearing is poor, the sole plate should have transverse timbers underneath it to spread the load.

Once the wall place, needle and top cleat and sole plate are in position the shore can be levered into place using a crowbar under the bevelled base of the shore. It is important to avoid excessive force as this could push the wall over. Dog irons can then be driven into the shore and sole plate timbers, with a timber cleat being spiked on to the sole plate. To prevent the shore from sagging, short struts can be nailed from the middle of the shore down to the wall plate, placed at right angles to the shore.

The shores should be placed at each end of a wall, set in about 600 mm and spaced at 4–5 m centres. Figure 10.23 shows a typical raking shore arrangement for a two-storey dwelling.

BIBLIOGRAPHY

Architects Journal (1949) Pynford method of underpinning, AJ Library of Information Sheets, 26 E2.

NHBC (2003) *NHBC Standards*, Chapter 4.2: Building Near Trees.

Parentis, E.A. and White, L. (1950 *Underpinning*, 2nd edn, Columbia University Press, New York.

Chapter 11
Contaminated land

11.1 CONTAMINATED SITES

There are prospective development sites world-wide which are contaminated, and which are a threat to health and safety, on the site and in areas surrounding the site. Such sites if used for residential housing need to be fully investigated and assessed in respect of what remedial measures are needed to restore the site to the level at which the residents' health and safety are ensured on a long-term basis.

Many sites contain toxic materials, often dumped illegally or placed there before there was any controlling body. There are sites that contain phosphorous wastes that can spontaneously ignite when exposed to the atmosphere, or which contain radioactive substances harmful to people and animals. Such sites are unlikely to be built over, as the cost of removing such toxic wastes would be very onerous even with the assistance of government grants. As a result of past industrial use many sites contain chemicals, acids or heavy metal contaminants. Not all the contaminants are hazardous to humans but they may affect plants or attack construction materials.

11.2 UK POLICY ON CONTAMINATED LAND

Current government thinking is to promote more development on 'brownfield' land and their present target is that 60% of all new development should be built on land that has had a previous usage. A 'brownfield' site is any site that has previously been developed.

Government estimates suggest that 4 million new homes are required by the year 2021 (EA Report No. 66 – 2000). A significant proportion of these new homes are likely to be constructed on or close to contaminated land.

Housing can be developed safely on such sites by adopting good practice in key areas such as:

- The initial assessment – desk study of the site
- Site investigation – very important

- Remediation selection
- Monitoring and validation following remediation.

In April 2000, new environmental legislation came into force in the UK regarding the remediation of contaminated land. Part IIA legislation was implemented into the 1990 Environmental Protection Act (by Section 57 of the Environment Act 1995). Its main objective was the identification and remediation of contaminated land in its current use and the apportioning of liability for remediation. Primarily local authorities and the Environmental Agency (EA) in the case of specific sites regulate this legislation.

The Environment Protection Act gives the following definition for contaminated land:

'**Contaminated land**' is any land which appears to the local authority in whose area it is situated to be in such a condition, by reason of substances in, on or under the land that:

(a) **Significant harm is being caused or there is a significant possibility of such harm being caused; or**
(b) **Pollution of controlled waters is being or is likely to be caused**.

'Harm' is defined in the Environment Act as harm to the health of living organisms, protected habitats, controlled waters and buildings or structures.

'Pollution' is defined as 'entry into controlled waters of any poisonous noxious or polluting matter or any solid waste matter' (DETR Part IIA 2000).

Part IIA gives local authorities greater powers of inspection and a statutory duty to concentrate on grossly contaminated sites with respect to current use that may pose a significant risk to human health, the environment and any controlled waters.

A risk management approach is now used in the UK to assess the pollutant linkages on a site, using a **contaminant → pathway → receptor** methodology (Fig. 11.1) and this underlies the 'suitable for use' approach. If there is no

Fig. 11.1 Pollutant linkage on sites.

pollution linkage between **contaminants**, **pathway** and **receptor** then there is no pollutant risk.

Receptors can be humans, animals, plants, or part of the environment, such as water resources, building structures and services.

Houses with gardens are the most sensitive end of the scale due to the potential exposure of children to contaminated soils through soil ingestion, dermal contact and vegetable consumption. Remediation that may be required for garden areas may not be required on land that will be used for industrial development. This is known as 'fit for purpose' or 'suitable for use'.

Thus it should be noted that land only needs to be remediated according to its intended end use. The purpose of the remediation should be to break the pollutant linkage by removing or treating the contamination, removing or blocking the pathway, or removing or protecting the receptor (EA Report No. 66 – 2000). As every site is unique, a risk-based approach should be carried out on a site-specific basis.

Assessing potential pollutant linkages should be carried out by formulating a site-specific conceptual model, as shown in Fig. 11.2. Development of the **conceptual site model** is given in Guidance Document BS 10175. The conceptual site model is not rigid in its design and it may be updated as new information comes to light or as the proposed development changes.

There is a misleading view that contaminated land is restricted to brownfield land, i.e. land where there has been previous residential, commercial or industrial development. It is possible to find agricultural land and other apparent greenfield sites, which are contaminated through the use of toxic substances or as a result of disposal of contaminated substances that fall into the definition of 'harm' in the Part IIA legislation.

Where land appears to be uncontaminated, it is prudent to check that no contamination is migrating onto or below the site from adjacent contaminated land. This could involve migration of contaminants through the soils or groundwater, via existing service trenches or by dust deposition (EA Report No. 66 – 2000).

Under Part IIA legislation, liability for the financial consequences of contamination rests with those who cause or knowingly permit contamination. This is known as '**the polluter pays principle**'. If the polluter cannot be found after reasonable enquiries, then the landowner may be responsible (EA Report No. 66 – 2000). Builders and developers need to be aware that in some circumstances they could be liable for remediation, particularly where land is 'sold with information' regarding known contamination.

The general public are becoming more aware of contamination issues via the media, local searches and through the Internet. It is therefore essential that the developer or builder understand the potential for liability with regard to cleaning up a contaminated site.

It follows therefore that an initial assessment and a basic site investigation by appropriate specialists is a crucial and worthwhile expenditure, carried out prior to purchasing the land. Key reports for obtaining regulatory approvals will include:

- The initial assessment, which will consist of a desk study and a walk-over survey
- Site investigation and risk assessment reports, backed up with a site conceptual model
- Remedial strategy, remedial options and method statements for the proposed remediation works
- Post remediation monitoring and independent validation reports
- Environmental monitoring reports.

As well as satisfying the statutory undertakings, the above reports should also provide the openness and transparency to reassure the house purchaser's professional advisers.

The vast majority of new homes are built annually by 19,000 builders registered with the National House Building Council (NHBC). Contaminated land cover has been included by the NHBC Building Control service since April 1999.

The NHBC Technical Standard Chapter 4.1 Managing Ground Conditions, provides guidance on geo-technical and contamination issues. The NHBC requires all registered sites to be assessed by a desk study and walk-over survey as part of the initial site assessment. They require this assessment be carried out by a 'suitable person' who has the required skills and knowledge in this field. The 'suitable person' must be able to determine when specialist advice and/or further detailed investigation is required. The initial assessment should be substantiated by a basic investigation in which trial pits should be excavated. The NHBC also requires the 'suitable person' to complete and sign the Contamination Statement form for the specific site, stating whether the site is uncontaminated (Part A) or contaminated (Part B) and to confirm that the procedures outlined in Chapter 4.1 have been carried out accordingly. The components of the initial assessment have been described earlier in Chapter 1 Site Investigation.

11.3 RISK ASSESSMENT

The Environment Agency and DEFRA are preparing a series of guideline values on the potential risks to human health to replace the prescriptive ICRCL trigger values. These new guideline values are to be known as the CLEA risk model (Contaminated Land Exposure Assessment).

Current land use

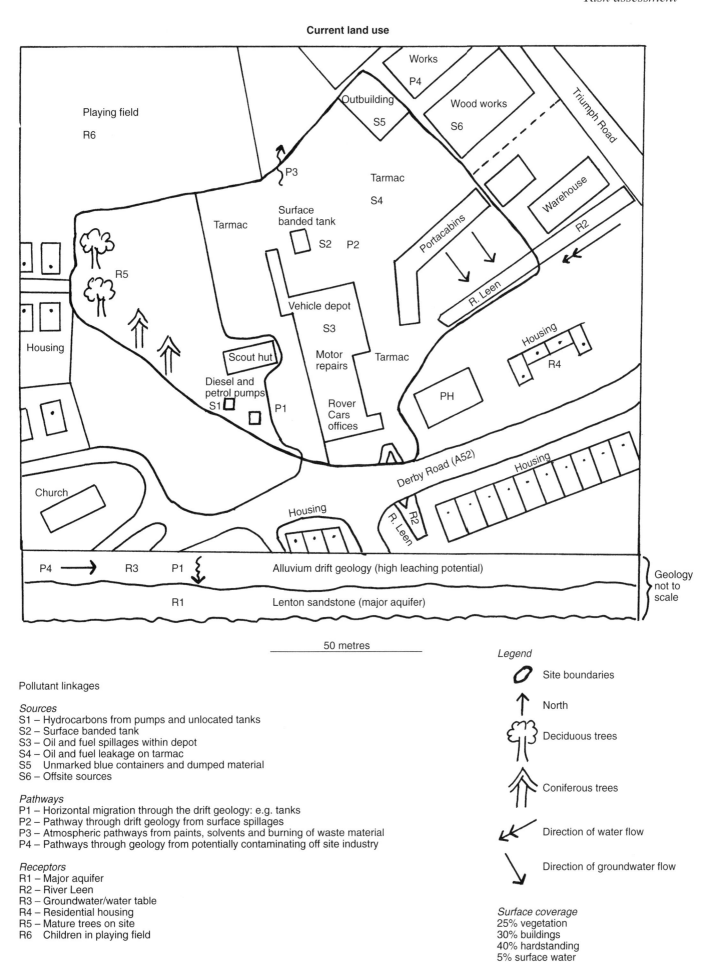

Fig. 11.2 Conceptual site-specific model for the Rover Garage, Derby Road, Woolaton.

Pollutant linkages

Sources
S1 – Hydrocarbons from pumps and unlocated tanks
S2 – Surface banded tank
S3 – Oil and fuel spillages within depot
S4 – Oil and fuel leakage on tarmac
S5   Unmarked blue containers and dumped material
S6 – Offsite sources

Pathways
P1 – Horizontal migration through the drift geology: e.g. tanks
P2 – Pathway through drift geology from surface spillages
P3 – Atmospheric pathways from paints, solvents and burning of waste material
P4 – Pathways through geology from potentially contaminating off site industry

Receptors
R1 – Major aquifer
R2 – River Leen
R3 – Groundwater/water table
R4 – Residential housing
R5 – Mature trees on site
R6   Children in playing field

Legend

⬭ Site boundaries

↑ North

Deciduous trees

Coniferous trees

Direction of water flow

Direction of groundwater flow

Surface coverage
25% vegetation
30% buildings
40% hardstanding
5% surface water

Risk assessment methodology has been developed to support the risk-based approach to managing contaminated land. Risk assessment should focus on the pollutant linkages on a specific site: whether it is on a specific contaminant, pathway or receptor, or all the pollutant linkages on a site.

Risk models are not all-encompassing because they focus on different risks, but they are common in that they focus on pollutant linkages. For example, LandSim, which has been developed by the Environmental Agency, focuses on risks to groundwater whereas CLEA uses a generic risk assessment for the protection of human health.

The cornerstone for all risk assessment is the conceptual site model (Fig. 11.2), for it provides a clear and concise basis for identifying pollutant linkages.

The following risk models are readily available and used by environmental consultants:

- Risk Human: Human health risk
- RBCA: Human health risk in relation to hydrocarbon contamination
- ConSim: Groundwater
- LandSim: Groundwater
- CLEA: Human health
- SNIFFER: Human health (ingestion and inhalation).

Risk assessment can apply to other elements on site, such as the risk of explosion from landfill gas or the risk to ecological systems.

CLEA: (Contaminated Land Exposure Assessment)

Risk-based guideline values have been developed by DEFRA to derive human health guideline values for exposure to contaminated soils (DETR CLR No. 10 – 2000). These will replace the previous ICRCL trigger values and are to be used as intervention values for the assessment of chronic risks to human health in relation to land use.

CLEA is a realistic model that estimates a range of human exposure to substances for the following three land uses:

- Residential (with or without plant uptake)
- Allotments
- Commercial/Industrial.

Soil guideline values (SGVs) for seven substances and four supporting documents were released in March 2002 by DEFRA and the Environment Agency. The first seven substances to be used in the CLEA risk model are:

- Arsenic (As)
- Cadmium (Cd)
- Chromium (Cv)
- Inorganic lead (Ca)
- Inorganic mercury (Hg)
- Nickel (Ni)
- Selenium (Se).

Other numerical criteria for polyaromatic hydrocarbons, phenols and cyanide will be issued later. DEFRA and the Environment Agency intend to publish reports for 55 contaminants, which will assist local authorities to determine the status of a specific contaminated site.

Soil guideline values are not a substitute for carrying out a risk assessment and where no SGVs are available, local authorities are advised to carry out a risk assessment using specific criteria.

The four supporting documents are:

- CLR 7
- CLR 8
- CLR 9
- CLR 10.

The CLEA model uses Monte Carlo simulation, which involves random computation reproduced many times for ten different exposure pathways to produce realistic exposure levels rather than maximum exposure levels. This results in a distribution of values rather than a deterministic single value.

The CLEA model is based on site-specific criteria, so the discrepancies and misinterpretation of the former ICRCL trigger values are replaced by a more realistic and robust methodology. For example, the soil properties used in the CLEA model are sandy, loam, clay, and organic soil. The soil type can affect the fate and transport of contaminants. Other criteria in the model that can be altered to match site conditions include:

- Exposure period
- Critical receptor
- Pathways
- Toxicological properties of contaminants.

When assessing the whole site, CLEA guideline values may not be sufficient to cover all the risks, for there may be receptors other than human health. CLR Documents 7–10 and Soil Guideline Values can be downloaded free of charge from the Environment Agency website at www.environment-agency.uk/subjects/landquality

RBCA (Risk Based Corrective Action)

This was initially developed by the American Standard for Testing Materials for the clean up of petroleum sites in the United States. It uses a tiered approach to calculate baseline risk levels and/or site-specific target levels for soil and groundwater remediation, based on information provided by the contaminated land specialists.

SNIFFER (Scotland and Northern Ireland Forum for Environmental Research Framework)

The SNIFFER model derives numeric targets to minimize the adverse human health affects of long-term exposure to soil contaminants. By working through a series

Table 11.1. Soil guideline values used in the CLEA model for sandy soil based on oral intake

Land use	Contaminant									
	As	Pb	Cd			Cr	Inorganic Hg	Ni	Se
			pH 6	pH 7	pH 8				
Residential with plant uptake	20	450	1	2	8	130	8	50	35
Residential without plant uptake						200			
Allotments				30		130	15	75	260
Commercial and industrial	500	750		1400		500	480	5000	8000

of formulae, a reference target in soil for a specific contaminant is obtained based on a tolerable daily intake (TDI)

A TDI is an estimate of a specific contaminant over a lifetime (based on 70 years) to be protective of human health. The exposure paths considered are ingestion of soil, consumption of home-grown vegetables, and inhalation of indoor and outdoor soil vapours, but **not** dermal exposure or inhalation of fugitive dust.

The SNIFFER methodology is used for sites contaminated by heavy metal and organic substances.

ConSim and LandSim

ConSim is a risk assessment model developed by the Environment Agency to assist in the impact assessment of groundwater pollution from contaminated land.

LandSim has been produced specifically for the assessment of landfill sites by the Environment Agency, and focuses on leachate generation and leakage rates from the site.

They are probabilistic models and do not take into account biological processes that can influence migration and bio-degradation.

Risc Human

Risc Human is the commercial programme for C-soil, which is the Dutch method for deriving generic values for human health. The model is deterministic like SNIFFER, with a point value being selected for each parameter. To run the Risc Human model, site-specific criteria must first be derived using the SNIFFER model.

The benefits of using risk assessment models are:

- They are site specific
- They provide consistency
- They are technically defensible
- They are transparent

- They can be cost-beneficial
- They contain best practice through a scientific approach.

These various risk models provide application of current best practice through a scientific approach. The transparency of the information generated provides the regulatory bodies with the necessary information for exposure risk on a site-specific basis.

The CLEA model and other risk assessment models will become the tools of choice for regulators, considering the risks to human health and the environment. The output results provide a more meaningful and justifiable risk than the old ICRCL guideline values and will be more likely to receive regulatory approval

Disadvantages

One model may not cover all the pollutant linkages. SNIFFER concentrates only on contaminants dispersed in soils and does not consider dermal exposure or risk from acute exposure. RBCA does not consider the vegetable consumption pathway and can be used only for organic contaminants, for example, contamination found at petrol station sites. There is also a degree of subjectivity when inputting data to run RBCA and Risc Human models in the measurement of skin exposed to a contaminant, as this will depend on the activity of the person.

11.4 INDUSTRIAL PROCESSES

Researching the former use of a site is vital in focusing the initial assessment. It will provide vital clues as to the type of contaminants that may be found and their approximate location on the site. Industry profiles, published by DEFRA in over 50 booklets detailing the activities that occurred and the likely contaminants that may have resulted, are most useful. They form a good starting point for any site investigation (Nathanial, P. 1999)

Contaminated land

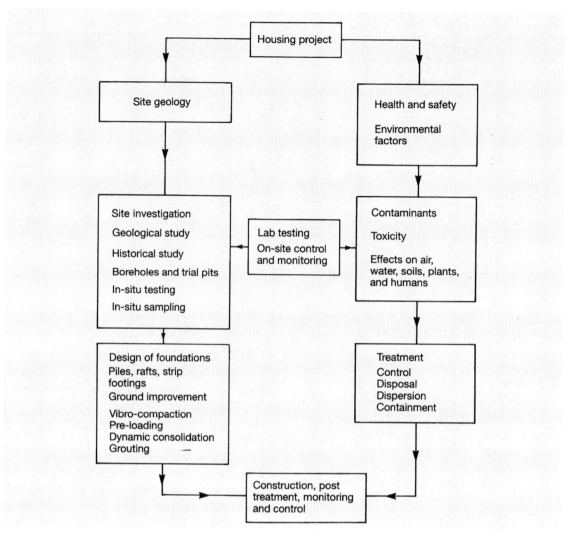

Fig. 11.3 Phased approach for redevelopment of a contaminated site for housing projects.

Fig. 11.3 shows a phased approach for redevelopment of a contaminated site for housing projects.

The following sites require investigation as they represent the 'Top 20' industrial processes likely to give rise to significant contamination:

- Landfill sites containing household or industrial wastes
- Old gasworks sites
- Timber treatment plants
- Iron casting works
- Iron, steel and coke works facilities
- Old shallow mineral workings, coal, fireclay, etc.
- Asbestos plants
- Scrap yards
- Old sewage treatment plants
- Metal smelting and metal refining
- Oil production
- Pharmaceutical industries
- Paint manufacture
- Leather works and tanning processes
- Galvanizing and metal treatment works
- Railway marshalling yards
- Explosives manufacturing
- Oil storage depots
- Chemical and pesticide plants
- Ship repair yards and dockland.

Having preconceived knowledge of the activities that occurred on site will focus attention during the walk-over survey. It will also influence the location of the trial pits and sampling points, and the specific chemicals to test for.

Guidance should be sought from BS 10175 – (2000) Code of Practice for the Investigation of Potentially Contaminated Sites. It can be very difficult to visualize former on-site activities without prior research. Sites that look inoffensive due to a change in land use may be hiding the true scale of contamination.

It is estimated that there are 360,000 hectares of contaminated land in the UK and a large proportion of this will be developed in the future (ENDS June 2000). The list from ENDS in Table 11.2 shows some of the sectors causing the largest estimated land investigation.

It must be remembered that not all brownfield land is contaminated and not all greenfield land is clean. Looking at the former history of the land and adjacent areas is

234

Table 11.2. Contaminated sectors and estimated site investigations (ENDS June 2000)

Industrial process	Number of sites
Petrol stations	120,000
Sewage works	800
Railway land	175
Scrap yards	12,000
Textile and dye works	11,200
Printing works	10,000
Gas works	2,000
Waste recycling treatment and disposal (including landfill sites)	5,500
Engineering works	43,305

essential in determining the risk to humans and the environment.

Remember: the consequences of no site investigation could be at least as expensive, and most probably much more so, than having one carried out.

11.4.1 Asbestos

Asbestos may present a risk to human health where there is a potential for asbestos fibres to be inhaled from damaged asbestos products (e.g. pipe lagging, asbestos cement sheets, insulating boards, sealants, and textiles) or from soil dust. It is estimated that 3,000 people die each year from asbestos-related deaths, which include asbestosis, mesothelioma and lung cancer (ARCA website, 10/2000).

Amosite (brown) and crocidolite (blue) asbestos products have been banned in the UK since 1985 but asbestos is widespread on land previously used for housing, commercial and industry (DOE Industry profile 1995). The introduction of new products containing chrysotile (white) asbestos has been banned in the UK since 1999 (British Asbestos Newsletter 35 – 1999). The wide and varied use of asbestos in industrial and non-industrial activities means that careful consideration is required when undertaking a site investigation. The presence of asbestos should be investigated on sites of former railway land, hospitals, textile works, cement manufacturers, and train carriage constructors' plants.

Asbestos surveys should be conducted by a specialist in this field, to ensure that the materials are properly identified and to provide recommendations as to necessary disposal. Asbestos is treated as special waste under the Special Waste Regulations 1996. Under these regulations all movements of asbestos waste have to be tracked using a consignment code until they reach a waste management facility that is licensed to handle asbestos waste.

On sites where existing buildings require asbestos to be removed, this removal must be carried out by a specialist contractor, using appropriate equipment. Under no

circumstances should other ancillary workers or the public be allowed into the affected buildings until a signed completion certificate has been issued.

11.4.2 Scrap yards

Scrap yards generally have had a succession of previous uses, often located on sites of former industrial use or wasteland unsuitable for development. Usually there is poor documentation of what has been received in scrap yards. Petrol and diesel contamination is common and chlorinated compounds are often encountered below the yard surface. Frag wastes are residuals left from crushing of cars consisting of asbestos, PCBs and oil spillages. The pollution generated from burning tyres is very toxic (Westlake 2000).

11.4.3 Sewage treatment works

A variety of solid, semi-solid and liquid wastes may be present in old sewage treatment plants or sewage farms. Raw sludge is generally the biggest contaminant and should be removed off site. Such sludge can contain a wide range of toxic and heavy metal concentrations, with raised concentrations in industrial areas.

Traditionally, solids and liquid waste were separated and spread on agricultural land. Extreme care must be exercised in its disposal onto land on which crops are grown for human consumption. Such toxics as cadmium and lead could be taken up into the food chain. There is also the potential for methane gas generation on former sewage works.

11.4.4 Timber manufacturing and timber treatment works

Timber preservatives are the largest cause of contamination; they include arsenic, copper chrome arsenates, phenols, and creosotes. Sources of contamination are often linked to old drainage systems such as sumps, soakaways and storage areas.

During the manufacture of timber products there is the potential for spillage of paints, stains and varnishes, which are volatile organic compounds (VOCs). Large volumes of adhesives, which include formaldehyde resins and polyvinyl acetate, are also used.

Resins are used in the manufacture of chipboard, joinery and veneer pressing. Containment bunds are now used during storage but these were rare in the past (Nathanial, P. 2000).

Waste in the form of dioxins and forans may have been buried on site as a result of incineration. Copper chrome arsenate (CCA), creosote and pentechlorophenol (PCP) should be included in a chemical testing suite when investigating such plants.

11.4.5 Railway land

Waste materials were often used to build up embankments. These could include ash from coal-fired engines and boilers, colliery waste, slag, and demolition rubble. Ash generally contains arsenic, lead, copper, nickel, sulphates and PAHs in excess of 50 mg per kg.

Contaminants from the operation and maintenance of rail tracks include oil, diesel and paint spillages, and combustible matter such as coal and sawdust is usually present (DOE Industry Profile 1995). Waste often occurs in the form of spent track ballast and wooden sleepers. Deposits of ash and coal with high calorific values can give rise to spontaneous combustion.

Chemical testing suites should include heavy metals, sulphates, ethylene, glycol, VOCs, TPHs, PCBs and asbestos.

11.4.6 Petrol stations and garage sites

Ever since supermarkets have been licensed to retail petrol, roadside filling stations have been closing down at a rapid rate. It has become very common to develop small housing estates on former petrol stations. Often the site is only large enough for up to five houses and this can affect the economics of the development.

The major consideration for the developer are the buried petrol tanks, which may still contain petrol or diesel products and will need to be made safe in accordance with the Local Fire Officer's requirements. Spillages and leakages will have resulted at the surface and seeped into the ground below.

It is also surprisingly common to have tanks that leak petrol and diesel into the ground. Such an occurrence happened on a site in Darlington in 1998, where petrol tanks from a service station adjacent to the site leaked into the ground below and spread a vapour flume over a large area of the proposed housing site. The petrol had leaked down into a band of sandy gravel 5 metres below the site and the only solution was to dig and dump all the contaminated material. This was a very costly process, paid for by the service station owners.

Chemical suites should always include tests for PAHs and VOCs. Other potential contaminants include paints, battery acids, solvents, anti-freeze, and brake fluids, particularly in areas of vehicle maintenance.

11.4.7 Gasworks sites

Generally the contaminants are toxic in a liquid or solid form. These result from coal by-products, coal tars, distillation liquids, ammonia, hydrogen sulphides, phenols, prussic acids, and spent oxides known as 'blue-billy'. Such sites generally cover a large area and the best method of investigation is to excavate plenty of trial pits on a closely spaced grid system at least 50 metres square initially,

followed up by additional infill pits to delineate any suspect areas encountered.

The first stage investigation should have regular sampling at 500 mm depth and the following contaminants should be checked out: arsenic, cadmium, lead, copper, cyanates, sulphates, toluene, cyanide, phenols, coal tars, and asbestos. Some of these compounds are carcinogenic and should be investigated very carefully, ensuring that proper precautions are taken in respect of the use of the correct protective clothing, boots and breathing apparatus.

It is also important to ensure that the contaminated soil from the trial pits is not allowed to spread or migrate to other cleaner areas of the site.

11.4.8 Metal smelting works

Usual contaminants are chromium, nickel, cyanides and cadmium. Generally such materials have to be removed off site to a special disposal facility.

11.4.9 Old mineral workings

Old mineral workings generally have concentrations of carbon dioxide and methane gas present. These can be a major problem if the workings are relatively shallow and do not have a substantial impervious cover of drift materials. As coal workings have been abandoned and surface operations closed down, the following conditions may prevail:

- When the mine water pumps are switched off, the mine galleries can flood. The subsequent rise in water levels can act like a bellows, pushing any mine gases up to the surface via fissures in the rock or up mine shafts. In addition the methane gases, being soluble, can be transported via the water to emerge long distances from their original source.
- As mines are closed their ventilation systems are shut down, resulting in an increase in the mine gas concentrations.

Where housing is to be built over suspected coal workings, it is prudent to have any known shallow workings check drilled to determine the seam depth and thickness and type of cover available. If the workings are still open and have less than 10 times seam thickness rock cover, they should be grouted with a mixture of PFA: cement on an appropriate grid system. Grouting up the workings will mitigate the potential for large volumes of gas to build up, enabling the gas levels at the surface to be more easily managed by gas-impermeable membranes.

During the drilling and grouting operations, gas sampling should be carried out from the drill holes using the method outlined in Fig. 11.4. This will enable the gas situation to be closely monitored.

Any shafts encountered should be grouted up from the base of the workings and capped off at rock head level

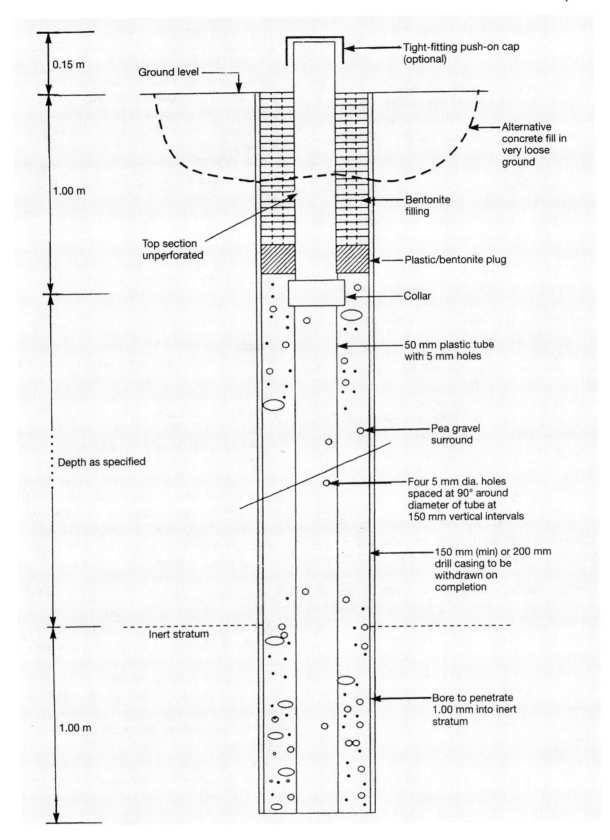

0.15 m

Ground level

Tight-fitting push-on cap
(optional)

Alternative
concrete fill in
very loose
ground

1.00 m

Bentonite
filling

Top section
unperforated

Plastic/bentonite plug

Collar

50 mm plastic tube
with 5 mm holes

Pea gravel
surround

Depth as specified

Four 5 mm dia. holes
spaced at 90° around
diameter of tube at
150 mm vertical intervals

150 mm (min) or 200 mm
drill casing to be
withdrawn on
completion

Inert stratum

Bore to penetrate
1.00 mm into inert
stratum

1.00 m

Fig. 11.4 Gas sampling borehole (Peckson and Lowe 1985).

with appropriate venting arrangements through the shaft cap.

Special foundations may be required together with gas-impermeable membranes below ground-floor slabs. These gas membranes should be underlain by a 300 mm layer of single-sized aggregate, with perforated vent pipes installed to vent any collected gas to the exterior of the building as shown in Fig. 11.9 and Fig. 11.10.

11.4.10 Toxicological effects of contaminants

11.4.10.1 Arsenic (As)

Occurrence of arsenic as a contaminant is quite common as it is associated with effluent, particularly sewage sludge and is a waste product from the production of coal gas at gasworks facilities. The burning of coal and ashes from smelting furnaces also contains arsenic. Industrial processes in the past involved the burning of vast amounts of coal and the domestic burning of coal has also produced widespread ash containing arsenic.

The primary exposure routes are inhalation and ingestion. Ingestion pathways of concern are through the consumption of vegetables and through the medical condition known as 'pica', whereby young children have a compulsion to eat soil. Acute exposure can lead to liver damage and kidney failure, while long-term exposure to inorganic arsenic can cause skin cancers (The Earth Shop.com–arsenic 2001).

Increasing the pH of the soil will decrease solubility and mobility of arsenic compounds and their availability for absorption by plant roots. Arsenic use in industrial processes is particularly widespread in industries and processes such as smelting, fungicides, pesticides, glass manufacturing, timber treatment plants, textiles, and dyeworks (Card 1995).

11.4.10.2 Lead (Pb)

Similar to arsenic, lead is found in sewage sludge, as a waste product from the smelting of ores and in coal combustion. A significant source of lead contamination in soils is from vehicle exhaust emissions. Industrial uses include batteries, ammunition, dyes, paints, and insecticides. Exposure routes are through ingestion and inhalation, which can result in cancers affecting the lungs and kidneys and which can also damage the brain.

Lead travels in the blood to affect the soft tissues and then moves into the bones (The Earth Shop.com–lead 2001). Lead poisoning tends to be bio-cumulative, which means that it can accumulate in the body.

11.4.10.3 Phytotoxic metals

Phytotoxic metals are a group of contaminants that can affect plant growth but are not generally hazardous to human health at low levels. These metals include boron, copper, zinc, and nickel. Components containing these metals can be hazardous to human health.

Boron: Found in motor-starting devices, electro-plating, manufacture of glass, and ore processing. Ingestion of boron compounds can lead to disorders of the central nervous system, depression and vomiting (Card 1995).

Copper: A variety of uses in the metal industry due to its use in forming alloys. Contamination derives from the effluent in the metal industries, pesticides and herbicides, and timber treatment works. Ingestion of copper compounds can cause nausea and vomiting, and high intake can result in liver and kidney damage.

Nickel: Used to form alloys with iron and used in the manufacture of batteries and electronic components. Sources of contamination are industrial effluents, fertilizers and sewage sludge. Compounds of nickel are acute toxins that mainly affect the kidneys; some are suspected carcinogens.

Zinc (Zn): Primarily used for galvanizing iron products and to make die casts. Sewage sludge contains high quantities of zinc and it is also common in flue dust and ash wastes from the burning of coal and smelting of non-ferrous metals. Zinc compounds are moderately toxic and can cause nausea and vomiting if ingested.

Generally speaking, metal toxicity and mobility are reduced with an increase in soil pH levels.

11.4.10.4 BTEX

Benzene, toluene, ethyl benzene and zylene are often grouped together and are referred to as BTEX constituents. They are commonly associated with contamination at petrol stations, coal carbonization plants and as a by-product from tars.

Benzene: A highly flammable colourless liquid with a sweet smell. Industrial processes are the main source of benzene particularly in the petroleum industry. Increased emissions result from the burning of coal and oil, vehicle exhausts and volatilization from service stations. It is also used in the manufacture of lubricants, dyes, and detergents. Benzene should always be tested at sites where coal or oil was burned and former petrol station sites.

Long-term exposure to benzene can cause blood problems, especially in bone marrow, and benzene is a known carcinogenic.

Toluene: A clear colourless liquid with a distinctive smell. It is produced in the manufacture of petrol and other fuels from crude oil and when making coke from coal. It is also used in the manufacture of paints, thinners, lacquers, adhesives, and rubber.

Acute doses of toluene can affect the brain, causing headaches, while memory loss and chronic dosages may permanently damage the brain (The Earth Shop.com–toluene 2001).

The combination of drinking alcohol and exposure to toluene can seriously damage the liver. The combined effect is known as synergism. Smoking also has this effect with many contaminants, exacerbating existing health problems.

Ethyl benzene: A colourless liquid that smells like petrol and volatilizes and burns easily. It occurs naturally in coal tar and petroleum but is also found in many man-made products including paints, solvents, inks, and insecticides.

Xylene: A colourless flammable liquid with a sweet odour that occurs naturally in petroleum and coal tars. It is primarily a synthetic chemical, which is produced from petroleum for use as a solvent and cleaning agent in the printing, rubber and leather industries. Its association with

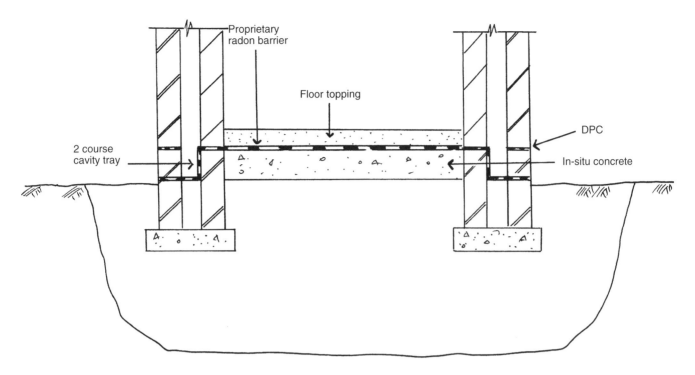

Fig. 11.5 Basic radon protection in house ground floor construction.

petroleum makes it a common contaminant at petrol station sites and should always be tested for.

Although not classed as a human carcinogen, short-term exposure to high levels of xylene can cause skin irritation and irritation of the eyes, difficulty in breathing and stomach ache.

11.4.10.5 Radon

Radon is a naturally occurring colourless and odourless gas that is formed from the radioactive decay of uranium. Long-term exposure to radon gas can cause lung cancer as a result of ingress through cracks and fissures in the ground and flowing into voids in buildings. Further research from the British Geological Survey (BGS) and the National Radon Protection Board (NRPB) has resulted in updated maps for the UK of areas where there is a potential risk from radon. Revised regulations from the Building Research Establishment came into existence in February 2000 for the protective measures to be incorporated into buildings in radon areas.

These are given in the Approved Document C and BRE Report 211 Radon Guidance on Protective Measures for New Buildings.

Risk from radon gas is higher in areas of igneous rock formations, which include Cornwall, Devon and Somerset, and the limestone formations of the Pennines and Northamptonshire.

Research at BRE, and comparison of radon exclusion techniques in other countries, has shown that radon can be kept out of buildings either by barriers in the ground

floor or by dispersing it beneath the building in such a way that it does not reach the occupants.

(BRE 1999)

The levels of protection from radon ingress have the following categories:

- Basic protection or geological assessment required to allow less
- Basic protection required
- Full protection or geological assessment required to allow basic protection
- Full protection or geological assessment required to allow less
- Full protection required.

Figure 11.5 illustrates the type of protection required to the ground floor construction of dwellings.

11.5 LANDFILL SITES

The dominant method of waste disposal by land filling in the UK is co-dispersal, whereby municipal and hazardous wastes may be deposited together. However, this system was to be phased out as a result of the EU Landfill Directive 1999 and replaced by a system where waste will be categorized. The categories will consist of hazardous waste landfill, non-hazardous waste landfill and inert waste landfill.

The Landfill Directive will require tighter controls on the disposal of wastes and will require landfill operators to flare gas and provide leachate management systems.

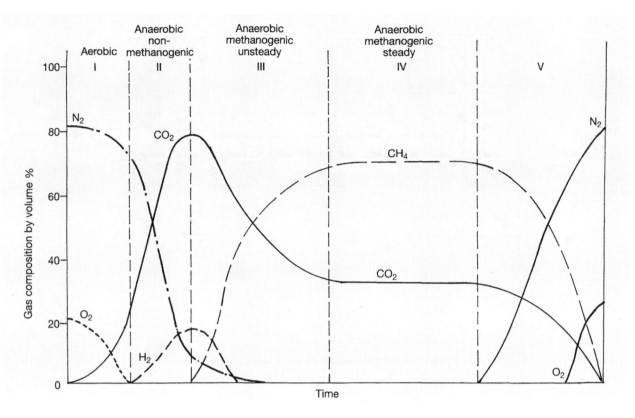

Fig. 11.6 Phases in landfill gas generation (schematic).

11.5.1 Gas migration

Domestic landfill sites present major problems for developers, both on the actual landfill site and on any site that is within close proximity to a landfill site. Landfill gas is produced during several phases of the breakdown and decomposition of biodegradable material under anaerobic conditions (Fig. 11.6).

Production of significant quantities of landfill gas may take from three months to a year to commence, but may continue for many decades. The composition of these gases vary with time but the main constituent parts are methane (CH_4) and carbon dioxide (CO_2).

Figure 11.6 (From Waste Management Paper No. 27 HM Inspectorate of Pollution 1989) shows the phases in landfill gas generation over time.

Phase 1 Aerobic decomposition of biodegradable materials. Entrained atmospheric oxygen is converted to carbon dioxide.

Phase II Anaerobic decomposition commences as oxygen is used up. Carbon dioxide concentration increases and some hydrogen is generated. No methane is generated at this stage.

Phase III Anaerobic methane generation commences and rises to a peak. Concentration of carbon dioxide declines: hydrogen generation ceases.

Phase IV Steady methane and carbon dioxide generation in proportions of 50–70% and 30–50% respectively.

Phase V Steady decline in generation of methane and carbon dioxide; gradual return to aerobic conditions.

Landfill gas can give rise to several hazards:

- It is flammable by virtue of its methane content. The explosive range for methane is between 5% (lower explosive limit) and 15% (upper explosive limit) by volume in air, and higher concentrations can be flammable.
- It is an asphyxiant by virtue of its methane and carbon dioxide content.

In view of these hazards, no housing development should be undertaken on a landfill site that is still generating gases in significant volumes. On such sites there is a high risk of spontaneous combustion, which can produce deadly carbon monoxide gases and in addition give rise to general ground instability.

There are old landfill sites which were filled prior to the 1965 Clean Air Act and which contain relatively low amounts of biodegradable matter. With adequate investigation and long-term gas monitoring it is usually possible to develop such sites provided suitable precautions in the construction of the dwellings are taken.

With housing developments, the main problems to contend with are found on those developments that are in close proximity to active landfill sites. The landfill gases produced can migrate in all directions, through the surrounding strata if they are fissured or permeable. The distance the gases migrate will depend on the permeability of the strata surrounding the landfill site, the volumes of gas being generated and the barometric pressure gradients. Other factors relating to the site geology, such as geological faults and fissures caused by past mining, can allow migrating gases to travel considerable distances from their source.

Old mine workings and mine shafts provide lateral and vertical conduits for gas to travel along. Old mine workings generally contain methane and carbon dioxide gases and these can often be misleading when surface monitoring is being carried out for landfill gas. If mine gases are suspected as being the prime cause it is possible to distinguish the mine gas from landfill gas by carbon dating tests.

Gas migration is influenced by several climatic factors, namely atmospheric pressure, groundwater levels or even the temporary effects of a snow-fall, which can act like a blanket over the landfill site and surrounding land, preventing the gas from venting at the surface.

Where housing is to be constructed within 250 metres of a known active landfill site, it is a condition of local planning authorities that gas monitoring must be carried out to check whether landfill gases are migrating onto the subject site. This planning condition usually takes the following form:

Prior to the development commencing, a contaminated site investigation and assessment shall be carried out to the satisfaction of the local planning authority, relating to the nature of all deposited materials within the adjacent tipping site. The investigation shall also take account of the possible migration of landfill gas. Any measures identified as required for safety reasons as a result of the investigations shall be agreed with the local planning authority in writing and shall be undertaken prior to the occupation of the dwellings.

11.5.2 Gas monitoring

Guidance on landfill gas monitoring adjacent to active-gassing landfill sites is to be found in Waste Management Papers No. 26 and 27 published by Her Majesty's Inspectorate of Pollution. The Approved Document Part C of the Building Regulations also stipulates that the maximum permitted trigger values of 1% (20% LEL) by volume for methane and 1.5% by volume for carbon dioxide must not be exceeded in the soils outside the landfill site or in dwellings on the proposed development.

The volume of gas migration can be determined by the installation of suitably positioned gas sampling pits or boreholes. Monitoring can also take place from perforated pipes inserted into backfilled trial pits.

There are various types of gas detection equipment on the market, such as the GMI Oxygascoseeker and the GMI Infra-red Carbon Dioxide Meter, which has a range from 0–100%.

Figure 11.4 illustrates such a gas sampling arrangement for use in a borehole. Figure 11.7 illustrates the best positions to install gas-monitoring points adjacent to a landfill site or in mining areas where shallow workings may be emitting gases.

The site should be monitored regularly over a long period, with any changes in barometric pressures being recorded. If significant gas migration is occurring from an adjacent landfill site, it will be necessary to introduce a gas management system in the form of relief wells, or provide a bentonite slurry trench around the landfill site to prevent the gas from migrating.

11.5.3 Carbon dioxide

The action level of 0.5% for carbon dioxide had long been considered to be too low. It was subsequently increased to 1.50% by HM Inspectorate in the 1992 Building Regulations changes, as high carbon dioxide levels can be found in virgin land as a result of aerobic and anaerobic decomposition of natural organic materials.

It has long been established that a more realistic value of 5% should apply and it is clear in the Approved Document Part C that, where levels of 5% are evident then specific design requirements must be incorporated in the design of the building's ground floor.

It is important to consider whether the carbon dioxide is being produced in sufficient quantities to migrate at these concentrations into a service duct or a building. In areas where elevated carbon dioxide concentrations are found in the ground, the risk may be considered to be negligible as there are insufficient sub-strata present to give rise to any other than small quantities of the gas.

It is essential that all deep drainage excavations and manholes are checked regularly for the presence of carbon dioxide. This gas is heavier than air and can build up in concentrated pockets in such situations.

Figure 11.8 illustrates a typical detail of a ground-floor construction to prevent landfill gas entering a building. The key gas protection measures should include:

- Provision of a well-constructed ground floor slab
- The installation of a low-permeability gas membrane of approved manufacture over the floor area of the dwelling. This membrane should extend across the external cavity wall to be fully effective
- Passive venting below the building – granular filled voids
- Passive venting below the building – under-floor void space
- Ventilation of confined spaces within the dwelling
- Minimum perforation of the ground slab by service pipes
- Sealing off all service entry points that pass into the walls or floor of the dwelling with approved sealants such as Bituthene MR top-hat patching or hot-poured bitumen sealants

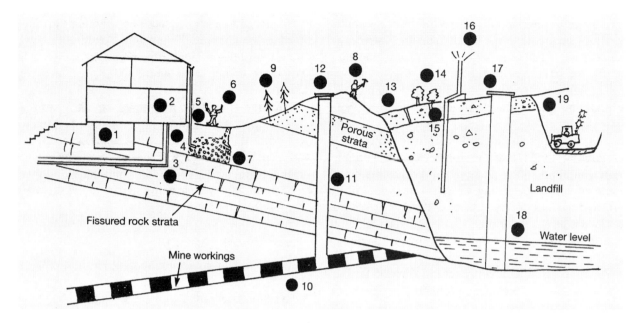

1 Gas in basements
2 Gas in buildings
3 Gas entry via services excavation
4 Gas entry via service pipe entry points
5 Gas in rainwater gutters and pipes
6 Odour nuisance from gas
7 Collection of gas by soak-aways
8 Gas in land drains
9 Vegetation damage (off site)
10 Migration from old mine workings

11 Gas up mine shafts
12 Gas from mine shafts
13 Gas from fissures in capping
14 Vegetation damage
15 Gas up boreholes
16 Efficiency of gas flaring
17 Gas in manhole wells
18 Gas pressure due to water level rising
19 Health and fire risks from gas in confined spaces or poorly ventilated areas on site

Fig. 11.7 Landfill gas monitoring requirements (after D.J.V. Campbell).

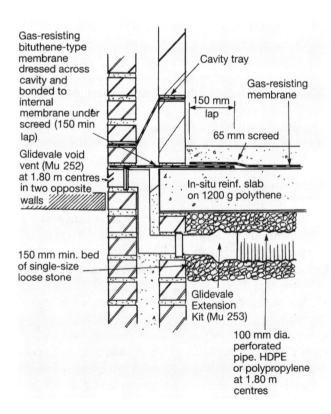

Gas-resisting bituthene-type membrane dressed across cavity and bonded to internal membrane under screed (150 min lap)

Glidevale void vent (Mu 252) at 1.80 m centres in two opposite walls

150 mm min. bed of single-size loose stone

Cavity tray

Gas-resisting membrane

150 mm lap

65 mm screed

In-situ reinf. slab on 1200 g polythene

Glidevale Extension Kit (Mu 253)

100 mm dia. perforated pipe. HDPE or polypropylene at 1.80 m centres

Fig. 11.8 Gas venting precautions to a suspended ground-floor slab.

• Ensuring that the granular bed surround around service pipes is interrupted with a bentonite or puddle clay seal prior to entering into the dwelling
• Active venting to the building
• Gas monitoring of installed measures with alarms.

Figures 11.9 and 11.10 illustrate the gas precautions to be used below raft foundations and voided suspended ground floors.

11.5.4 External measures

(a) Slurry trenches

The use of slurry trenches is increasing as a means of effectively controlling gas migration from landfill sites. The method of conventional clay filled trenches has now been superseded by bentonite and cement slurry mixtures. Though the raw materials are more expensive, installation is quicker and the end result is a less permeable trench. Hydrating the bentonite with water and then blending in the cement forms the cement slurry. The cement slurry

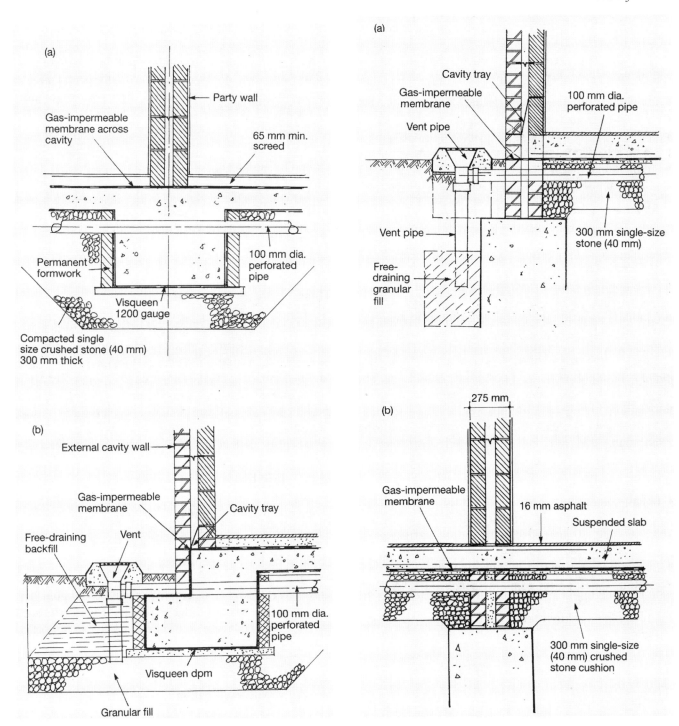

Fig. 11.9 Gas-venting precautions to raft foundation.

Fig. 11.10 Gas-venting precautions to suspended in-situ ground floor.

is placed into the trench as excavation proceeds on a continuous basis. The slurry itself gives support to the deep trench sides and sets to a firm consistency within two days. It usually achieves the consistency of a firm to stiff clay and its permeability may be as low as 10^{-9} m/s. The minimum recommended trench width is 600 mm and depths up to 8 metres can be achieved using traditional excavating plant. Greater depths require specialized plant equipment.

It is important to ensure that any slurry trench is keyed into the underlying low permeability stratum, and on-site supervision is critical and essential if good results are to be obtained.

(b) Gas venting trenches

Generally placed between the landfill site and the development site, these consist of excavated trenches filled with large single-size hard materials such as bricks or stones. A 100 mm diameter perforated pipe is installed with vertical risers at 25–30 metre centres. The top surface of the trench should be sealed off with compacted clay or impervious

Contaminated land

membrane to reduce the ingress of surface water into the trench fill.

Questions that need to be considered for ground gas investigations (after Gregory 2000) are:

- What gases are present and where?
- What are the gas concentrations?
- What is the gas source?
- Who is liable for the gas?
- What are the surface emission rates?
- What are the rates of gas production?

Identifying sources of methane gas:

- Landfill operations
- Mine workings
- Wetlands (peat)
- Main sewers and water service pipes
- Gas mains
- Other: compost, fly-tipping, cemeteries, buried carcasses, dung heaps and cesspits.

Apart from carbon dioxide and methane, it is also important to be aware of hydrogen sulphides, carbon monoxides, hydrogen and volatile organic compounds (VOCs).

Further guidance can be obtained from the following publications.

CIRIA

Methane and Associated Hazards to Construction – CIRIA Doc. No. 79 – 1992
Measurement – CIRIA Doc. No. 131 – 1993
Protecting Development from Methane – CIRIA Doc. No. 149 – 1995
Interpreting Gas Monitoring. CIRIA Doc. No. 151 – 1995
Risk Assessment. CIRIA Doc. No. 152 – 1995

Non-statutory guidance

Waste Management Paper 26a Landfill Completion 1993
Waste Management Paper 26b Landfill design, Construction and Operational Practice – 1995
Waste Management Paper 27 Landfill Gas – 1991
Waste Management Paper 4 Licensing of Waste Management Facilities – third edition – 1994
BRE Report No. 212 – 1991 Construction of New Buildings on Gas Contaminated Land

11.6 DESK STUDY

When a site is suspected of containing contaminants likely to present a hazard to any housing development, a desk study should be commissioned in accordance with BS 5930.

As with any site investigation, the time spent studying the site history and geology is money well spent. Once the past history has been established the range of contaminants may be narrowed down, so reducing the cost of the fieldwork and laboratory testing. The desk study is therefore extremely cost effective. It generally consists of:

- Local geological study
- Industrial history of the site
- Mining investigation
- Site reconnaissance.

11.6.1 Local geological study

Examination of the site geology should establish information such as depth to rock-head and the position of any geological faults or ground fissures. Fissures often occur in rock measures that overlie known coal-mining areas. These faults and fissures can provide pathways for methane migration from old abandoned coal workings or mine shafts.

11.6.2 Industrial history of the site

Contaminants in the ground will generally be high if there has been past industrial processes on the proposed development site. They are likely to constitute the largest single hazard to a housing development.

This scenario is more likely to occur on redevelopment sites in urban areas where the Victorian industrialists deposited their waste products from local industries such as iron smelting, metal works, and cotton and woollen mills.

11.6.3 Mining investigation

Both past mining and future mining can give rise to problems on a site and a thorough search should always be made for such activities. This check should not restrict itself to coal workings as many other minerals, such as fireclay, sandstone, chalk and gypsum, were obtained by shallow mining methods.

11.6.4 Site reconnaissance

This should be carried out to determine visually whether there is any evidence of contaminants likely to pose a risk to development. This reconnaissance should also relate to areas around the subject site, as streams flowing through adjacent sites can pick up contaminants and deposit them off site.

11.7 SITE INVESTIGATION

Once the desk study has been compiled and evaluated, it should be possible to determine the size and scope of the site investigation that will be required. This will most likely consist of trial pits and boreholes, together with long-term monitoring of ground waters and gas instrumentations.

11.7.1 Trial pits

These can be excavated using a JCB or a Hymac and should be deep enough to enable full identification of any fills, contaminated wastes or other toxic substances. If the trial pits exceed 1.20 metres depth they should not be entered unless they are properly shored up in accordance with health and safety requirements.

Groundwater levels should be recorded and groundwater should be sampled for testing. Any evidence of buried tanks should be recorded and investigated further. The trial pits can be monitored for methane or other gases during the excavations. On backfilling the trial pits, perforated plastic pipes can be inserted with a gravel surround to enable long-term gas monitoring to be continued. Soil and fill samples should be taken at regular intervals for laboratory testing.

11.7.2 Boreholes

These are usually required if the depths of fill exceeds the reach of the excavating plant. They are also required if old coal workings are suspected at a shallow depth. Rotary core drilling is the usual method adopted when drilling through rock strata. For soft ground and filled sites, shell and auger boring is used and samples can be obtained for laboratory testing by using a split mandrel.

The boreholes can be lined and used for gas monitoring and groundwater checks if they are properly prepared.

11.7.3 Testing for toxic gases

The type of investigation and degree of monitoring carried out will depend on the level of risk to the site from contamination by methane, carbon dioxide or solvents. These risks depend on the following factors:

- **Proximity** of the proposed development to any landfill operation. Where the distance is less than 250 metres the site must be monitored for gases, which may be migrating from the landfill operation
- The **local geology** beneath the site and beneath the landfill operation must be established, as this can have a major effect when methane is present
- The nature of the **materials** that have been deposited in the landfill site or other filled areas on the subject site

- The **time period** over which the fills were placed, which should include a start date and a completion date
- If a **landfill tip** is being checked it is useful to determine whether the tip was being managed and operated under a proper licence and that materials deposited have been properly capped off and sealed
- The level of **groundwater** if present
- Any evidence of past **mine workings** below the site.

Once these factors have been assessed the testing for gases can consist of:

- Probing any excavations or boreholes with a gas monitor
- Checking gas readings when barometric pressures are both low and high. These checks should be carried out very frequently to be effective.

11.7.4 Chemical analysis

Once the site investigation has been completed and the samples have been analysed in the laboratory, another factor, the **environmental significance** of the results, can be examined. The course of action required may depend on the intended usage of the land. For housing developments the risk assessment will need to be carried out using the appropriate model.

11.7.5 Safety

All investigation work and remedial measures should be carried out in a controlled manner so that the contaminants are contained and the safety of site personnel and the public are guaranteed. This will require the specialist firm to identify problem areas and inform site personnel during the works programme.

Some of the hazards may be physical, such as abandoned tar wells or lagoons. Other hazards, such as carbon dioxide, methane or flammable organic residues, may be encountered during deep trench excavations. Great care should be exercised when backfilling old tar wells and underground tanks, and any liquids present should be removed totally prior to filling in.

11.7.6 Conclusion

Developers or engineers who fail to carry out or initiate adequate site investigation of contaminated land run the risk of being charged with negligence. The investigations should be specifically designed to relate to the past history of the site, the geology of the area and the type of development being considered. The investigation should be carried out in accordance with BS 5930 1981 and BS 10175 1988.

The object of a site investigation is to obtain information. The quality of the investigation will be determined by the skills of the specialist firms and the equipment used. False economies at the preliminary stages could result in huge expense at a later stage. After sites have been made good and built over, some allowance should be made for monitoring to confirm that the measures taken have been effective.

Where the concentrations of contaminants are not sufficiently high to warrant removal off site it is generally considered advisable to introduce a 200 mm minimum thickness of dolomite or crushed limestone to act as an alkaline break layer to prevent potential upward migration of water-soluble metal contaminants. A suitable depth of clay or topsoil should then be placed to prevent root contact from trees or shrubs; this is usually 700 mm to 1000 mm thick. In domestic gardens the existing 'soils' are usually removed for a depth of 1,000 mm and replaced with an alkaline break layer at least 200 mm thick with approximately 800 mm of clean neutral or alkaline soils above. The excavated 'soils' may be re-used in public open spaces or landscaping works.

Phyto-toxic guidelines

The following group of contaminants are not normally hazardous to human health in the concentrations shown, but they are phyto-toxic, i.e. they will affect plant growth.

Contaminant	Guideline value mg/kg
Boron	3
Copper	50
Nickel	20
Zinc	130

The phyto-toxic effects of the above contaminants may be additive and the soil pH value can influence the phyto-toxic effects.

Case Study 11.1

It is proposed to develop a former dye works in Halifax with a mixture of commercial and residential properties. Because of the known contaminants associated with its former use, a comprehensive site investigation has been called for by the local planning authority. This investigation is to be carried out following demolition of the existing buildings and subsequent removal of the ground-floor slabs.

The site investigation report will be required to outline recommendations for treating fills that may be classed as hazardous and also to designate which areas of the site will be best for residential development.

INVESTIGATION

Prior to the investigation commencing a desk study was carried out and a site-specific conceptual model was developed (Fig. 11.11). This enabled the engineers to delineate the most likely areas to carry out trial pits and chemical sampling.

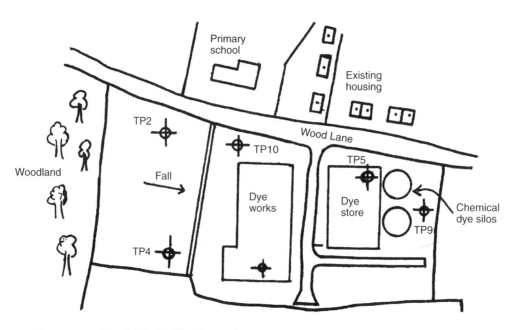

Fig. 11.11 Site-specific conceptual model for Halifax dye works.

SAMPLING

Samples sent to the laboratory for chemical testing were analysed for contaminants likely to be harmful to residents when the site is developed. These tests showed concentrations of three main contaminants, arsenic, cadmium and nickel.

Sample location no.	Contaminant	Concentration mg/kg
TP 2	Arsenic As	16
TP 4	Arsenic As	53
TP 6	Cadmium Cd	1100
TP 9	Cadmium Cd	850
TP 10	Nickel Ni	200
TP 5	Nickel Ni	3750

ASSESSMENT OF RESULTS

The various values were simulated using the CLEA 2002 software program, doing two simulations for arsenic, cadmium and nickel. This simulation showed how the soil concentrations can vary according to the specific nature of the site ground conditions and the proposed end use for the development.

The primary consideration in the initial site assessment and any proposed remedial measures must be:

- To identify and assess any hazard resulting from the possible ingestion or inhalation of, or any direct skin contact with, the site fill
- The phytotoxicity of the surface fills.

TOXICITY OF FILL MATERIALS TO POTENTIAL HOUSING OCCUPANTS

Nickel

Nickel and its compounds are recognized as having a significant role in the aetiology of eczema on the human skin. Ingestion of nickel may exacerbate eczema in those people already affected. Normal soils contain up to 500 mg/kg.

Cadmium

Cadmium compounds can be extremely toxic if ingested. They accumulate in the kidneys, causing kidney damage.

Arsenic

The toxicity of arsenic compounds in soils varies with the acidity of the soils. Occurrence of arsenic as a contaminant is quite common because it is associated with effluent, particularly sewage sludge, and is a waste product from the production of coal gas at gasworks facilities. The burning of coal and ashes from smelting furnaces also contains arsenic. Industrial processes in the past involved burning vast amounts of coal, while the domestic burning of coal has produced widespread ash containing arsenic.

The primary exposure routes are inhalation and ingestion. The ingestion pathways of concern are through the consumption of vegetables and through the medical condition known as 'pica', whereby young children have a compulsion to eat soil. Acute exposure can lead to liver damage and kidney failure and long-term exposure to inorganic arsenic can cause skin cancers (The EarthShop.com–arsenic 2001).

Increasing the pH of the soil will decrease solubility and mobility of arsenic compounds, and their availability for absorption by plant roots. Arsenic use in industrial processes is widespread, particularly in the following industries and activities:

Smelting, fungicides, pesticides, glass manufacturing, timber treatment plants, textiles and dye-works (Card 1995).

CLEA Outputs (based on CLEA 2002 software program)

Arsenic

Health criteria value	Index dose
Entry route	All exposures
Site use	Residential
Soil type	Sandy
pH	4
Soil organic matter	5%
Human database	Female
Averaging time	1–6 years
Calculate background exposure	No
Run simulation and optimize	5000 iterations
Soil concentration	19.26 mg/kg

Arsenic

Health criteria value	Index dose
Entry route	All exposures
Site use	Commercial
Soil type	Clay
pH	9
Soil organic matter	5%
Human database	Female
Averaging time	Working life 16–59 years
Calculate background exposure	No
Run simulation and optimize	5000 iterations
Soil concentration	552.84 mg/kg

Cadmium

Health criteria value	Tolerable daily intake
Entry route	Oral
Site use	Residential

Case Study 11.1 *(contd.)*

Soil type	Sandy
pH	4
Soil organic matter	5%
Human database	Female
Averaging time	1–6 years
Calculate background exposure	No
Run simulation and optimize	5000 iterations
Soil concentration	0.016 mg/kg

Cadmium

Health criteria value	Tolerable daily intake
Entry route	Oral
Site use	Commercial
Soil type	Clay
pH	9
Soil organic matter	5%
Human database	Female
Averaging time	Working life 16–59 years
Calculate background exposure	No
Run simulation and optimize	5000 iterations
Soil concentration	1842 mg/kg

Nickel

Health criteria value	Tolerable daily intake
Entry route	All exposures
Site use	Residential
Soil type	Sandy
pH	4
Soil organic matter	5%
Human database	Female
Averaging time	1–6 years
Calculate background exposure	No

Run simulation and optimize	5000 iterations
Soil concentration	263.36 mg/kg

Nickel

Health criteria value	Tolerable daily intake
Entry route	All exposures
Site use	Commercial
Soil type	Clay
pH	9
Soil organic matter	5%
Human database	Female
Averaging time	Working life 16–59 years
Calculate background exposure	No
Run simulation and optimize	5000 iterations
Soil concentration	9188 mg/kg

Conclusions

It is clear from the specific site conceptual model that the most serious contamination is within the confines of the previous factory building and the chemical storage silo area. These areas should therefore be designated for commercial/industrial development with residential development in the remaining areas.

Consideration will need to be given to the phyto-toxic effects of the nickel concentrations in the commercial areas, as the fills are acidic. This will probably require removal of some of the fills and replacement with clean soils.

BIBLIOGRAPHY

Bardos, P. (2000) MSc Contaminated Land Management Course Notes, Remediation module, University of Nottingham.
British Asbestos Newsletter Issue No. 35 (1999).
Building Research Establishment (1999) *Radon: Guidance on Protective Measures for New Dwellings*, Report No. 211, third edition.
Campbell, D.J.V. (1981) *Landfill Gas Production and Migration*, Environmental Safety Centre, AEA Harwell.
CARD Geotechnics *Contaminative Substance Datasheets*, July 1995.
CIRIA (1987) *Building on Derelict Land*, Vol. 3, Appendices A–E, Construction Industry Research and Information Association, London.
Department of the Environment, Transport and the Regions (Feb. 2000) Draft DETR Circular, Environmental Protection Act 1990: Part IIA Contaminated Land, London: Department of the Environment.
Department of the Environment Industry Profile, *Asbestos Manufacturing Works* (1995).
Department of the Environment Industry Profile, *Railway Land* (1995).
Department of the Environment, Transport and the Regions and Environment Agency (2000) *Guideline Values for Contamination in Soils*, Contaminated Land Research Report GV series, London.
Department of the Environment, Transport and the Regions (2000) *The Contaminated Land Exposure Assessment Model (CLEA)*: Technical Basis and Algorithms, Contaminated Land Research Report No. 10, London.
Environment Agency (2000) *Guidance for the Safe Development of Housing on Land Affected by Contamination*, Report No. 66.
Gregory, B. (2000) MSc Contaminated Land Management Course Notes, Landfill Gas module, University of Nottingham.
Ground Engineering (July 1998) Wash and Scrub Up, pp. 16–19.
Harris M., Herbert S. and Smith M. (1995) *Remedial Treatment for Contaminated Land, In-situ Methods of Remediation*, Volume IX, CIRIA Special Publication 109, pp. 74–81.

GLOSSARY OF TERMS RELATING TO CONTAINATED LAND

Aerobic	In the presence of oxygen
Anaerobic	In the absence of oxygen
Controlled landfill	Where waste materials are deposited in an orderly planned manner at a site licensed under the Control of Pollution Act 1974
Gas migration	The movement of gases from the wastes within a landfill site to surrounding strata or emission into the atmosphere
Impermeable	Used to describe materials which have the ability to resist the passage of fluid through them. The materials can be natural or synthetic-based
Inert waste	These are wastes which will not physically or chemically react or undergo biodegradation within the landfill mass
Landfill	Waste materials deposited in accordance with the Waste Management Paper No 26 and Control of Pollution Act 1974
Leachate	The result of liquid seeping through landfill waste and by so doing extracting substances from the deposited wastes
LEL	Lower explosive limit: the lowest percentage by volume of a mixture of flammable gas with air which will propagate an explosion in a confined space, at 25°C and at atmospheric pressure
Monitoring	A process of measuring for gas using portable instruments and analysis of samples to provide information for assessment of site conditions
Perched water	An accumulation of water at a level above that of the adjacent water table
Permeability	A measure of the rate at which a fluid will pass through a medium
Putrescible	Describes a substance that is readily decomposed by bacterial action giving rise to offensive odours
UEL	Upper exposure limit: the highest percentage by volume of a mixture of flammable gas and air which will propagate a flame at 25°C and atmospheric pressure
Venting (active)	The removal of landfill gases by forced extraction from within the landfill mass
Venting (passive)	The natural movement of landfill gases from a landfill site into the atmosphere, assisted by porous drainage media
Venting trench	A trench containing porous granular material of uniform size to permit the passive venting of landfill gas

HM Inspectorate of Pollution (1986) *A Technical Memorandum for the Disposal of Wastes on Landfill Sites*, Waste Management Paper No. 26, HMSO, London.

HM Inspectorate of Pollution (1989) *The Control of Landfill Gas*, Waste Management Paper No. 27, HMSO, London.

Nathanial, P. (1999) *An Introduction to Contaminated Land Management*, Seminar notes.

Nathanial, P. (2000) MSc Contaminated Land Management Course Notes, Industrial Processes Module – Timber Products Manufacturing, University of Nottingham.

NHBC (1999) NHBC Standards Chapter 4.1 *Land Quality – Managing Ground Conditions*.

Peckson, G.N. and Lowe, G. (1985) *Methane and the Development of Derelict Land*, London Environmental Supplement No. 13.

Scotland and Northern Ireland Forum for Environmental Research (1999) Communicating Understanding of Contaminated Land risks. SNIFFER Project No. SR97 (11)F.

The EarthShop.com http:// www.eco-usa.net/toxics/arsenic

The EarthShop.com http:// www.eco-usa.net/toxics/lead

The EarthShop.com http:// www.eco-usa.net/toxics/toluene

Wilson, D.C. and Stevens, C. (1981) *Problems Arising from the Development of Gas Works and Similar Sites*, UKAEA, Harwell.

Wolfe and White (1997) *Principles of Environmental Law*.

Index

acidity classification 9, 10
acidic conditions in contaminated land 238, 246–8
acids in groundwater 10
active mining 100, 107
activity of clay soils 76
active pressures (Rankine) 150–1
adhesion on piles 48–9
Adits 96, 111
aerial photographs 3
aerobic 240, 249
alignment of piles 53
allowable bearing pressures
   on cohesive strata 8–9
   definition of 67
   on granular strata 84–7
   for limiting settlement 71–2
allowable loads on piles 179
alluvial soils 3
anaerobic 240, 249
analysis of circular slips 137–8
anchor piles 50
angle of internal friction μ 137
angle of shearing resistance of retained materials 137
anhydrite mining 95
argillaceous rock 183
arsenic 232–3, 235–6, 238, 247
artificial cementation 202
asbestos 235

basements
   damp-roofing of 169
   under-pinning adjacent to 209
bearing capacity factors
   in cohesive strata 69
   in granular strata 85–6
bell pits 95
benzene 238
Berezantsev, V.G. 48, 90
Bishop, A.W. 137
bored piles
   in cohesive strata 46–7, 49, 169
   design of 47, 90
   in granular strata 89–90
   methods of installation 48
   underpinning with piles 216
boreholes
   records of 4–6
   rotary flush 99, 245
boron 238

British Coal 3, 21, 100–1
Broms, B. 48, 90
building on wooded sites 122–4
bulking factors 111

calcareous sands 82
cadmium 232, 235, 247
capping off mine shafts 97–8, 111
Casagrande
   plasticity chart 9, 77–8
   test apparatus 77
chalk mining 12, 95
clasp system 102
classification of soils 78, 81–2
Claymaster 33, 123, 124–5, 127, 131, 135, 214
clays
   desiccation of 33, 74, 121, 125
   swelling of 75, 123
climatic variation 119–20
cohesionless soils
   allowable bearing pressures 84–6
   cased piles in 90
   piles in 90–93
   settlements in 87
   skin friction in 90
cohesive soils
   allowable bearing pressures 71
   cold storage building on 67
   moisture movements in 74–5
   piles in 47–8
   settlements on 72–3
compaction of fills 181, 183
contaminated land
   contaminated sites 229
   carbon dioxide 240–1
   landfill gas 239–41
   old mine workings 236
   testing for toxic gases 245
   timber yards 234–235
copper 238
crib walls 149–50
crown holes 111, 176

desktop study 1–2, 244
desiccation of clay soils 118, 121
differential settlements
   in filled ground 34–35, 175–7
   in mining subsidence areas 102–3
dilatancy 82

Index

disused wells 44
drained cavity construction 169, 172
drainage services 126
driven piles
    set calculations 49, 93
    testing of 49, 93
Dutch formulae 94
dynamic consolidation 198–201
dynamic piling formulae 92–93

Earth pressures
    active 150–151
    passive 152
effective stress analysis 139
end bearing of piles
    in cohesive strata 48
    in granular strata 48, 90
excavations
    on steep slopes 146–7
    in underpinning 213–14

faults 111
factors
    bearing capacity 68–70
    influence line 74
Fellenius 139
field dry density 9
filled ground
    chemical wastes in 188, 235–6, 238
    foundations on 175–8, 179
    piling in 178
    stiff raft foundations on 178
    surcharge loading of 201

gabion walls 149
gap graded gravels 81
gas works sites 236
gas monitoring 241
geological memoirs 3
geo-textile walls 150
ground improvement
    chemical injection in 202–3
    by dynamic consolidation 198–200
    by surcharge loading 201–2
    testing of 192, 201
    by vibro-compaction 193–8
groundwater 2, 67
groundwater lowering 86
grouting
    in ground improvement 202–3
    in mine shafts 97
    in mine workings 95, 99–100
gypsum mining 14, 95
gypsum solution features 14

hazardous gases on landfill sites 240, 247–8
human health risks 232, 246

immediate settlements 68
industrial processes 233–6, 238
influence factors 74
in-situ testing of soils 4, 181, 184, 192–3
ironstone mining 5, 95

joints in mining subsidence areas 103, 109–10
Jumikis 139–140

lateral displacements 67
lateral loads on retaining walls 150–1
limestones 11
liquid limit 76
load testing
    in ground improvement 190, 192–3
    of piles 50–3
    using plate bearing 89
    zone tests 192–3
leachate 249
lead 238

Mackintosh probe 82
metal smelting works 236
Meyerhof, G.G. 69–70
mine shafts 14
Mineral extraction
    bell pits 95
    mine shafts 96
    pillar and stall 111
Mine workings
    grouting old workings 99
    void migration in 99
Mohr circle stress diagram 9–12
Moisture content 4, 8
Movement joints 109

Negative skin friction 49, 179
Net bearing pressures 67
NHBC Chapter 4.2 33–5, 113–15, 118–20, 121–2, 124, 133
nickel 238, 247

opencast coal workings 177
outcrops 99, 111

particle size distribution 9
pad and pier foundations 44
passive resistance 162
peak particle velocity 199
peat 8, 33
Perth penetrometer 82
pH values 9, 10
piezometers 4, 146
phyto-toxic 238, 246
piled foundations
    bored 47, 90
    in cohesive strata 118
    continuous flight auger 90
    design of 50–3
    driven 50–3
    shaft resistance 48
pile testing
    re-drive 93
    static 50–3
pillar and stall coal workings 97, 111
plastic limit 4, 77
plasticity index 4, 76, 113, 117, 119, 130
plate bearing tests 89, 190–2
pollution 229

pore-waterpressures 137, 182
proof rolling 34

raft foundations
    design 33–44
    irregular shaped buildings 104
    on filled ground 178, 189, 198
    in mining subsidence areas 103
    NHBC Chapter 4, 2, 119
radon gas 239
railway land 236
railway tunnels 3
re-drive test on piles 93
reinforcement bar areas 32
reinforced strip foundations 32
retaining walls
    dampproofing to 169–70
    gravity type 152
    pocket type 156
    reinforced cavity type 157–8
    reinforced earth type 150
    reinforced concrete design to BS 8002 163–8
risk assessment 230–3
rotational slip in cohesive strata 138

safety 245
salt extraction 13
sandstone mining 95
scrap yards 235
settlement of foundations
    on cohesive strata 67
    on granular strata 67–8, 176
serviceability limit state 163
sewage treatment works 235
shape factor λ 70
shear vane test 4, 71
shoring 226
shrub planting 121
sink-holes 12
site investigation procedure 4
site testing of filled ground 181–2
skip tests 192
slurry trenches 242
slope stability
    Bishop's rigorous method 137
    Fellenius 139
    Jumikis 139–40
soft spots 23
solution features 11

solution brine mining 13
split spoon sampler 83
standard penetration test
    in cohesive strata 4
    correction factor 84
    in granular strata 9, 83
steel sheet piling 150
strip foundations in active mining subsidence areas 107
suspended ground floor construction 179–80

testing soils
    in laboratory 8–10, 76–7
    in the field 4, 83, 89, 181, 190–2
timber treatment works 235
toluene 238
toxological effects of contaminants 238
tree heights 114–15
tree identification 115–17
tree species 113
trench fill foundations
    in filled ground 27, 32
    to NHBC Chapter 4, 2, 123
trial pits 4, 21, 236, 245
tri-axial tests 77–8

underpinning 206, 209, 212–14
undisturbed samples 4, 9
undrained shear strengths 71, 77–78
undrained tri-axial test 8
ultimate bearing capacity of cohesive strata 67, 69
ultimate limit state 163

vertical stress distribution 70
vibration 198–9
vibro-compaction 186–98
vibro-flotation 187
vibro-replacement 186
void migration 97–9

walk-over survey 1
water demand of trees 114–15
well pointing 86
widened reinforced strip foundations 28
wild brine extraction 13

xylene 238

zinc 238
zone testing 193